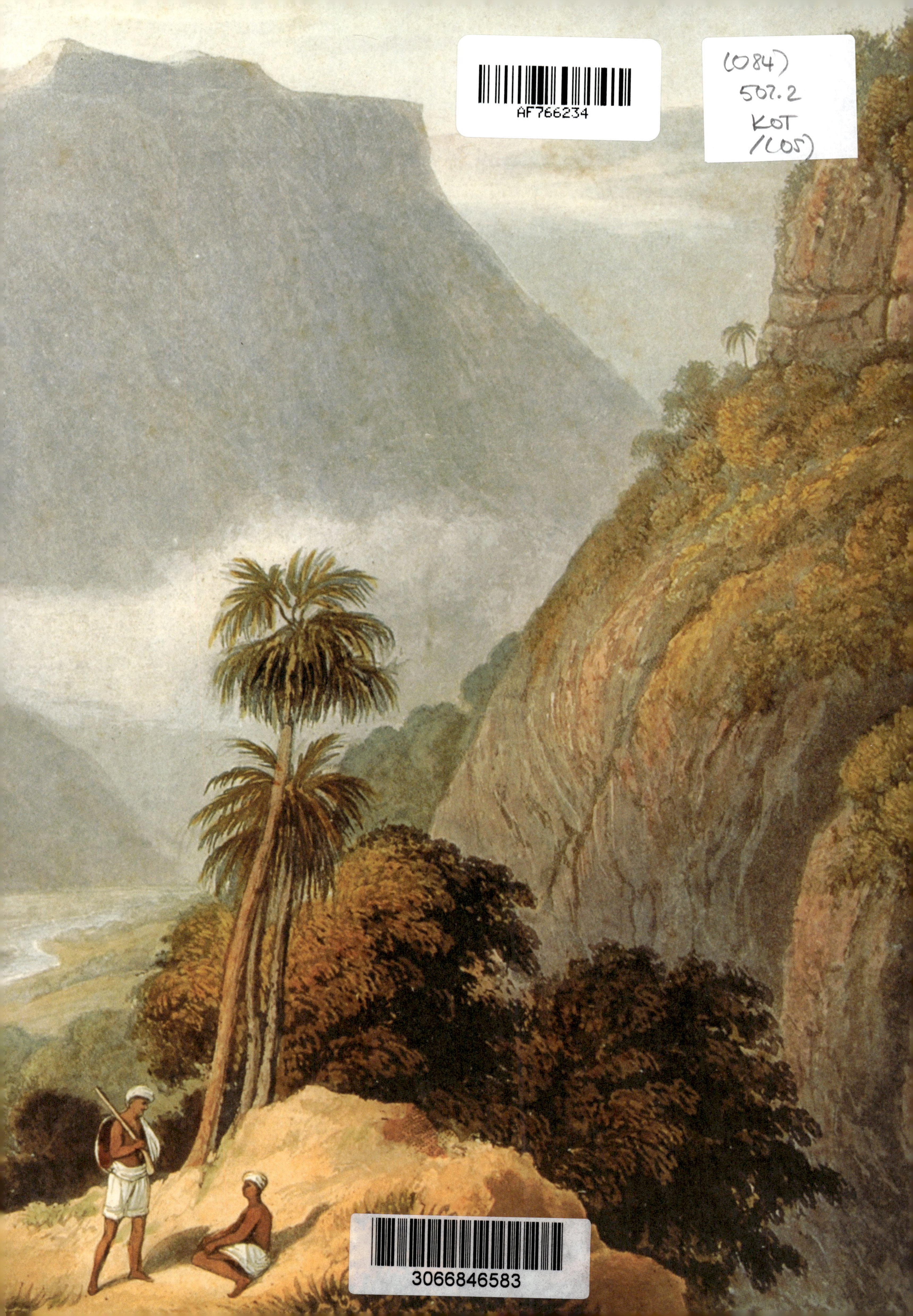

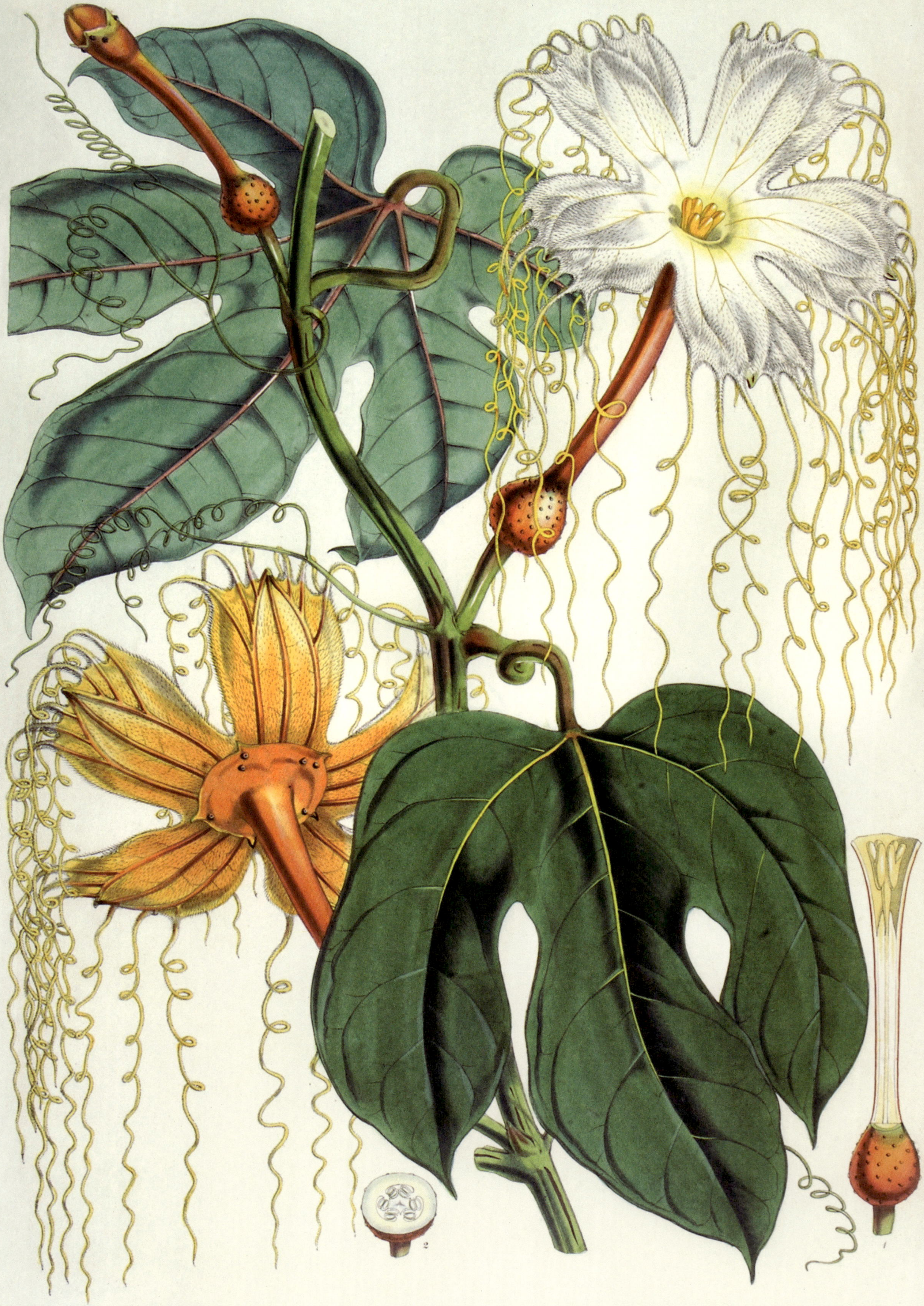

Hodgsonia heteroclite

Courtesy
Kavin C. Parikh, Parikh Foundation

LIVING JEWELS *from* THE INDIAN JUNGLE

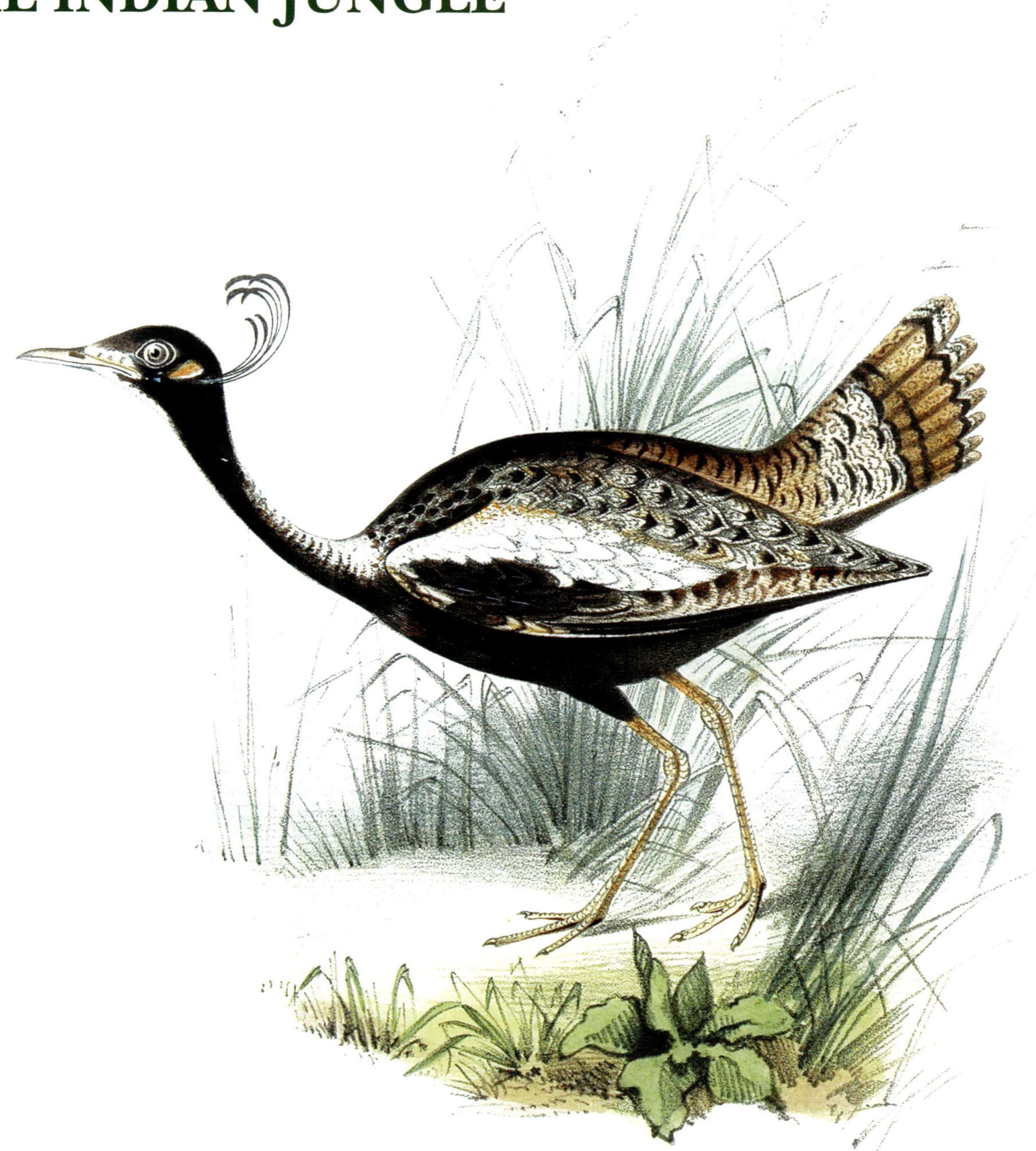

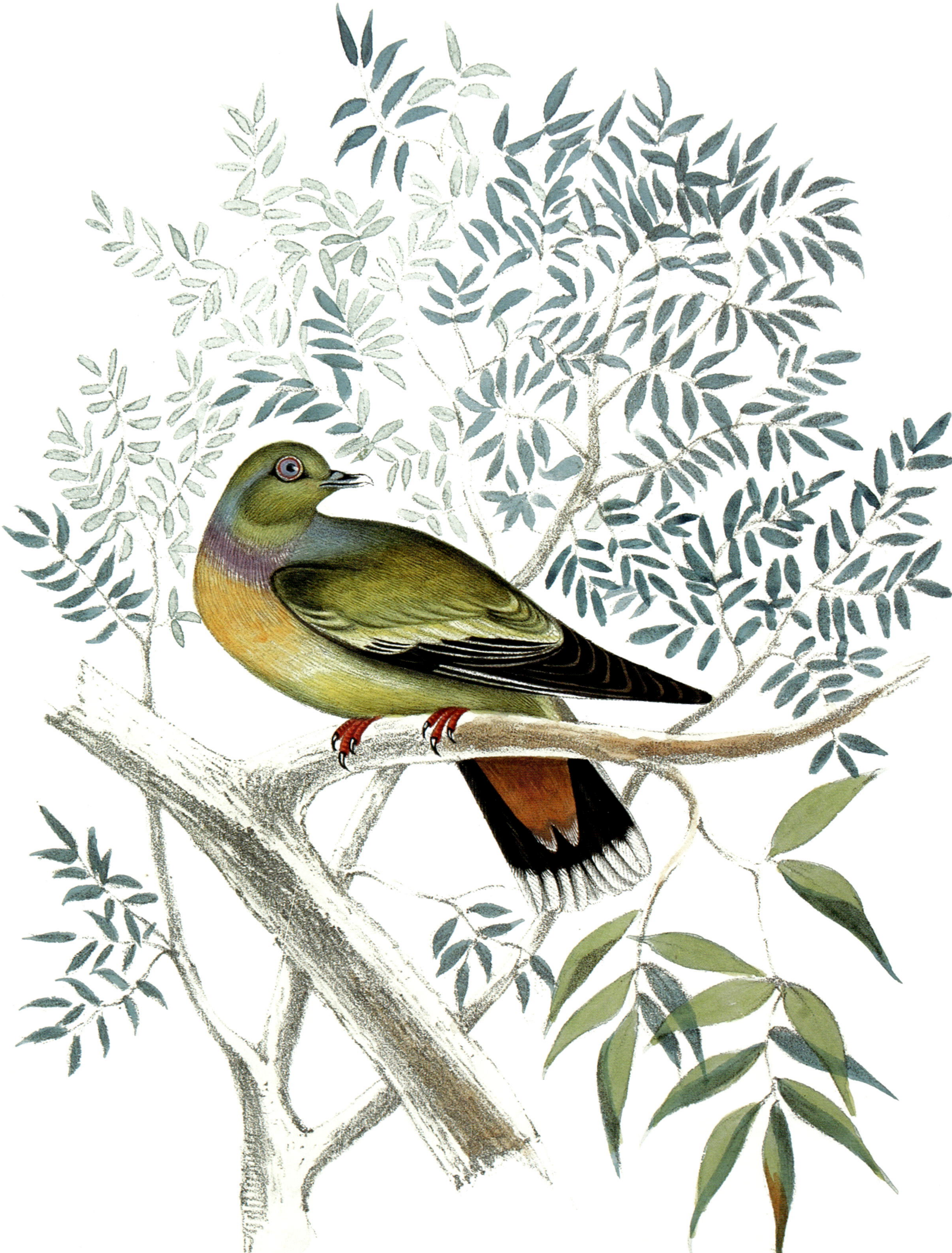

Treron bicincta

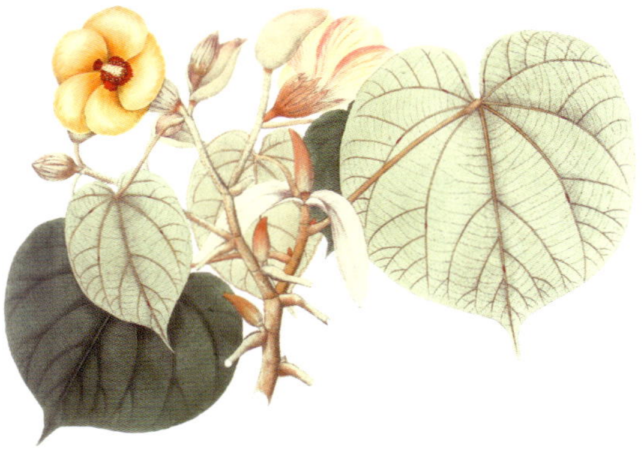

LIVING JEWELS *from* THE INDIAN JUNGLE

edited by

Ashok S. Kothari and Boman F. Chhapgar

BNHS INDIA CONSERVING NATURE SINCE 1883
125 Years

OXFORD
UNIVERSITY PRESS

Souvenir Sub-committee Ashok S. Kothari (Chairman), J.C. Daniel, Asad R. Rahmani, Rishad Naoroji, B.F. Chhapgar, the late Naresh Chaturvedi, Isaac Kehimkar, Vibhuti Dedhia, Nirmala Reddy (Librarian/Convenor)

Technical Guidance M.R. Almeida, Ranjit Manakadan, Varad Giri, S. Balachandran, Vibhu Prakash, Deepak Apte, Saraswathy Unnithan, Swapna Prabhu, Vidyadhar Ogle, Chandrakant Lattoo, Vithoba Hegde, Nikhil Bhopale, Vinod Patil

Library Sub-committee Ashok S. Kothari (Chairman), Rishad Naoroji (Co-Chairman), B.F. Chhapgar, J.C. Daniel (Hon. Secretary), E.A. Kshirsagar(Hon. Treasurer), Asad R. Rahmani (Director), Doreen D'Sa, Monica Bose, Soonoo Taraporewala, Virendra Widge, Nirmala Reddy (Librarian)

Library Staff Nirmala Reddy (Librarian), Tarendra Singh (Library Assistant), Sadanand Shirsat (Library Attendant)

Publicity and Publications Vibhuti Dedhia (General Manager – Publications), J.P.K. Menon (Systems Manager), V. Gopi Naidu (Design Manager), Atul Sathe(Public Relations Officer), Sonali Vadhavkar (Publications Officer), Priyanka Iyer (Publications Assistant)

Chief Coordinators M.G. Mathews (Administrative Manager), Divyesh Parikh (Head – Central Marketing Department), Sachin Kulkarni (Manager – Projects), Deepali Chaubal (Executive PA to the Director), Shalet Alva (Secretary), Sonu Bellani (Membership Officer), Rutika Paradkar (Officer – Front Office), Santosh Mhapsekar (Officer – Sales), Sunil K. Ghanvilkar (Officer – Dispatch), Vasant R. Naik (Administrative Officer), Suresh M. Shetty (Liaison cum Office-Assistant)

Accounts Varsha D. Chalke (Chief Accountant), Shabnam Shaikh, Darshana Patil, Ajit Majgaonkar, Sandeep Dhuri, Manisha Dahiphale, Swapnil Mistri

Additional Assistance Uma Pratap Singh, Ravi Tavde, P. Ranga, Sunder P. Ranga, Shantaram Karambele, Parshuram Parvate, Kamlesh Hasaye, Shyam Jadhav, Parasnath Jaiswal, Vijaya Das

Photography Dharmendra Mistry

ISBN: 0-19-806465-9
Price: Rs. 1600.00

Produced by **Marg Publications** for the Bombay Natural History Society
Text Editor Rivka Israel
Editorial Savita Chandiramani, Gayatri W. Ugra, Arnavaz K. Bhansali
Design Concept, Jacket, and Preliminary Pages Kirti Trivedi
Layout Vidyadhar R. Sawant
Production Gautam V. Jadhav
Data Entry Rajkumari Swamy

Printed at Thomson Press (India) Ltd., Navi Mumbai 400708.

Published by J.C. Daniel, Honorary Secretary,
The Bombay Natural History Society, Hornbill House,
Dr. Sálim Ali Chowk, Shaheed Bhagat Singh Road, Mumbai 400023
and copublished by Manzar Khan, Oxford University Press, YMCA Library Building,
Jai Singh Road, New Delhi 110001.

Oxford University Press, Walton Street, Oxford OX2 6DP
Oxford, New York, Auckland, Bangkok, Buenos Aires, Cape Town, Chennai, Dar es Salaam, Delhi, Hong Kong, Istanbul, Karachi, Kolkata, Kuala Lumpur, Madrid, Melbourne, Mexico City, Mumbai, Nairobi, Sao Paulo, Shanghai, Taipei, Tokyo, Toronto

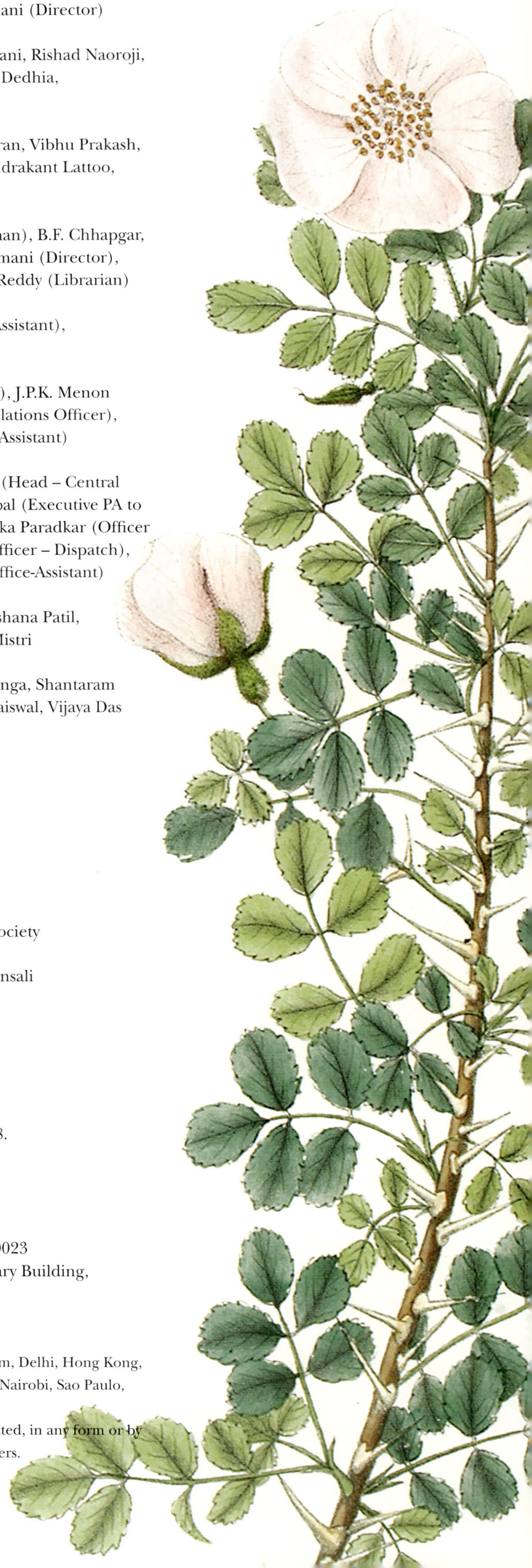

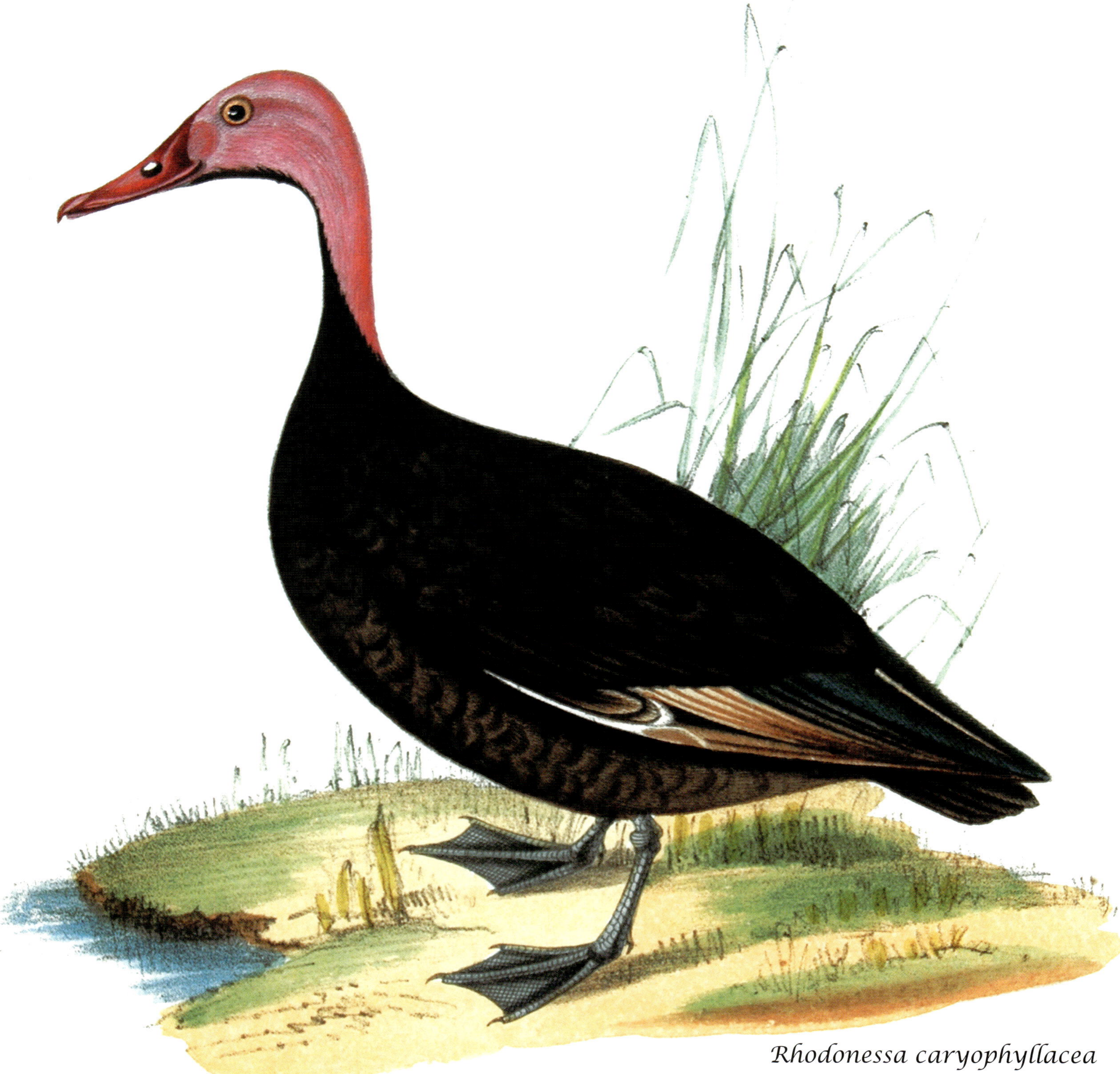

Rhodonessa caryophyllacea

This book is dedicated to
the Pink-headed Duck,
which we may never see again

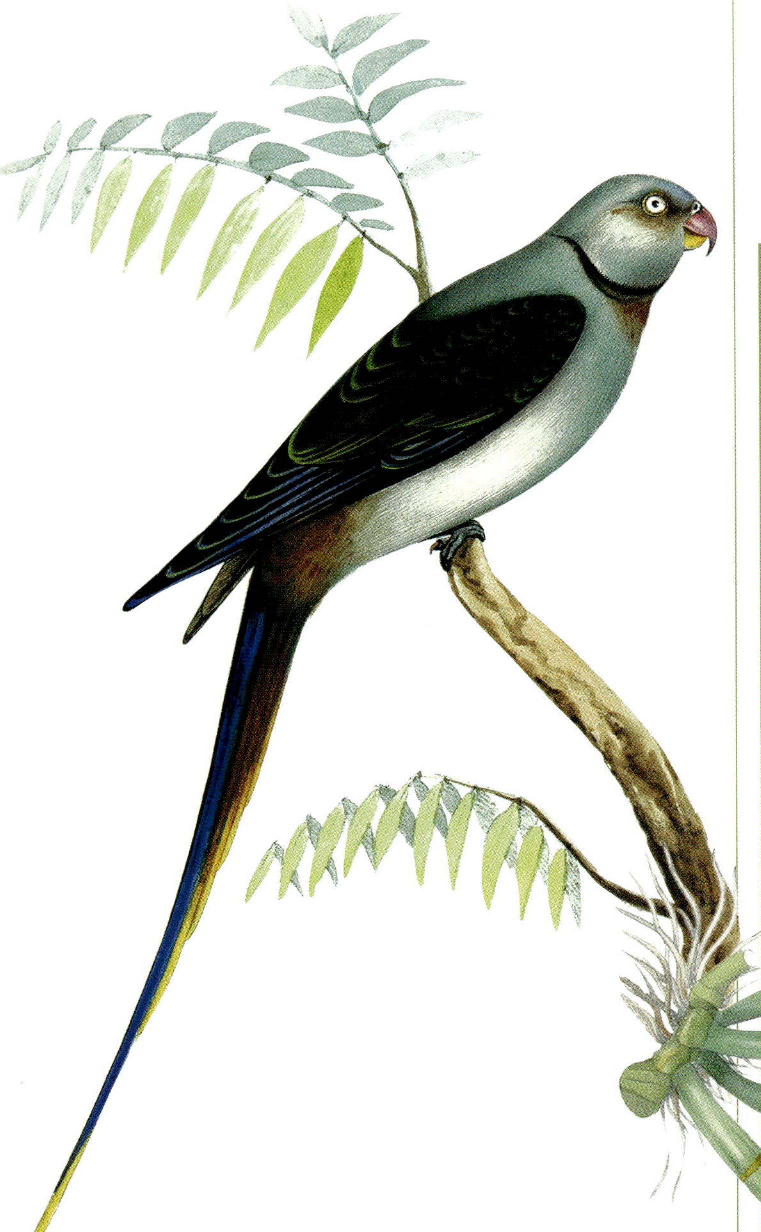

Contents

ARTICLES

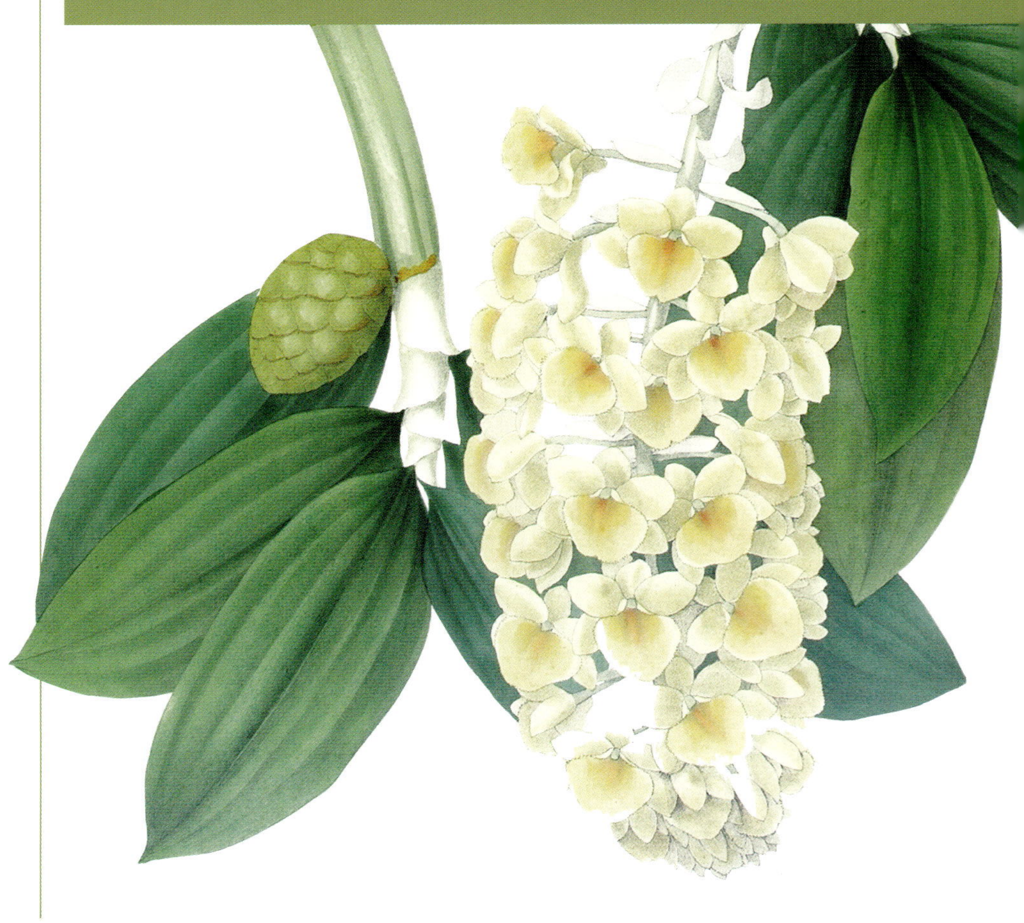

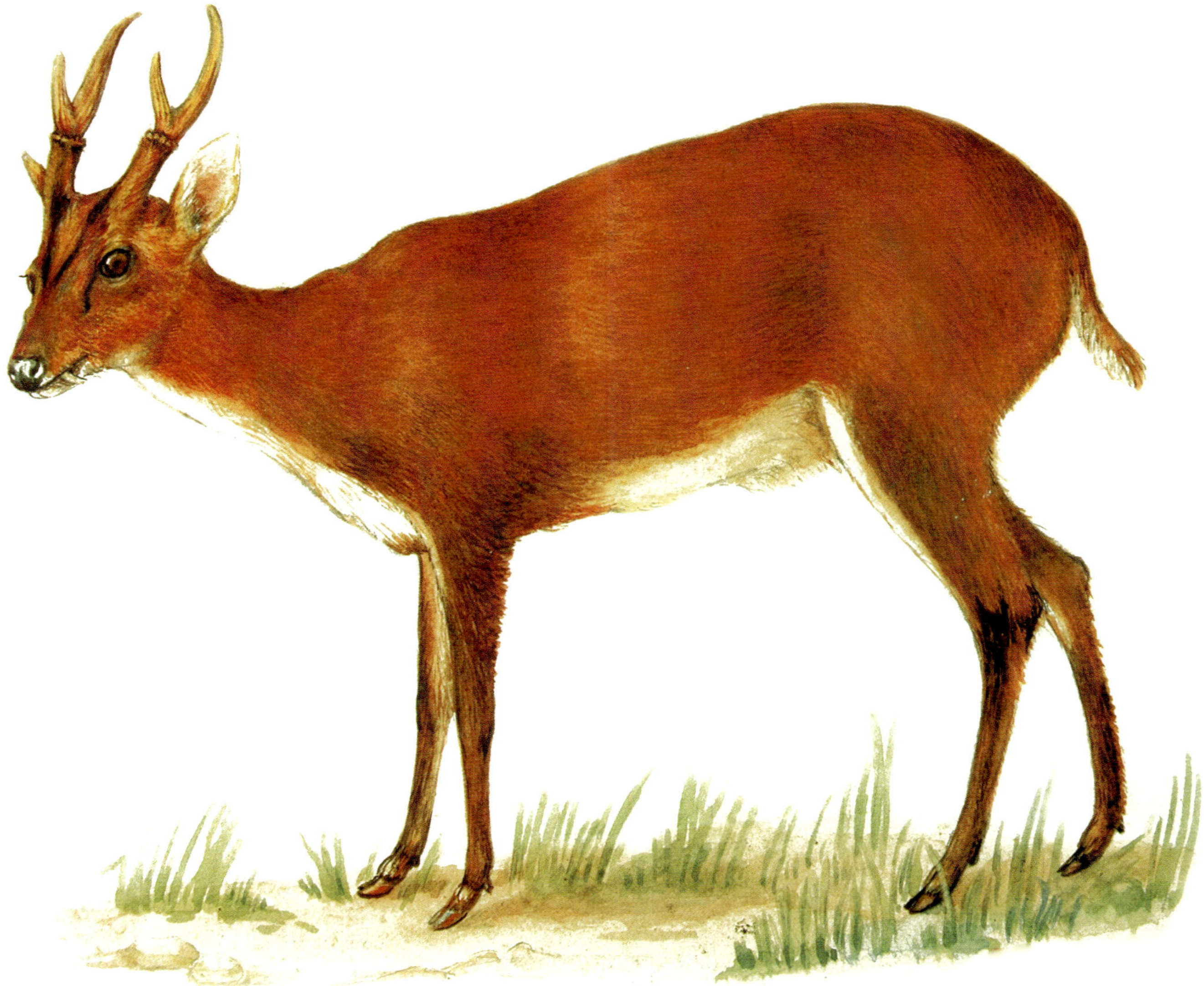

INTERLUDE

NOTES from the *Journal of the BNHS*

Colour Plates

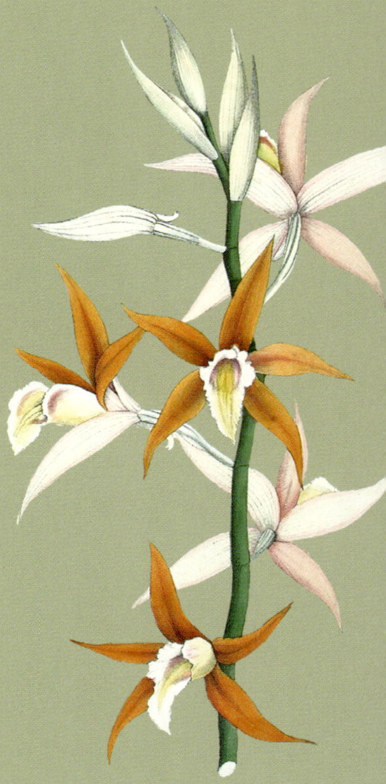

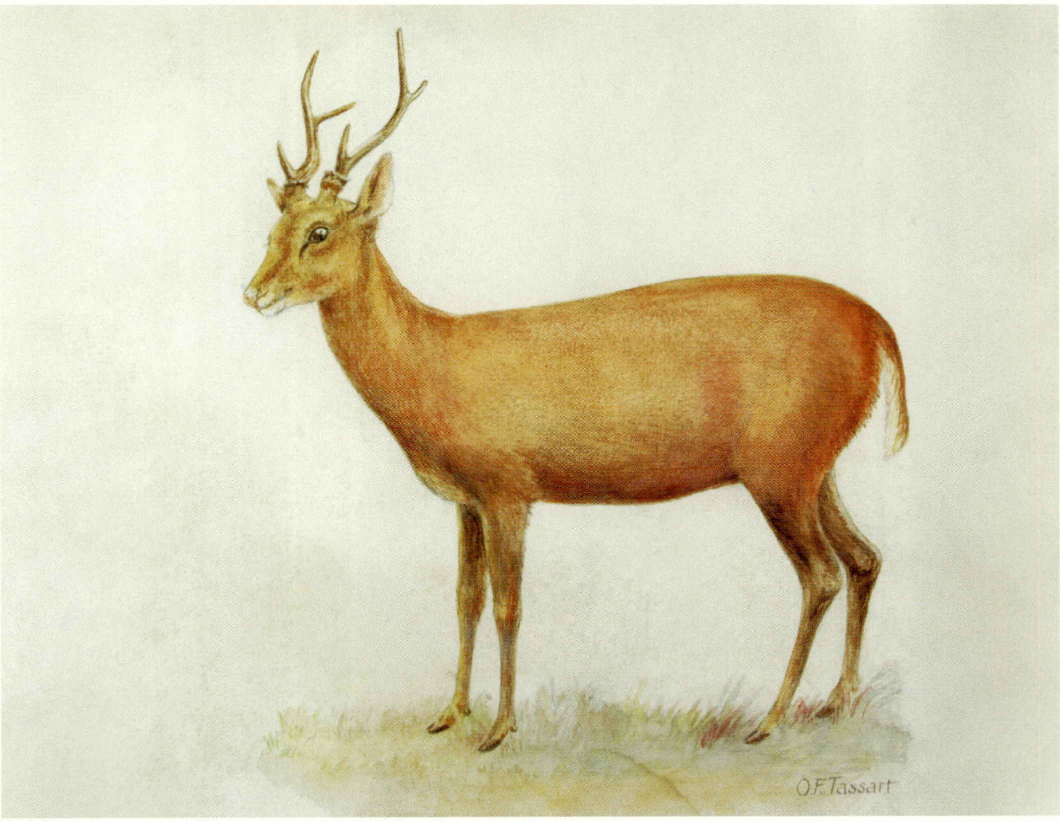

Hog Deer
Axis porcinus

Painting by O.F. Tassart, from *JBNHS* Vol. 37, No. 4 (Supplement, *The Wild Animals of the Indian Empire and the Problem of their Preservation*, Part II).

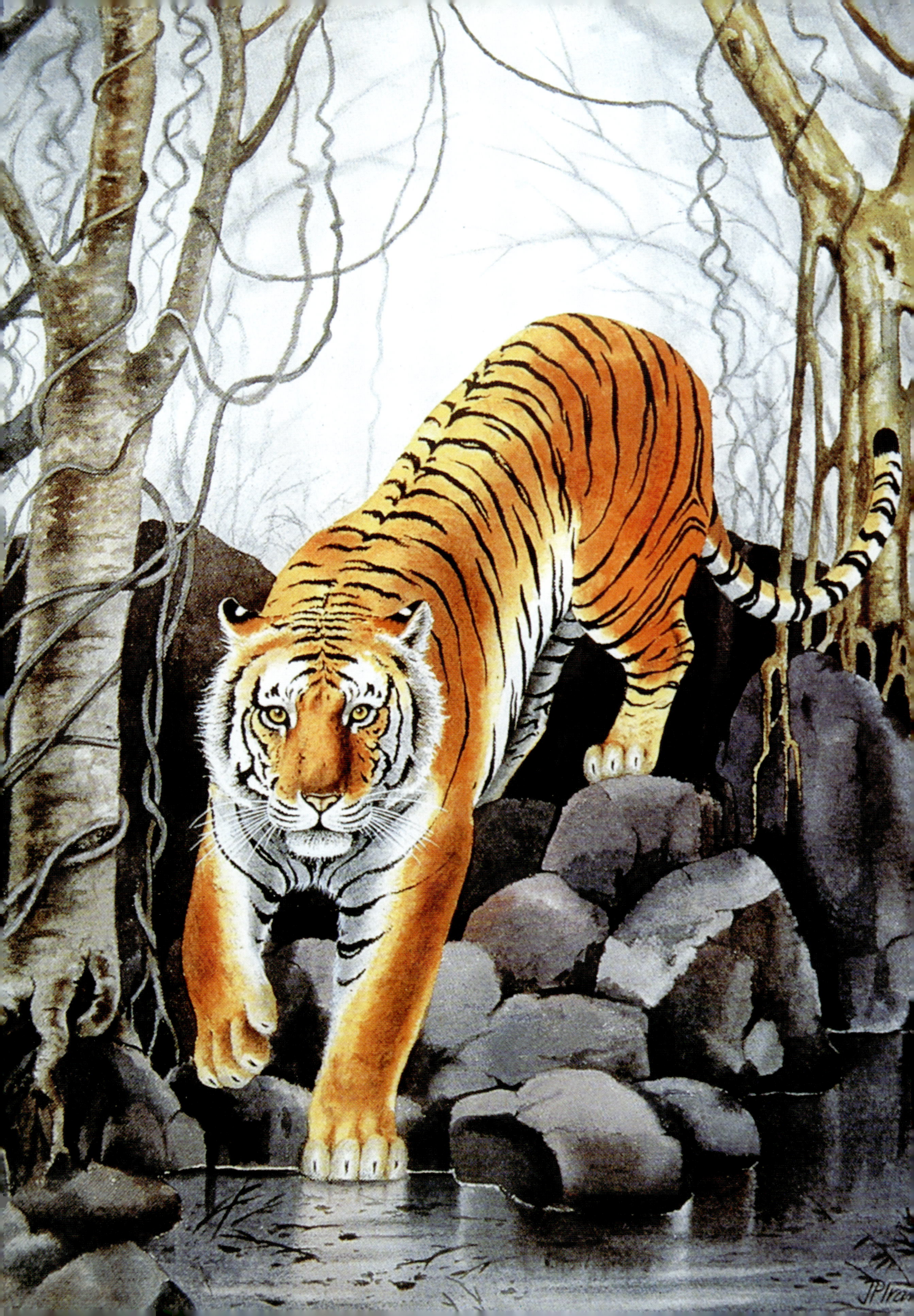

Foreword

Tiger
Panthera tigris

Painting by Jamshed P. Irani. Reproduced by kind permission of Larsen & Toubro.

This book celebrates the completion of 125 years of existence by the Bombay Natural History Society. During this period, members have helped the Society in many ways. Some have offered monetary assistance; others have given their time and rendered voluntary service. Still others have donated books from their personal collections. There are many such, but two immediately spring to my mind: Mr. F.V. Evans, of the famous pre-Independence department store Evans Fraser situated in Bombay (now Mumbai) and Mr. Van Ingen, of the equally illustrious taxidermy firm of the same name in Mysore that was responsible for mounting many of the animals displayed in museums and in maharajas' drawing rooms all over India.

Life then was quite different. Unlike today's fast-track, frenetic race against time trying to meet deadlines, life in those days was lived at a slow and leisurely pace, with plenty of time to indulge in hobbies and pastimes. One favourite pastime was, of course, big game hunting. And the animal of choice was predominantly the tiger. We should not forget that, in those days, as "Fleetwood" so succinctly put it, tigers were "plentiful as blackberries".

There were an estimated 40,000 tigers in India at the beginning of the 20th century. Our colonial masters have been held responsible for the drastic decrease in the tiger population. Had this been true, the tiger population should have bounced back to normalcy after 1947, when the British left India. However, the statistics after Independence, particularly those of the last decade, tell a different, grim and sordid tale. India had 3,805 tigers in 1999, 3,642 in 2002, and 1,411 in 2006. The tiger is extinct in large portions of its historical habitat. In 28 tiger reserves, 251 sq. km of moderately dense forest were lost between 1997 and 2002, and in 21 out of the 28 reserves, forest cover within 10 km of the periphery of the sanctuaries dropped by 124 sq. km over the same period. Further, over 94 sq. km of tiger habitat within protected areas has been lost between 1997 and 2002. To add to the crisis, rampant poaching has played havoc with tiger populations. The shameful reports on the complete disappearance of the tiger from Sariska and Panna National Parks are alarming portents.

The tiger has been a victim of over-publicity. Clubbed together with the catch-words "environment" and "conservation", the tiger has become a handy stepping-stone for some to climb on the bandwagon of publicity. While this helps to spread the message, what is urgently required is political will and action in a broader perspective. Despite the drastic decrease in numbers, the tiger is not the most threatened animal in India today. Even among mammals, the Asiatic Lion (now confined to Gir) and the Gangetic River Dolphin are far fewer in number. Yet not many people speak out for them.

As is best exemplified by Jim Corbett, many hunters later relinquished the rifle, and replaced shikar with more humane options. Their skills honed by years of studying the habitat and behaviour of their quarry, they made copious notes, which were printed in books. Thus many "shikar" books contain a mine of information on Indian wildlife in those days. Today many "shoot" animals on camera; these photographs are more educative and impressive than the lifeless trophies on the walls of colonial bungalows. In pre-camera days those endowed with artistic talent made beautiful paintings of wildlife, especially birds. Many of these paintings were also printed in books. These books, already expensive when they were published, today are priceless collectors' items eagerly sought by connoisseurs.

Dr. Ashok S. Kothari and Dr. B.F. Chhapgar, editors of the Society's earlier books *Sálim Ali's India* and *Treasures of Indian Wildlife,* have once again teamed up to produce this "jewel" of a book on India's natural heritage. I sincerely hope it also motivates young India to protect them.

B.G. Deshmukh, I.A.S. (Retd.)
President

Mumbai, September 2009

Preface

This is our third venture in compiling a book on Indian Natural History comprising vintage paintings and excerpts from books, magazines, and journals, some of which go back to the late 18th and early 19th century. Encouraged by the success of our earlier books, *Sálim Ali's India* and *Treasures of Indian Wildlife*, we were prompted to prepare this volume to celebrate the 125th year of our Society's existence.

The BNHS Library has a wonderful collection of old books, thanks in large part to the munificence of donor members. Among these are *A Century of Birds from the Himalaya Mountains* (1832) and *The Birds of Asia* (1850–73) by John and Elizabeth Gould, *Fifty Illustrations from Indian Ornithology* by T.C. Jerdon (1844), *Oriental Memoirs* by James Forbes (1812–13), *Scenery, Costumes and Architecture Chiefly on the Western Side of India* by Captain Robert Melville Grindlay (1826), *Curry and Rice* by Captain George Francklin Atkinson (1859), and *Oriental Field Sports* by Captain Thomas Williamson (2nd edition 1819).

Reviewers of our earlier books had commented that we had given disproportionate weightage to birds and tigers. We agree, but there are reasons for this.

Birds are non-controversial; one cannot hate a bird (not even the ubiquitous, pesky crow!). They make up by their abundance what they lack in size. One does not have to go into the deep forest to watch them. You can see them from your window if there is a tree or bush nearby. And finally, like butterflies and coral reef fishes, many of them are brilliantly, even gaudily, coloured. Thus they make excellent subjects for artists.

Tigers may not be so non-controversial. After all, they have been feared for centuries as predators, cattle-lifters, and even man-eaters. They are not the biggest or mightiest among the beasts of the jungle; elephants and rhinos come in that class. But they have a charisma all their own. No one who has seen a tiger walking silently, effortlessly, along a jungle trail, muscles rippling under its beautifully striped coat, can fail to be hypnotized by it. Compare this with the ungainly gait and the wrinkled grey skin of the elephant or the equally drab thick skin of the rhino hanging in loose folds like a suit stitched by an apprentice tailor. Elephant, rhino, and other large mammals inspire awe in their own right, but the tiger is the darling of naturalists.

Some readers may have also wondered why we did not include plants in the last two volumes. This was simply because we did not have suitable matter – or so we thought. Last year, in the Library, we chanced upon a parcel wrapped in newspaper and tied carefully with string. On opening it, we were thrilled to find a cache of veritable jewels: N. Wallich's *Plantae Asiaticae Rariores* (1832), J. Forbes Royle's *Illustrations of the Botany and Other Branches of Natural History of the Himalaya Mountains and of the Flora of Cashmere* (2 volumes, 1839), and J.D. Hooker's *Illustrations of Himalayan Plants* (1855). So this time we have been able to amply redress the balance by including a large number of botanical plates, and we are sure that readers will appreciate them.

Rudyard Kipling, one of the most famous "Anglo-Indian" poet-writers of all time, is familiar to all of us as the author of *The Jungle Books.* He was born in Bombay, where his father Lockwood Kipling was Principal of the Sir J.J. School of Art. He imbibed Indian culture and traditions from his long stay in India, and his *Jungle Books* in particular illustrate his deep love and understanding of the Indian jungles through the adventures of Mowgli and animal characters like Akela the wolf, Baloo (bear), Bagheera (panther), Sher Khan (tiger), Hathi (elephant), Chil (kite), Nag (cobra), and Dharzi (tailor bird). To provide some comic relief after some of the hair-raising accounts of the first section, we have included one of Kipling's *Just So Stories,* which lampoons the rhino. Incidentally, Pestonjee Bomonjee, the baker in the story, was in real life a student in Lockwood Kipling's art and sculpture classes.

Preparing this book has given us immense pleasure. We hope that readers will get as much pleasure from reading it.

Dr. Ashok S. Kothari
Dr. Boman F. Chhapgar

Mumbai, September 2009

Acknowledgements

It is a major task to thank all those who have helped and guided us in this endeavour. Their suggestions have helped give final shape to this volume. Three people who are not in this world today – Mr. Sorabjee Godrej, Mrs. Vijaya Deshmukh, and Mr. J.S. Serrao – provided constant moral support and inspiration when we first began our explorations of the Library's rich collection 15 years ago (from the time of *Sálim Ali's India*). Our President, Mr. B.G. Deshmukh supported the idea of this volume as he had done for the previous ones. Director Dr. Asad Rahmani provided necessary help at every stage and also made valuable suggestions. Vice President Mrs. Pheroza J. Godrej as always provided generous guidance. Members of the Advisory Committee (Mr. J.C. Daniel, Mr. E.A. Kshirsagar, Mr. Rishad Naoroji), Ms. Vibhuti Dedhia, the late Mr. Naresh Chaturvedi, Mr. Isaac Kehimkar, and Ms. Nirmala Reddy also made important contributions. Dr. Ranjit Manakadan, Dr. M.R. Almeida, Dr. Swapna Prabhu, and Mr. Varad Giri readily provided necessary information. Solid assistance was given by Mr. M.G. Mathews, Mr. Sachin Kulkarni, Ms. Deepali Chaubal, Ms. Shalet Alva, Ms. Rutika Paradkar, Mr. Vasant Nayak, Mr. Santosh Mhapsekar, Mr. Sunil Ghanvilkar, and Mr. Suresh Shetty. Mr. Vasant Naik also worked hard in getting sponsors for the plates. The Executive Committees of 2006–08 and 2008–10 supported the project. The Library staff in particular has provided excellent and dedicated support. Ms. Nirmala Reddy, our sincere Librarian, Mr. Tarendra Singh, Library Assistant, and Mr. Sadanand Shirsat, Library Attendant worked very hard. Tarendra and Sadanand have many years of experience of library work behind them. They helped at every stage, keeping the required books and information ready for us. Mr. Dharmendra Mistry once again competently photographed the paintings.

Our sponsors have been particularly generous with their help. The Palanpuri community, and particularly diamond merchants, have always extended financial support whenever approached. They are Mr. Kavin Parikh, Mr. Harshad Mehta, Mr. Arunbhai Mehta, Mr. Kamlesh Jhaveri, Mr Pareshbhai Mehta, Mr. Ramnikbhai Shah, Mr Dineshbhai Shah, Mr Arvindbhai Shah, Mr Jitendra Bhansali, Dr. Vikram Mehta, Mr. Ashish K. Mehta, Mr. Dilipkumar Lakhi, Mr. Suresh Kothari, Mr. Raj Shekhar Parikh, Mr. Premal Parikh, Mr. Satish Choksey, Mr. Chhotalal P. Shah, Mr. Piyush Kothari, Mr. Devang Shah, Mr. Nayan Kothari, and Mr. Zubin Mehta. Well-wishers of BNHS, Ms. Malini Thadani, Head, Group Communications and Corporate Sustainability, Mr. Unmesh Brahme, Senior Vice President, Corporate Sustainability, and Ms. Vandana Bhansali, Associate Vice President of HSBC were the first to support this project. This time we received support from Mr. Suresh Pai, Executive Vice President, Mr. Paul Abraham, Chief Operating Officer, Ms. Sunita Prabhu, Chief Manager, and Ms. Hetal Pandya, Deputy Manager of IndusInd Bank. Various other banks and corporate houses also gave generous support in spite of the economic recession.

Several individuals gave wholeheartedly for this publication. A BNHS member and ex-Principal of Bombay International School, Ms. Dhun Panthaki made an extremely generous donation. It is not possible to personally thank all the other donors here, but it must be said that the task of enriching this book with rare prints and interesting articles was completed without any financial burden on the Society thanks to their generosity. Mr. Rajshekhar Parikh, Founder President of Share & Care, New York, which has helped many institutions, enquired often about the progress of the volume despite delicate health. Among others Hansa Kothari, Hardik Kapadia, Rajal Kapadia, Sheetal Shah, and Mamta Mehta extended valuable help.

Most of the prints reproduced here were available to us thanks to the late Mr. F.V. Evans who donated his treasure trove of books to the Society in the 1920s. The volumes are well preserved and not exposed to the public eye, except on rare occasions. Over the years, Ms. Dilnavaz Mehta of Rare Finds, Mumbai and Mr Anjan Gupta of Messrs Ranjan Gupta & Co., Kolkata helped us in our search for rare and useful books for the BNHS Library.

Finally, we thank Marg for their continued support and for giving priority to this project. Ms. Radhika Sabavala, General Manager, and her dedicated team have played a key role in the making of this book. Ms. Rivka Israel and Dr. Gayatri Ugra had suggested that we include flowering plants in this book. Dr. Ugra was Publications Officer in BNHS for a long time. Rivka with her deep knowledge, experience, sincerity of purpose, and cheerful disposition, has been involved in various stages of producing this book. The Designer, Professor Kirti Trivedi, infused new ideas in the presentation of pictures and text. He was ably supported in production and layout by Mr. Gautam V. Jadhav and Mr. Vidyadhar R. Sawant.

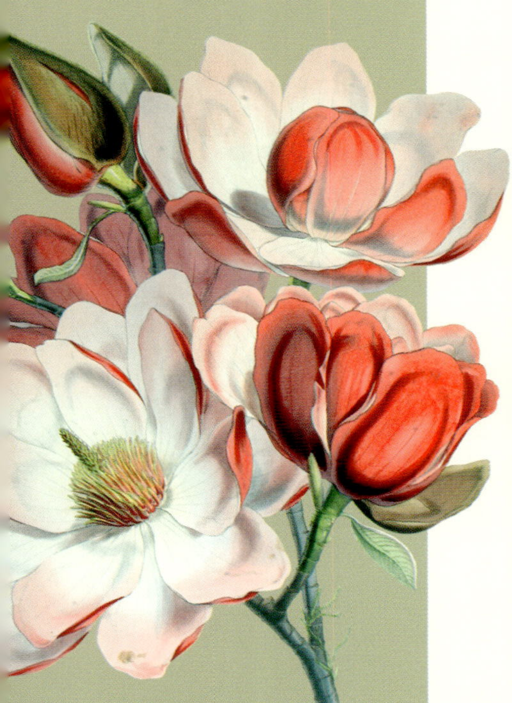

Discovery of *Amherstia nobilis*

Nathaniel Wallich

THE first notice I had of the existence of this magnificent tree was at Rangoon, in August 1826; when Mr. Crawfurd favoured me with some dried unopened flowers and a leaf of it, together with the following account, taken from a statement of his visit to Martaban, contained in the Calcutta Government Gazette of the 29th May of that year.

"Around the hill" at Kogun, on the Saluen river in the province of Martaban, "is a garden belonging to a neighbouring monastery, in no very good order. The only plant in it, which struck us as remarkable, was a tree about twenty feet high, abounding in long and pendulous panicles of rich geranium-coloured blossoms, and long and elegant lance-shaped leaves. It is of the class and order of Diadelphia Decandria, and too beautiful an object to be passed unobserved even by the uninitiated in botany. Handfuls of the flowers were found as offerings in the cave before the images of Buddha."

Captain Hitchens, of the Honourable Company's Madras military service, was good enough to give me a promise at the time, that he would if possible obtain for me, against my arrival on the Tenasserim coast, some more dried specimens; which he in part performed. In the mean time I accompanied the British Envoy to Ava; but neither while we stayed there, nor during our progress on the Irawaddi, could I procure the slightest additional information concerning the tree. It was not even known by name at the capital of the Burma empire.

The subject of these pages is so interesting that I venture to make one more extract respecting it, from my official report of a journey on the river Saluen, undertaken in March 1827, in order to examine the site and capabilities of the Teak-forests in that direction.

"In about an hour I came to a decayed Kioum (a sort of monastery), close to the large hill of Kogun, distant about two miles from the right bank of the river, and twenty-seven from the town of Martaban. I had been prepared to find a tree growing here, an account of which had long ago been communicated to me by Mr. Crawfurd, and which I had been fortunate enough to meet with for the first time a week ago at Martaban; nor was I disappointed. There were two individuals of this tree here; the largest, about forty feet high, with a girth, at three feet above the base, of six feet, stood close to the cave; the other was smaller, and overhung an old square reservoir of water, lined with bricks and stones. They were profusely ornamented with pendulous racemes of large vermilion-coloured blossoms, forming superb objects, unequalled in the Flora of the East Indies, and, I presume, not surpassed in magnificence and elegance in any part of the world. I call this tree *Amherstia nobilis*. The Burmese name is *Thoka*. Neither the people here nor

Amherstia nobilis, popularly known as the Tree of Heaven, or the Pride of Burma, is considered by many the most beautiful tree in the world. When the tree is in bloom tree-lovers flock to behold its beauty. In Mumbai there are seven individuals of this species, of which three are at the Jijamata Udyan, Byculla, and two in the garden of the Bombay University Fort campus. A tree-lover Mr Kamlesh Jhaveri has successfully raised two trees at his bungalow garden in Juhu.

Tree of Heaven, Pride of Burma, Orchid Tree
Amherstia nobilis
Family: Fabaceae

The illustration shows a branch with a leaf and a raceme, much reduced; and a leaflet, natural size.

From N. Wallich, *Plantae Asiaticae Rariores*, 1832.

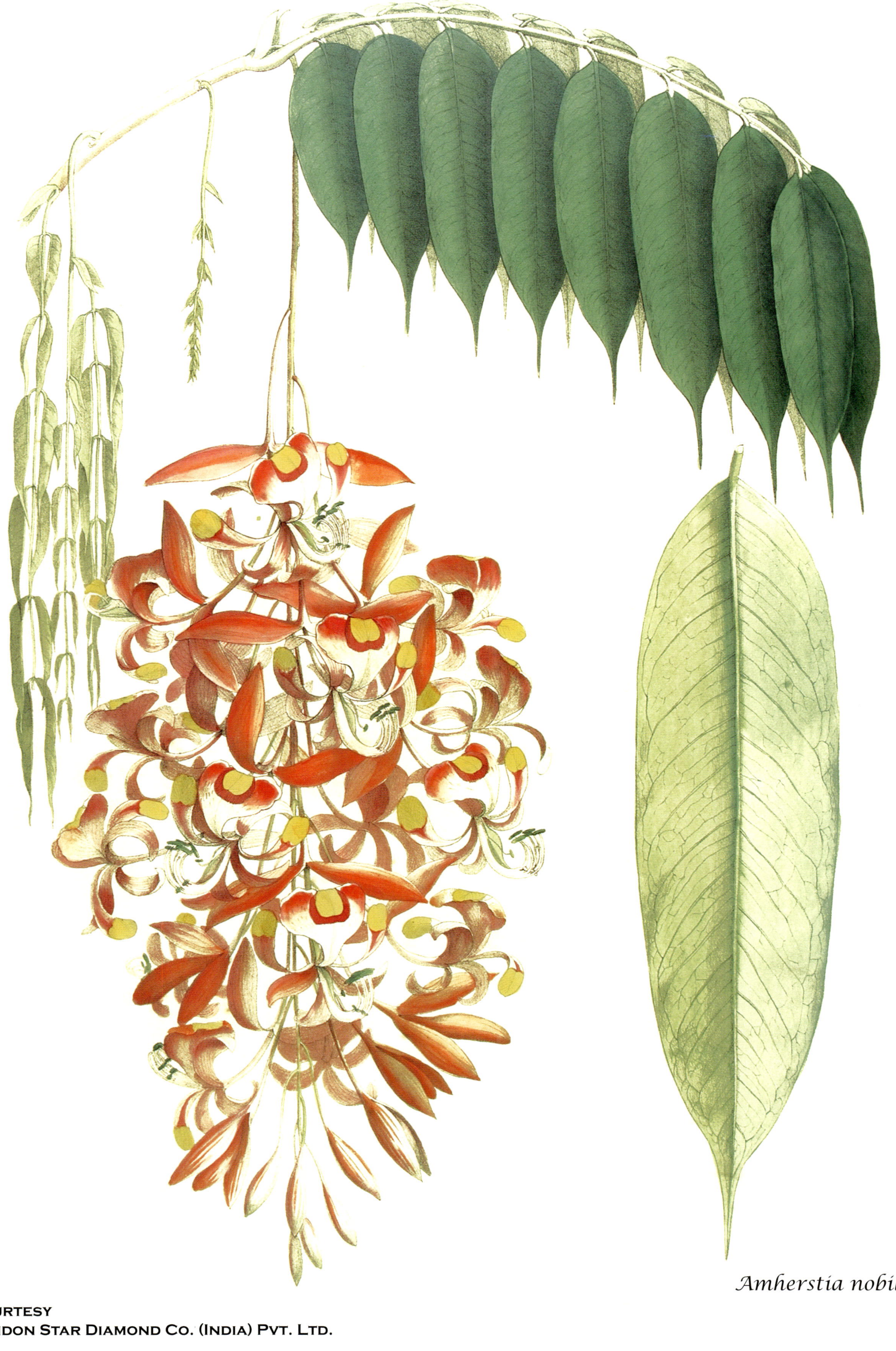

Amherstia nobilis

COURTESY
LONDON STAR DIAMOND CO. (INDIA) PVT. LTD.

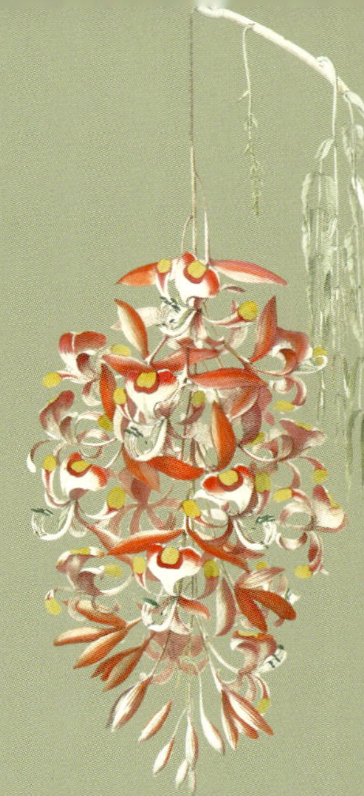

at Martaban could give me any distinct account of its native place of growth; but there is little doubt that it belongs to the forests of this province. The ground was strewed, even at a distance, with its blossoms, which are carried daily as offerings to the images in the adjoining caves. Round the spot were numerous individuals of *Jonesia asoca* [now *Saraca asoca*] in full blossom, inferior in beauty only to those trees; also several Nagesurs (*Mesua ferrea*). It is not a little remarkable, that the priests in these parts should have manifested so good a taste as to select two sorts of trees as ornaments to their objects of worship, belonging to a small but well-marked and extremely beautiful group in the extensive family of leguminous plants. At Martaban I found a third member of the same group, also of great beauty, *Bauhinia variegata* planted close to the *Amherstia.*

There can be no question that this tree, when in full foliage and blossom, is the most strikingly superb object which can possibly be imagined. I have very great satisfaction in naming it after the Right Honourable Countess Amherst and her daughter Lady Sarah Amherst, the zealous friends and constant promoters of all the branches of Natural History, especially Botany, who, after a residence of nearly five years in India, during which they performed an arduous and extensive journey to the lofty regions of western and northern Hindustan, spending many weeks among the mountains near the Himalaya, at an elevation of 10,000 to 12,000 feet, have returned to England with a large and very interesting collection of preserved specimens of plants, gathered and excellently preserved by their own skill and industry.

The leaves are ample, pinnate, and of a dark green colour; while young they are glaucous-purple and hang down loosely, together with the tender shoots to which they are attached. The leaflets are large, oblong, tapering into a most slender point, very glaucous underneath, and furnished with slender prominent rib and nerves, the latter uniting towards the margin in a most elegant manner. The flowers are numerous and very large, scentless, of a brilliant vermilion colour, diversified with three yellow spots; they are arranged in gigantic ovate bunches, pendulous on their long peduncle, and partially hidden beneath the profuse and elegant foliage. I have only seen one single pod, which was unripe; it was sword-shaped, flattened, with one of the margins thick and ribbed, seven inches long, pointed, dark red, containing five or six imperfect seeds.

By the assistance of my people I was able to procure some layers for the botanic garden at Calcutta; two of these I attempted to bring with me to England, but although they produced fresh shoots soon after their embarkation, they perished early during the voyage.

The genus is nearly allied to *Heterostemon* of Desfontaines, but differs in the large valvate floral bracts, five-petalled irregular corolla, ten diadelphous stamens, and finally in having all the anthers polliniferous. Together with that genus, *Jonesia, Bauhinia,* and some others having a distinctly tubular calyx, it may perhaps be fitly separated into a distinct order, to which I would propose the designation of *Bauhinieae.*

From Nathaniel Wallich, *Plantae Asiaticae Rariores, or, Descriptions and figures of a select number of unpublished East Indian plants,* 1832.

The Glory Lily

K.R. Kirtikar

Gloriosa superba (Linn.)
Marathi – *khadyanag*
(Natural Order – Liliaceae)

THE plant is an annual climber, wild in hedges and jungles; often cultivated in gardens for its beautiful flower.

ROOT – Tuberous, fleshy, budding from the convexity above; tubers cylindrical or flattened, in some parts sometimes slightly twisted, 4 to 9 inches long; about an inch thick. Substance internally white, mealy, juicy, with a slight acrid odour. The external appearance of the tuber is brown. This is due to the thin paper-like epidermis which encloses it at all parts of the bulb, except the growing point, which latter looks like the eye of a potato. The epidermis is easily removed by the mere handling of the bulb, and is fragile. The old bulb, as the plant matures, shrivels up, and gradually throws out a new bulb at right angles which terminates in a growing point. It is from this growing point that the future plant sprouts out. The rootlets are chiefly confined to the part of the bulb directly opposite to where the stem is situated.

STEM – Usually a single stem arises from the bulb. It may branch off almost immediately after it leaves the tuber. The stem is put down as 10–20 ft. by Hooker (*Flora Br. Ind.*), but is often much longer than that in the jungles, where it is seen in its natural condition. When the stem branches off immediately on leaving the tuber underground, three or four slender cylindrical green sprouts shoot up above the ground, thus making it appear that they are separate plants from one and the same bulb; but it is not so. One bulb, one plant is the rule invariably in a number of bulbs I have carefully examined. The stem is nearly cylindrical, and though herbaceous, is tough and interspersed with woody fibres in close-packed bundles. The medulla of the stem is slightly pithy and full of watery juice when fresh; hollow when the plant dries up and the pith shrivels up. At each insertion of the leaf, which is arranged in pairs in an opposite manner, the stem deviates from the median line, thus giving the whole plant an angular or regular wavy appearance.

LEAVES – Sessile or short petioled; the inferior ones opposite and oblong; the superior ones single, ovate-lanceolate; the other leaf replaced by the flower-stalk, when the plant is in flower. The length of the leaves varies from six to eight inches; the greatest breadth is often as much as two inches. *Venation,* parallel-nerved. Mid-rib prominent,

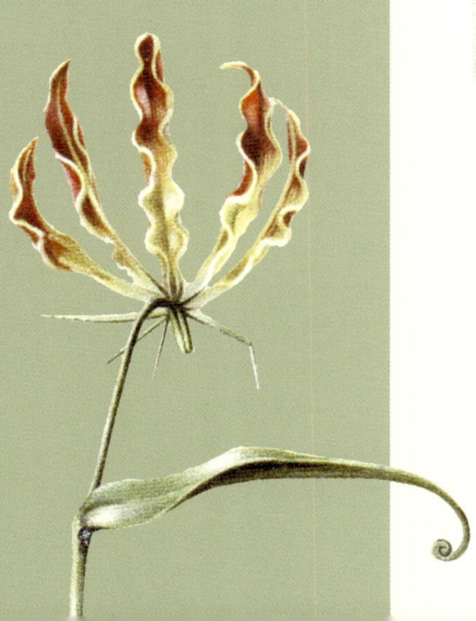

terminating into a stiff spiral coil, which resembles a tendril. This is a marked characteristic of the plant, which, so far as I know, has no parallel in the Vegetable Kingdom.

FLOWERS – Solitary, on long slender variously curved peduncles, 4–6 inches long; deflexed, often giving the flowers with their reflex petals the appearance of a slender-winged butterfly poised in the air. It is this that gives the gorgeous flowers their unique appearance, and shows off their rich, varied, and ever-varying hues.

PERIANTH – Of six petals, persistent. *Flower-buds* conical or pyramidal, bright green.

PETALS – 3–4 inches long; linear-lanceolate, narrow, undulate or crisped; bright green with a pale yellow margin when fresh opened; bright yellow up to a third from their insertion, and scarlet up to tip when full-blown; gradually converting into rich scarlet, with golden-yellow margin and mid-rib when the ovary is maturing into fruit and growing in size. The flower retains its beauty for several days. The petals are persistent, and don't fall off even after the plant has dried. They fall off long after the fruit and seeds are mature. It is this that makes it so appreciable as a forest and garden beauty. The petals, which are reflexed when fresh-open, assume a spreading aspect as they grow older.

FRUIT – A capsule 2 inches long; an inch broad when fully developed; coriaceous; septicidal.

SEEDS – Numerous, subglobose, in dense double rows in each of the three carpels of the capsule. *Testa* spongy white when young, covered with a brilliant rich scarlet mucilaginous coat when the fruit matures.

REMARKS – *Gloriosa superba* is essentially a monsoon plant; it begins to sprout out in the early part of the rainy season, and dies soon after the rains. The bulb remains dormant throughout the cold weather. It flowers and bears fruit from July to September. Should the plant be cultivated as a garden beauty or ornamental trellis-plant, it is best to repot or replant it in February or March, or at the latest in April, when the terminal bud of the tuber begins to sprout out. Dalzell and Gibson call it "Buchnag". This is a mistake. Buchnag is the root of *Aconitum ferox,* which is different in form and appearance. There is no doubt, however, that the tubers of this plant are adulterated with the tubers of the real Buchnag. Dalzell and Gibson also call the plant "*Kalawee*". This is evidently a misprint, and meant for the Marathi synonym "Kala-lawee". The colour of the flowers depends a great deal on the amount of the sun or the intensity of it, to which the flowers are exposed as they are maturing and expanding. The specimen from which our drawing is made was obtained from a plant which flowered under a persistent cloudy sky; the scarlet colour is therefore not quite so deep or bright as it might be, and very often is.

THE POISONOUS PROPERTIES

This plant is one of the nine secondary poisons mentioned by Hindu writers.

Glory Lily
Gloriosa superba
Family: Liliaceae

1. Flowering branch of the plant
2. Root, bilobed; vertical portion being half the shrivelled tuber of the current year; the horizontal portion being the new tuber from the terminal eye of which at the tip of the part marked 3, the future plant will sprout out: from the convexity of the angle rises the green stem of the plant dividing into four branches immediately on escaping from the bulb; from the under surface of the angle are shown numerous rootlets
4. Half-mature fruit cut across about the middle
5. Half-mature capsular fruit *in situ*

From *JBNHS* Vol. 7, No. 4

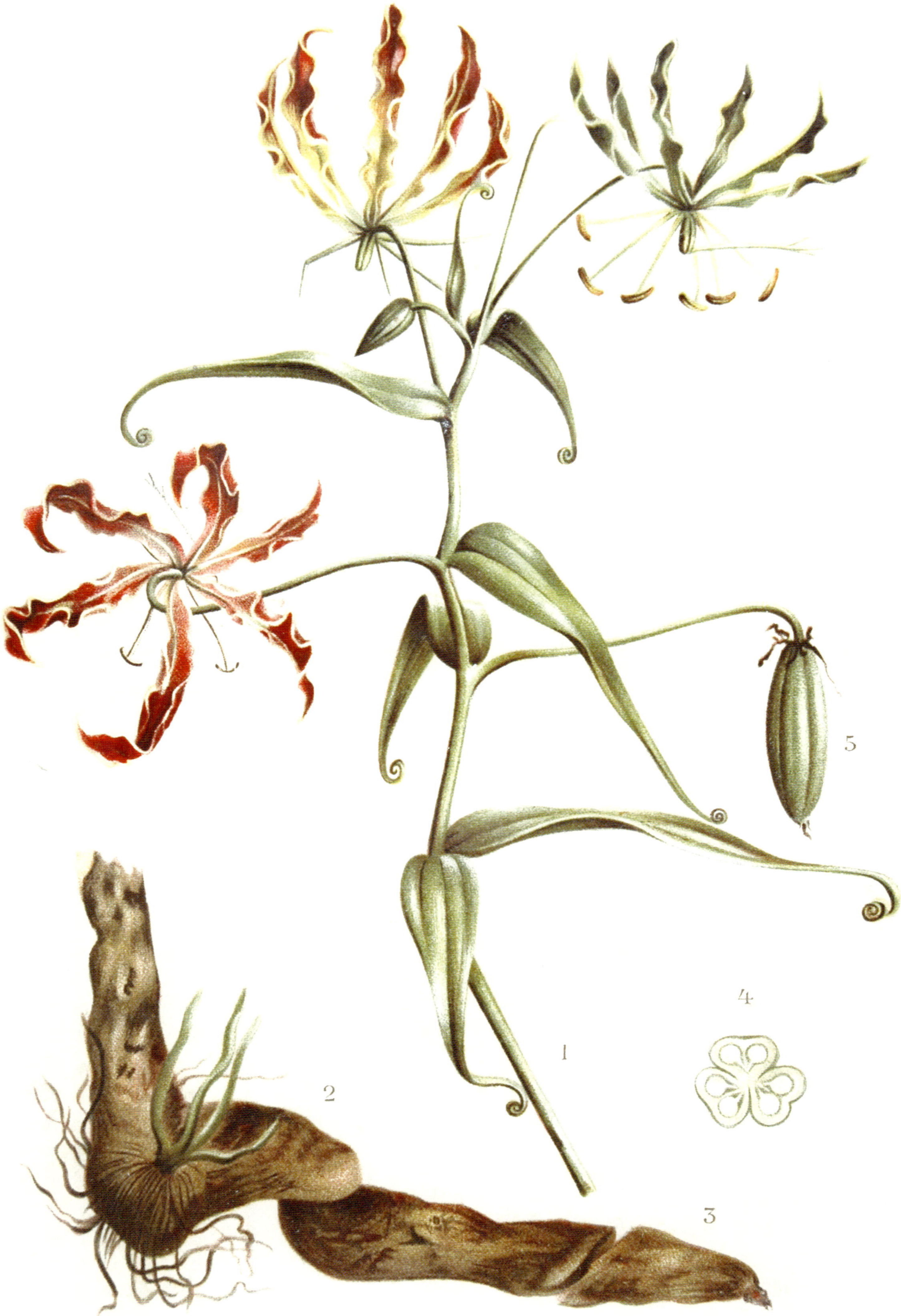

Gloriosa superba

In memory of Mr. Naresh Chaturvedi,
former Curator, Bombay Natural History Society

They are given as follows in *Nighanta Ratnakar* (Vol. III, p. 255):– 1. Euphorbia neriifolia. 2. Calotropis gigantea. 3. Gloriosa superba. 4. Abrus precatorius, red and white. 5. Nereum odorum. 6. Strychnos nux-vomica. 7. Datura alba and fastuosa. 8. Jatropha curcas. 9. Papaver somniferum (its product opium).

Very few cases of poisoning are on record, but the plant is well known all over the country as a virulent poison, although it is said to be useful in fevers as an antiperiodic. Dr. Norman Chevers mentions two cases (*Medical Jurisprudence,* p. 284), in one of which the effects are stated to have been those of a Narcotico-irritant poison; it is not stated what poisonous symptoms were observed in the other. In the one 1/10 tola and in the other 2 tolas of the root caused death. Brigade-Surgeon Lyon mentions (*Medical Jurisprudence,* p. 454) a case reported in the *Indian Medical Gazette,* 1872, p. 153, in which the powdered root (quantity not mentioned) produced the following effects:– "Symptoms of poisoning appeared in half an hour, and were:– retching, violent vomiting, spasms and contortions of the body, with fearful racking pain; from time to time there were short intervals of relief, followed by recurrence of the same symptoms. Death took place in four hours. The *post-mortem* appearances were congestion of the brain and its membranes with extravasations of blood. The lungs, liver, and kidneys were all deeply congested. The gastric mucous membrane showed signs of inflammation. The peritoneal covering of the fundus of uterus (unimpregnated) was also found inflamed."

Ancient Hindu writers agree in attributing violent emetic properties to the root; it is also said to cause abortion, and as such prescribed by Hindu physicians for expelling retained after-births. Dr. Dymock, in citing the researches of Warden, says that there are two resins and a bitter principle in the root. Warden names this bitter principle *superbine,* and considers it identical with that of *Urginea scilla,* the ordinary Squill. The bitter active principle of Squill, says Dr. Lauder-Brunton, is a glucoside Scillitoxin or Scillain. Squill is classed by Lauder-Brunton among stimulants of the Cardiac muscle, when moderate doses are given. When, on the other hand, larger doses are given, it acts as a "Cardiac poison" (p. 276, *op. cit.*). In such cases, that is to say, where the dose is large, "the stage of stimulation is followed by one of peristaltic action and final arrest in Systole." It is a pity the case quoted by Dr. Lyon gives no information regarding the *post-mortem* condition of the heart, although the condition of every other organ is noted. It will be useful to note this condition in all future observations with a view to verify the opinion of Warden. In excessive doses, "the operation of Squills," says Dr. Waring (*Therapeutics,* p. 489), "is that of an acro-narcotic poison, 24 grains having proved fatal." Squill is known to cause nausea; a small dose may act with extreme violence. I have already said that the plant is well known among the Hindus as possessing emetic properties. The conjecture of Warden is in my opinion based on a striking similarity of the physiological effects of the two plants on the human body.

From K.R. Kirtikar, "The Poisonous Plants of Bombay", *JBNHS* Vol. 7, No. 4 (1898).

Meconopsis simplicifolia
Family: Papaveraceae

This plant is found in the Darjeeling and Sikkim Himalaya ascending to 4,000 to 5,000 m in Chemthang and Gulchi La and also in northern Sikkim. It is a fairly common upper alpine species from central Nepal to southeast Tibet, 3,300–4,500 m. It has long, narrow leaves tapering at both ends, irregularly toothed, with bristly hairs. The plant flowers in June–July; the purple to sky-blue blooms are solitary, borne on long, bristly stalks arising from a rosette of leaves. They have 5–8 obovate petals, 2–4 cm long. This is the national flower of Bhutan. The capsule is erect, narrow-oblong, 5 cm. In Hooker's words, "This is the most beautiful and conspicuous of all the alpine flowers of Sikkim and is very common in rocky and gravelly places at [4,000 m] elevation and upwards, where it expands its delicate blossoms in May, exposed to violent winds and snow storms of these inhospitable regions. It was originally discovered by Dr Wallich's collectors in central Nepal, but has not been found further west in the Himalaya."

From J.D. Hooker, *Illustrations of Himalayan Plants,* 1855.

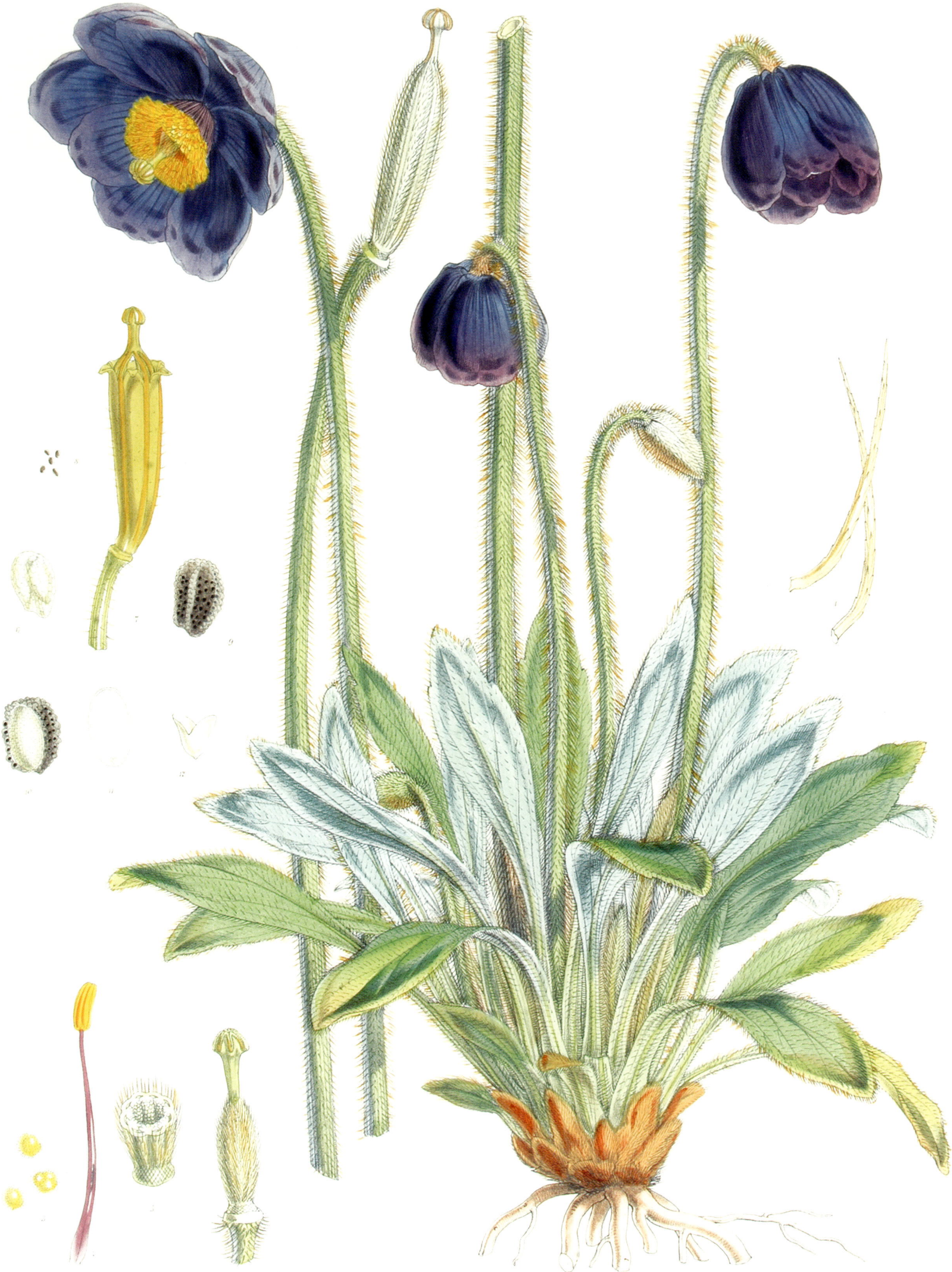

Meconopsis simplicifolia

COURTESY
GODAVARI BIOREFINERIES LTD.

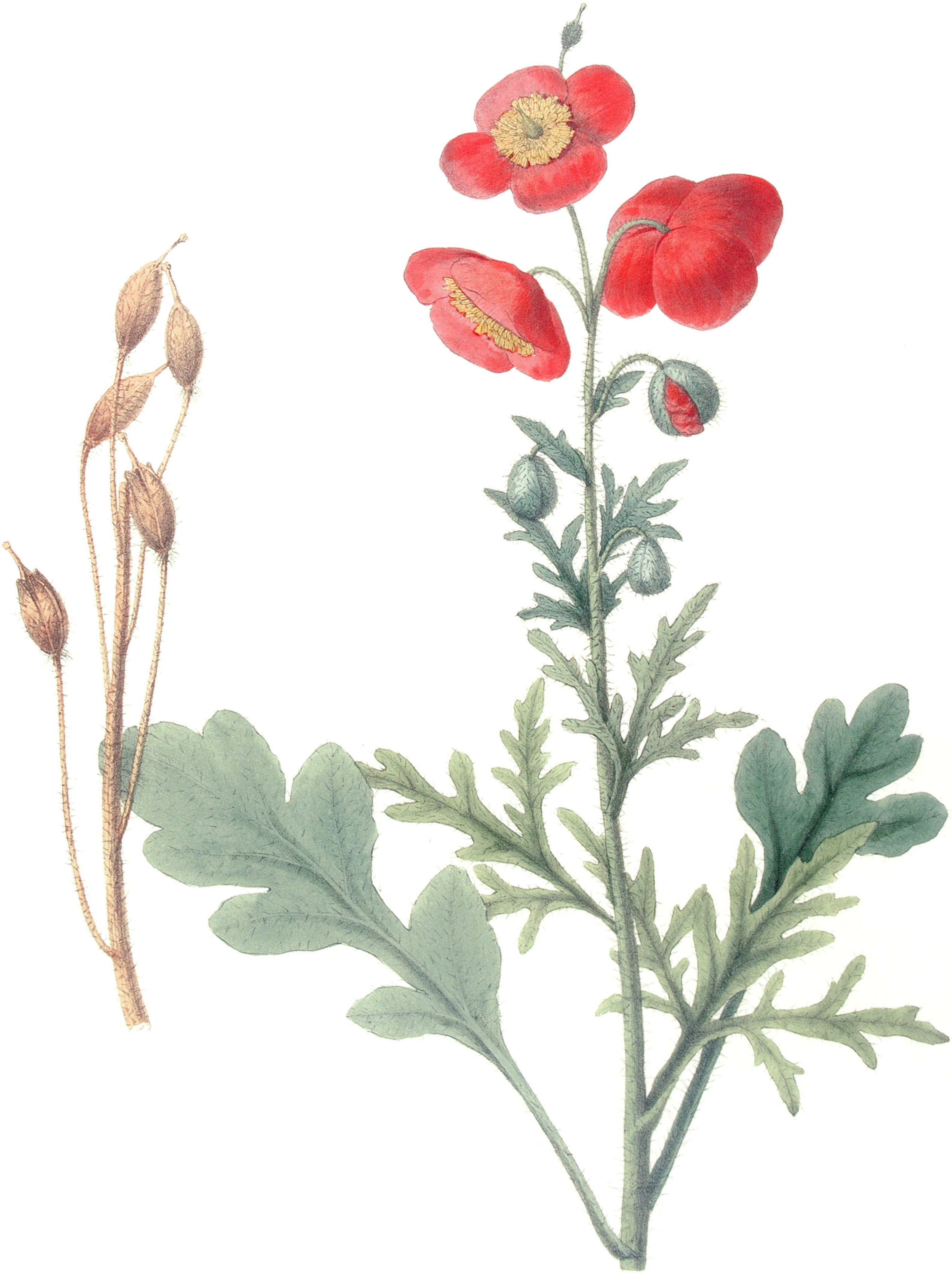

Meconopsis aculeata

COURTESY
GODAVARI BIOREFINERIES LTD.

Blue Poppy
Meconopsis aculeata
Family: Papaveraceae

An alpine perennial, this is considered one of the most beautiful Himalayan plants. A species of the western Himalaya, in India it is found in Himachal Pradesh, Uttarakhand, and Kashmir (on the hills above Sonmarg and most of the high passes and hills, including the area surrounding Amarnath), at 3,000–4,000 m on open rocky situations. The leaves are deeply and irregularly lobed. The lobes are usually rounded, toothed, and widely spaced. Leaves and stems have bristly hairs. The flowers are seen from July to September. They are numerous, large, usually bright blue but very rarely (as shown here) purplish-red or red, 5–7 cm across. There are usually 4 petals, round to obovate. The fruit capsule is oval to oblong with bristly hairs. The roots of the Blue Poppy are said to possess narcotic properties and are considered poisonous. The entire plant is used in Tibetan medicine, mainly to help in the healing of fractures and reducing the inflammation around them, and to treat pain in the upper part of the body – especially around the thoracic cage. Over-harvesting has put pressure on the wild population of this plant.

From J. Forbes Royle, *Illustrations...*, 1839.

Mass-flowering of "Karvi" Plants

J.S. Serrao

For the Naturalist 1992 is the year of *Karvi*. The plants have mass-flowered in the first week of September. It is rather late for the bloom to commence. August/September should have been peak months for the flowering, were the rains not to play truant. The rather late flowering tempts one to hopefully expect the bloom to continue till around November this year.

Karvi is the Marathi name for the plant. For science it was once *Strobilanthes callosa*. The plant was brought to the attention of science by Nees, a resident Britisher of Bombay in the 1800s. Nees was a shipbuilding engineer at the Naval Establishment at Bombay's present Lion Gate docks. However, by hobby he was a botanist. But *Karvi* differed from the genus *Strobilanthes* found in India. So the late Fr. H. Santapau, S.J. of St. Xavier's College, who later became the Director of the Botanical Survey of India created a separate genus of its own for the plant by latinising its Marathi name *Karvi* into *Carvia* and coined in 1954 a new combination to name the plant as *Carvia callosa*.

The life-history of *Karvi* is interesting. Each year with the commencement of the *true* SW Monsoon rains, the dormant plants show signs of coming to life. No amount of pre-monsoon showers induce them to do so. With the end of the rains the plants shed their leaves by November/December. By January the plant is almost nude. Its bare dead-looking stems, either upright or prostrate, can be seen all round. When the monsoons arrive on the dot, by July/August the plants attain a height of 9 to 10 feet. This life-cycle is repeated year after year for seven consecutive years. Dormancy of the plant continues till the following southwest monsoon season. In the eighth year the plants explode into their mass-flowering. Invariably always an individual plant flowers here and there in the seventh year – a precursor of the mass-flowering which has to follow in the eighth. So too, some stray individuals flower in the ninth year, dropping the curtain on the mass-flowering that took place in the eighth year.

The flowering of *Karvi* has always been a matter of controversy. It is generally believed that plants come into mass-bloom once in 8 to 12 years. The villagers in Thane maintain that they mass-flower once in three years. This view was supported by the late Major K.R. Kirtikar, a retired army surgeon, who intensely worked the Thane area botanically during the close of the last and the early years of the present centuries.

Flowering Broadcasts

Careful records maintained in Bombay and its environs since the start of the

J.S. Serrao, a keen naturalist and ornithologist with vast knowledge on varied subjects, worked as Librarian at Hornbill House. He initiated nature walks in and around Mumbai for members of the Society. In 1992 he took active part in the Karvi Festival. In 2000 he was eagerly waiting for Karvi to bloom again but fell seriously ill and did not recover. Recently during the monsoon of 2008 Karvi plants in the Western Ghats were in full bloom, the slopes covered with pinkish-purple flowers. BNHS organized Karvi trails for its members and the public at its CEC land in Sanjay Gandhi National Park.

twenties of this century, show that *Karvi* mass-bloomed in 1929, 1944, 1952, 1960, 1968, 1976, 1984. The current year's flowering follows in the eighth period since 1984. Seemingly for some reason hitherto unknown, mass-flowering between the years 1929 and 1944 went unrecorded. Perhaps it either escaped the attention of Bombay naturalists, or the plants failed to phenologically react. Mass-flowering heralds the end of the life-cycle of now standing *Karvi* plants. Their place will be taken up by the seeds the current year's flowering broadcasts. These seedlings will mass-flower in 2000 to usher in the 21st century.

Karvi is a wild plant. Geographically it is a denizen of the hill slopes of Madhya Pradesh, parts of Gujarat, thence down the Maharashtra Ghats, and inhabits vast stretches in the Konkan and North Kanara. The plants reach their southernmost peninsular limit about the Nilgiris. It prefers the moist deciduous forests, and tends to form dense gregarious patches in such areas.

It is an unforgettable sight to see *Karvi* in mass-flower. The beautiful purplish blue flowers, tinged with a pink or rosy hue, break the drab monotony of our barren hillsides, and add beauty and colour to our wilderness. The nearest place for the Bombay man to see and enjoy this natural beauty happens to be the Borivli National Park, preferably in the easily accessible parts of the park around the Film City at its south end. So brace up fellow naturalists to have a look at the year's mass-flowering of *Karvi*, and please do not wait for flowering in 2000. I look forward to meeting some of you among my old acquaintances and friends as I hobble about on my 76-year-old legs in the park enjoying the beauty and colour of these flowering plants.

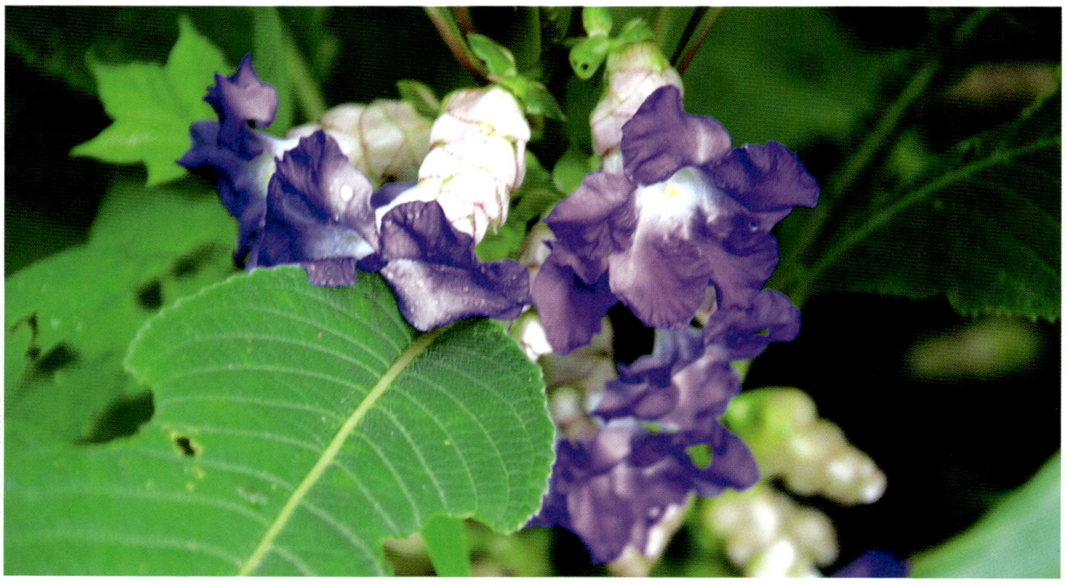

Ashok S. Kothari

From *Afternoon Despatch & Courier*, October 2, 1992.

Slender-billed Scimitar babbler
Xiphirhynchus superciliaris

Resident in the Himalaya from Nepal east to Arunachal Pradesh, and Bangladesh, this bird frequents bamboo thickets and dense undergrowth in thick forest. It is a dusky rufous-brown with dark grey head, whitish brow, and grey-streaked white throat. Its beak is long and curved. A shy bird it usually keeps hidden in thick undergrowth, probing for insects and ants among fallen leaves, moss, and bark crevices, but also ascends into flowering and fruit trees for nectar and berries. In April–July it builds a large globular nest on or close to the ground.

From T.C. Jerdon, *Illustrations of Indian Ornithology*, 1847.

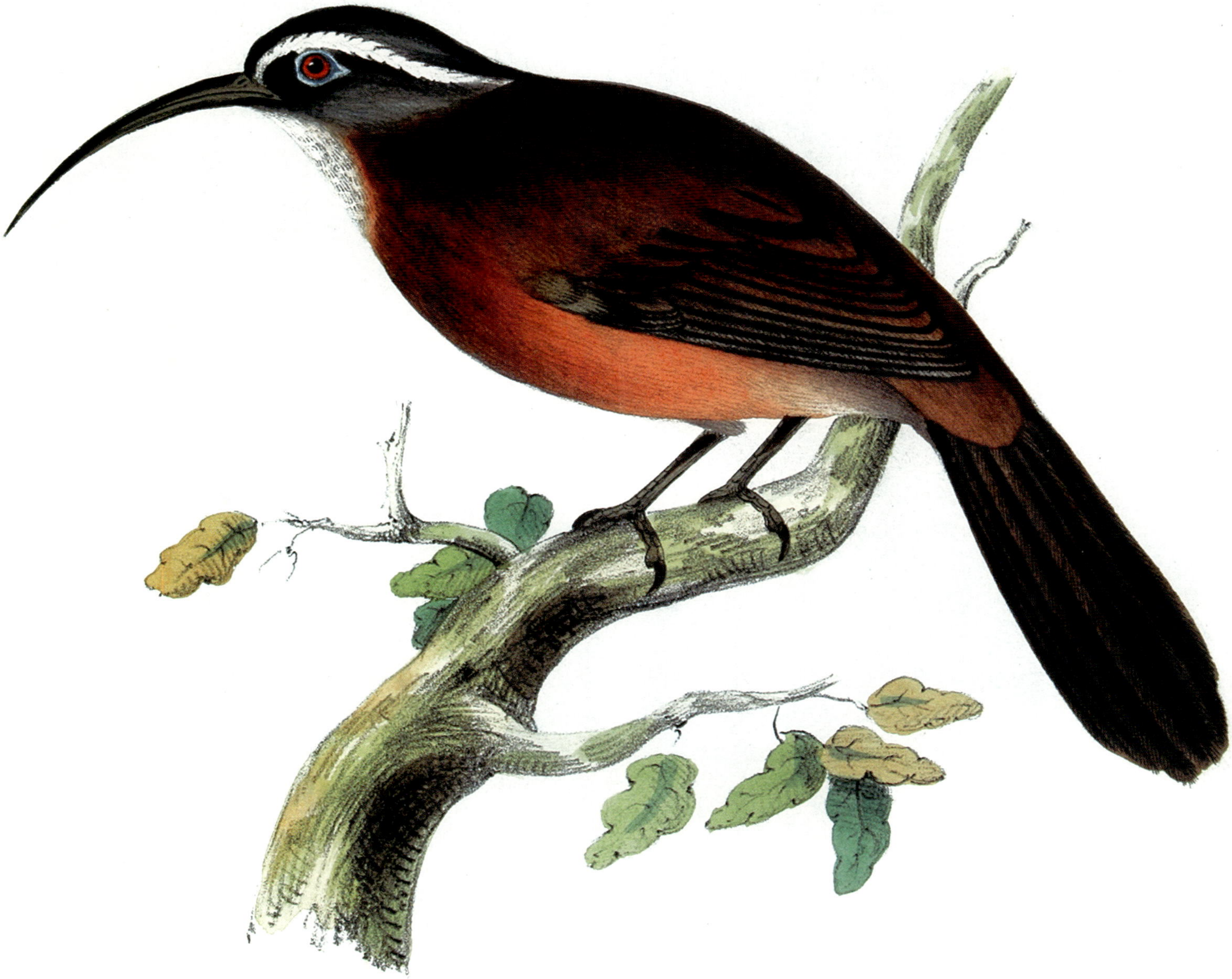

Xiphirhynchus superciliaris

COURTESY
INDUSIND BANK

Emperor Jahangir's Elephant Hunt in the Panch Mahals

Manekshah S. Commissariat

At Dohad, on the frontiers of Gujarat and Malwa, Jahangir celebrated, on March 10, 1618, "in joy and thanksgiving", the commencement of the 13th year of his accession to the throne. On the following day, the ceremonies associated with his fifty-first birthday, according to the lunar year, were carried out by the weightment of the Emperor against gold and silver, and other festivals. After a halt of several days, the party proceeded to the hilly country surrounded by jungle, some 10 *kos* away from Dohad, for the elephant-hunt to which the Emperor had been looking forward ever since he first entered Gujarat. The grazing-ground of the elephants was in the forests at the foot of a hill called "Rakshas Pahar" or demon hill, by some thought to be the great hill of Pavagadh in the Panch Mahals. On all sides the jungle had been surrounded by a large body of horse and foot, and a ring-fence arranged in the manner known as *qamurgha* among the Mughals. Outside the jungle, a wooden platform was put up on a tree for the Emperor, and seats were provided on other trees for the nobles. Two hundred decoy male elephants and many female ones had been got ready, and on each of them two *mahuts* or drivers were seated with strong nooses. These were to drive the wild elephants from the jungle towards the royal presence so that he might witness the hunt. Unfortunately, when the men placed round the ring-fence entered the jungle, the chain was broken owing to the thickness of the forest and the hills and hollows, and the circle of the *qamurgha* did not remain perfect. The wild elephants in bewilderment rushed in every direction, but about a dozen of them appeared on the side where the Emperor was seated. The tame elephants were now brought under requisition and with their help several of the wild ones were captured. Of these two were very handsome in shape and with "perfect marks", and these were named by Jahangir as "Pavan Sar" and "Ravan Sar".

Leaving the Darogha and the head huntsman to carry on further operations during the remainder of the season, Jahangir retraced his steps from the forest. Four months later, when he was again at Ahmadabad, the officers in charge of the hunt waited on him with the news that altogether 185 elephants had been caught, 73 males and 112 females. This large haul was the result of the combined operations of the imperial huntsmen and fauzdars, and the huntsmen of Prince Shah Jahan (*Tuzuk*, II, 5, 24). It seems no scientific measurements were taken, in the generations that followed, for the preservation of elephants in the Dohad forests, for today the region is almost destitute of these animals.

From Professor Manekshah S. Commissariat, *History of Gujarat*, Vol. II, 1957.

In the 17th century the Panchmahals in south Gujarat had thick forests with a large population of elephants. The abundance of wildlife in the Panchmahals is mentioned by James Campbell in his *Gazetteer of the Bombay Presidency* (26 vols.), Vol. III: *Kaira & Panchmahals*, 1879, reproduced in *Treasures of Indian Wildlife*, pp. 112–13.

Michelia cathcartii
Family: Magnoliaceae
Synonym: *Magnolia cathcartii*

This tree is found in Meghalaya, Nagaland, Sikkim, West Bengal, Myanmar, southwest China, and Vietnam. According to Hooker, "This is a very common tree on the outer range of Sikkim Himalaya, [from 1,500 to 2,000 m], beyond which elevation it rarely ascends. It is conspicuous in April from the abundance of blossoms with which in some years the branches are covered appearing as if snowed upon; however the trees flower much more freely at some seasons than others. The leaves are only partially shed in winter, the new ones being put forth during or immediately after flowering in April and May. It has hitherto been found nowhere but in Sikkim, and bears the name of Mr. Cathcart, around whose residence at Leebong, near Darjiling, some fine trees stood; these were about [20 m] high and had straight trunks and rather short branches. The wood is good and used for household purposes by Bengali carpenters, who give it the name Champa, which is commonly applied to several other species of Magnoliaceae."

From J.D. Hooker, *Illustrations of Himalayan Plants*, 1855.

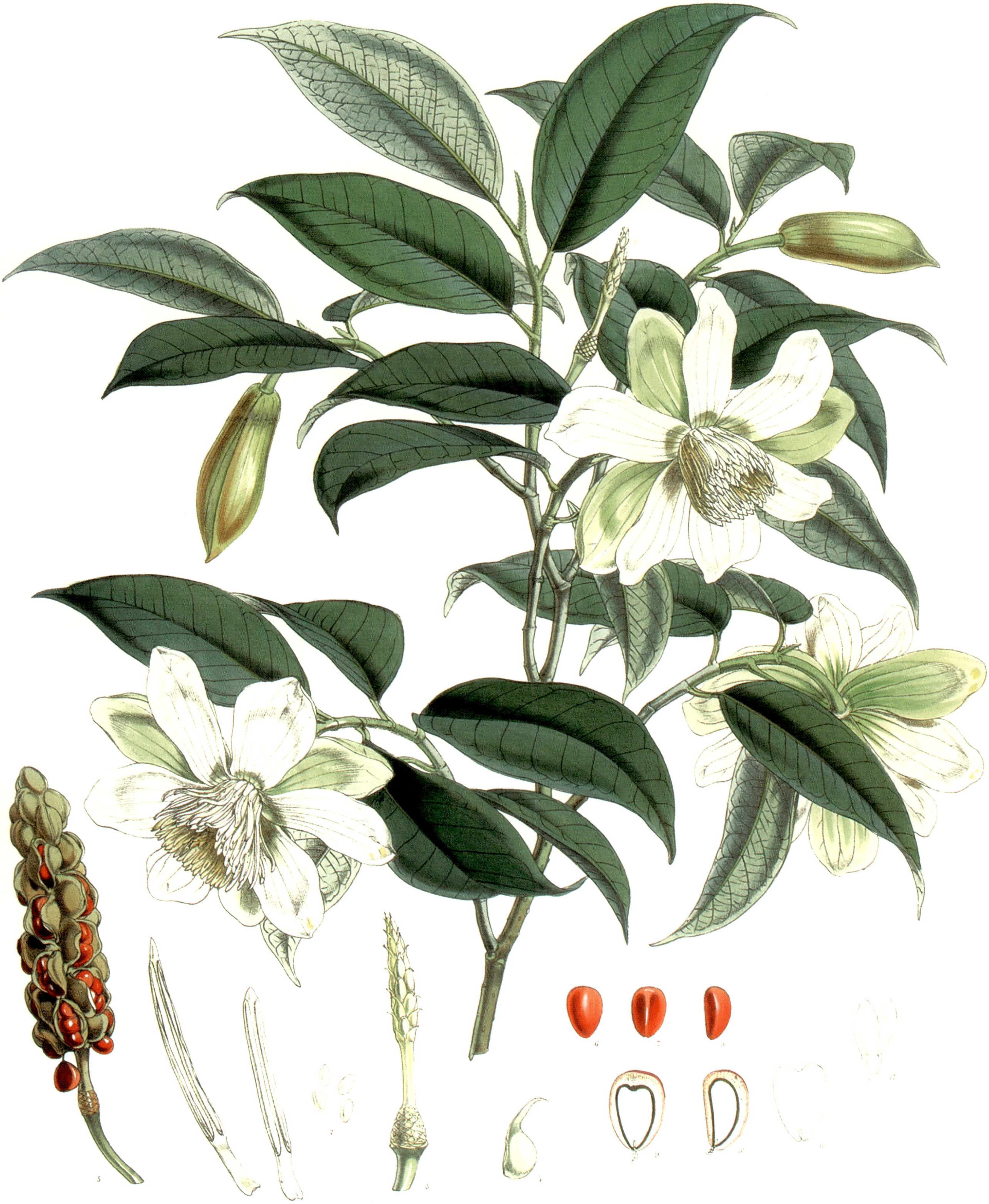

Michelia cathcartii

COURTESY
ROSY BLUE (INDIA) PVT. LTD.

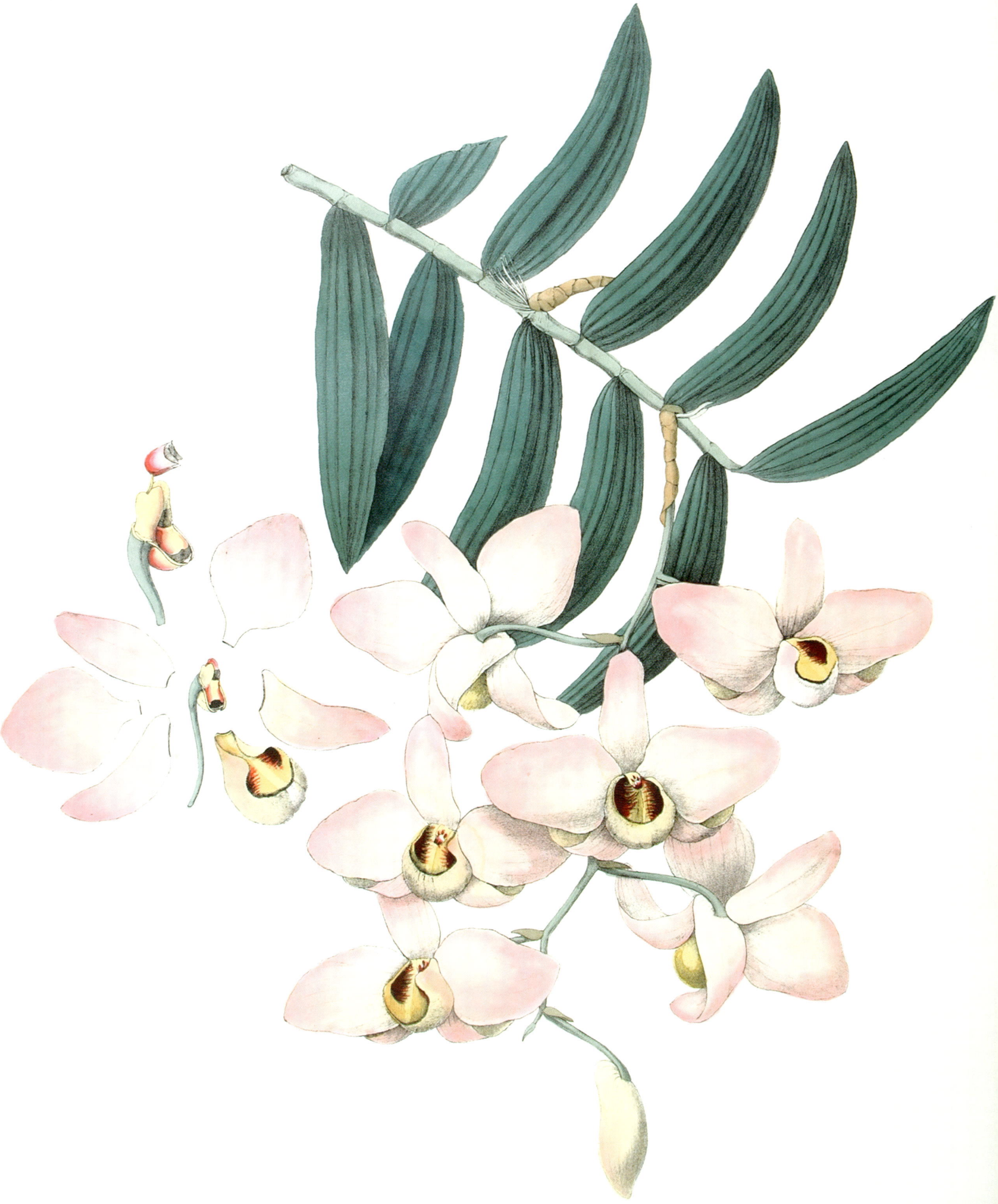

Dendrobium moschatum

COURTESY
KANSAI NEROLAC PAINTS LIMITED

Musky-smelling orchid
Dendrobium moschatum
Family: Orchidaceae
Synonyms: *D. cupreum*, *D. pulchellum*

This large-sized, evergreen, epiphytic orchid is found at altitudes of c. 1,300 m in India's Himalaya – Uttarakhand, Sikkim, Assam, Meghalaya, Arunachal Pradesh, Manipur, and Nagaland – and in Jharkhand (Chhota Nagpur), Orissa, Bangladesh, Nepal, Bhutan, China, Myanmar, Laos, Thailand, and Vietnam. Cylindrical pseudo-bulbs and lance-shaped leaves are borne on arching pendulous stems; the stem is thickest at the middle, attenuated towards the base and apex. The inflorescence arises from axillary nodes in the apical part of old stems, and is pendulous with 7–15 flowers. The flowers appear in May–July and are short-lived; they measure 8 cm across, and are musky-scented, creamy buff flushed with rose, the lip forming a globular or pear-shaped yellow pouch. The sepals have reddish tips and reddish or orange veins, and are narrower than the petals.

From N. Wallich, *Plantae Asiaticae Rariores*, 1832.

An Elephant Saves a Shikari from the Jaws of a Lion

Captain Godfrey Charles Mundy

A DAY'S march brought us into the territories of the Khytul Rajah, which are divided by a small stream from the Patialah dominions. The effects of the efficient government of the latter province are plainly visible in the great superiority in cultivation possessed by it over the Khytul country, whose Rajah, a mere boy (cousin of the king of Patialah), is ruled entirely by his mother.

Two marches brought us to Pewur, through a country which, with the exception of sufficient cultivation round the villages for the bare existence of the inhabitants, presents one vast sheet of wild jungle, abounding in game. Two mounted shikkarees of the Rajah brought a couple of wild hogs into camp which they had killed, and three others arrived from Colonel Skinner, to assist us to find a lion between this place and Hansi. Of these animals there were formerly great numbers in the surrounding wilderness: but from the zeal of English sportsmen, and the price put upon their heads by Government, the royal race of the forest – like other Indian dynasties – is either totally extinct, or has been driven farther back into the desert.

By *crack* sportsmen the lion is reputed to afford better sport than the tiger: his attack is more open and certain; a peculiarity arising either from the noble nature of the jungle king, or from the country which he haunts being less favourable for a retreat than the thick swampy morasses frequented by the tiger. Colonel Skinner relates many interesting anecdotes of lion-hunts, with the exploits and narrow escapes of the horsemen of his corps, who always accompanied the line of elephants into the jungle on these occasions.

A gentleman of our party had, perhaps, as perilous an adventure with one of these animals, as any one of the former; he having enjoyed the singular distinction of lying for some moments in the very clutches of the royal quadruped. Though I have heard himself recount the incident more than once, and have myself sketched the scene, yet I am not sure that I relate it correctly. The main feature, however, of the anecdote, affording so striking an illustration of the sagacity of the elephant, may be strictly depended upon.

A lion had charged my hero's elephant, and he, having wounded him, was in the act of leaning forward in order to fire another shot, when the front of the howdah

In the 19th century, lions existed in many parts of India. This incident took place in Punjab, between Patiala and Hansi. Captain Godfrey Charles Mundy (1814–60) describes how a trained elephant, at the instance of its mahout, hammers the lion with the branch of a tree and saves the shikari. Captain Mundy, author and soldier, was decorated in 1825–26 as aide-de-camp to Lord Combermere, Chief of the Indian Army, at the siege and storming of Bharatpur.

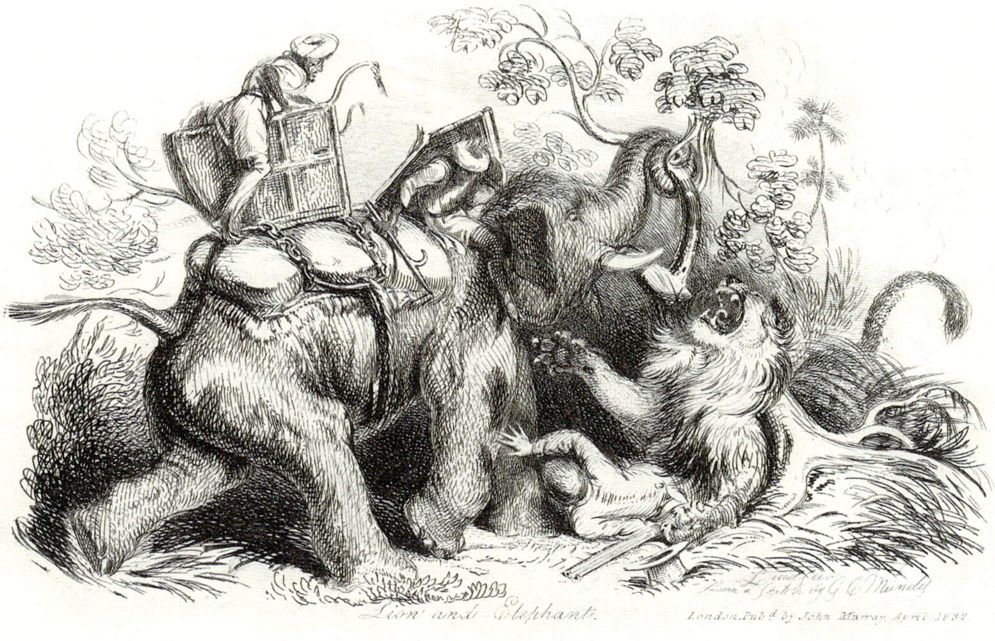

Lion and Elephant.

suddenly gave way, and he was precipitated over the head of the elephant into the very jaws of the furious beast. The lion, though severely hurt, immediately seized him, and would doubtless shortly have put a fatal termination to the conflict, had not the elephant, urged by the mahout, stepped forward though greatly alarmed, and grasping in her trunk the top of a young tree, bent it down across the loins of the lion, and thus forced the tortured animal to quit his hold!

My friend's life was thus preserved, but his arm was broken in two places, and he was severely clawed on the breast and shoulders. The lion was afterwards slain by the other sportsmen who came up.

From Captain Mundy, *Pen and Pencil Sketches, being the Journal of a Tour in India,* in two volumes, Volume I, 1832. (Dr. Ashok S. Kothari Collection.)

Pink Tulip Tree
Magnolia campbellii
Family: Magnoliaceae

According to Hooker, "This superb tree which forms so conspicuous a feature in the scenery of Darjiling is chosen by Dr. Thomson and myself to commemorate the services of Dr. Campbell, a resident of Darjiling. Dr. Griffith discovered the tree in Bhutan. It is a large forest tree abounding on the outer edges of Sikkim", at elevations of 2,500–3,000 m and also in Darjeeling, Assam, Meghalaya, Bhutan, Nepal, southwest China, and northern Myanmar. Hooker also states that the trunk is straight, often 25–30 m high and 4–6 m in girth, covered with black bark; the wood is soft and almost useless. The flowers are produced abundantly in April, at the end of branches, when the tree is yet perfectly leafless. They vary from white to deep rose colour or almost crimson, and in size from 15–25 cm; the scent is faint. In May the tree is in full leaf and the fruits ripen in October. Young plants have perfectly glabrous leaves; those of older trees are more or less silky on the outer surface. Once a staple tree in the forest, it is much diminished due to its constant use as firewood and as timber for planks and tea-boxes.

From J.D. Hooker, *Illustrations of Himalayan Plants,* 1855

IN LOVING MEMORY OF RAGHUNATH KASHINATH MITRA OF NAIROBI
FROM DOROTHY & SHEELA, AND ALAKA RAGHUNATH NADKARNI

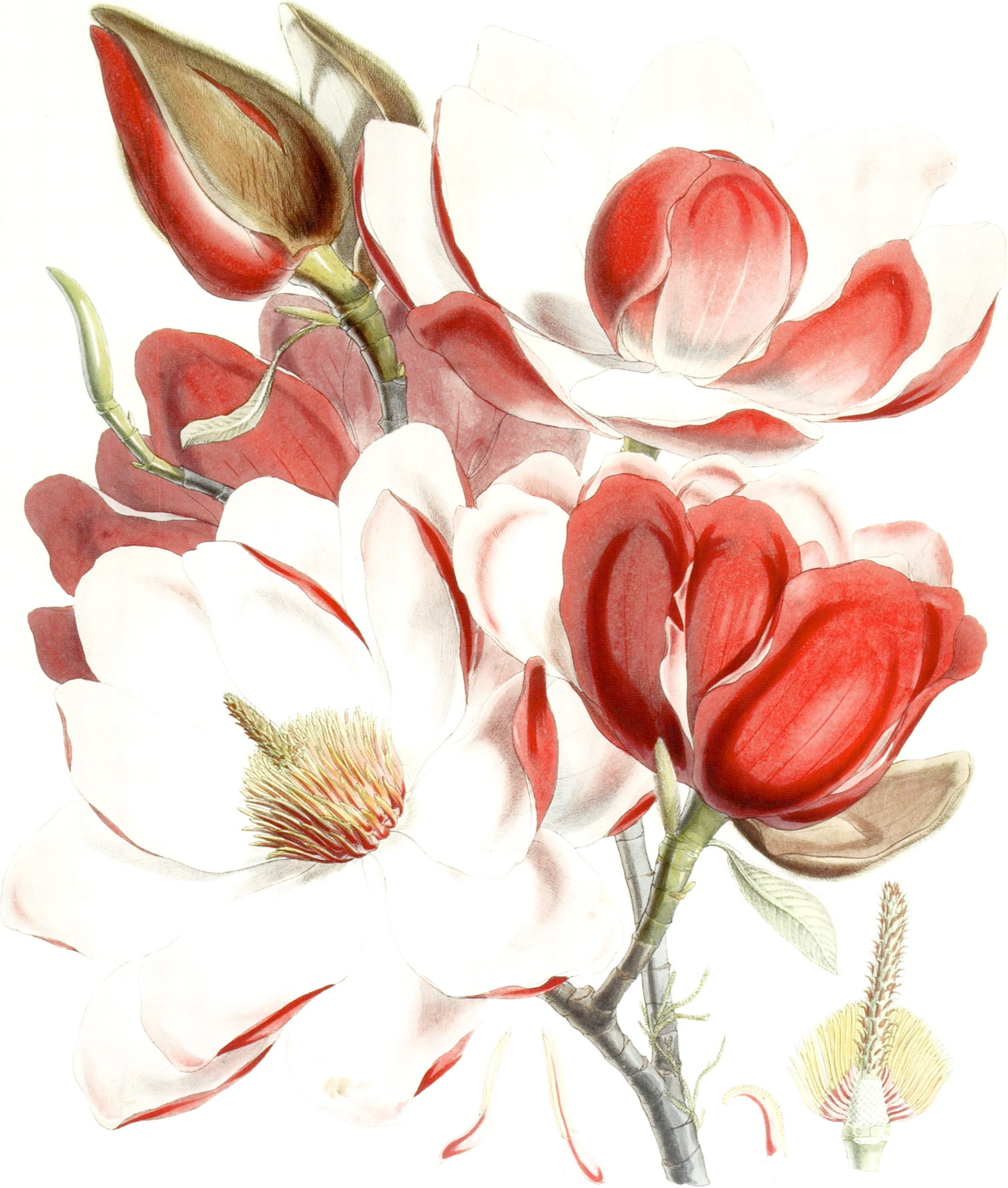

Magnolia campbelli

N MEMORY OF MANJULA DHIRAJLAL SHROFF
– RUDRA AND WILLIAM CONNAL

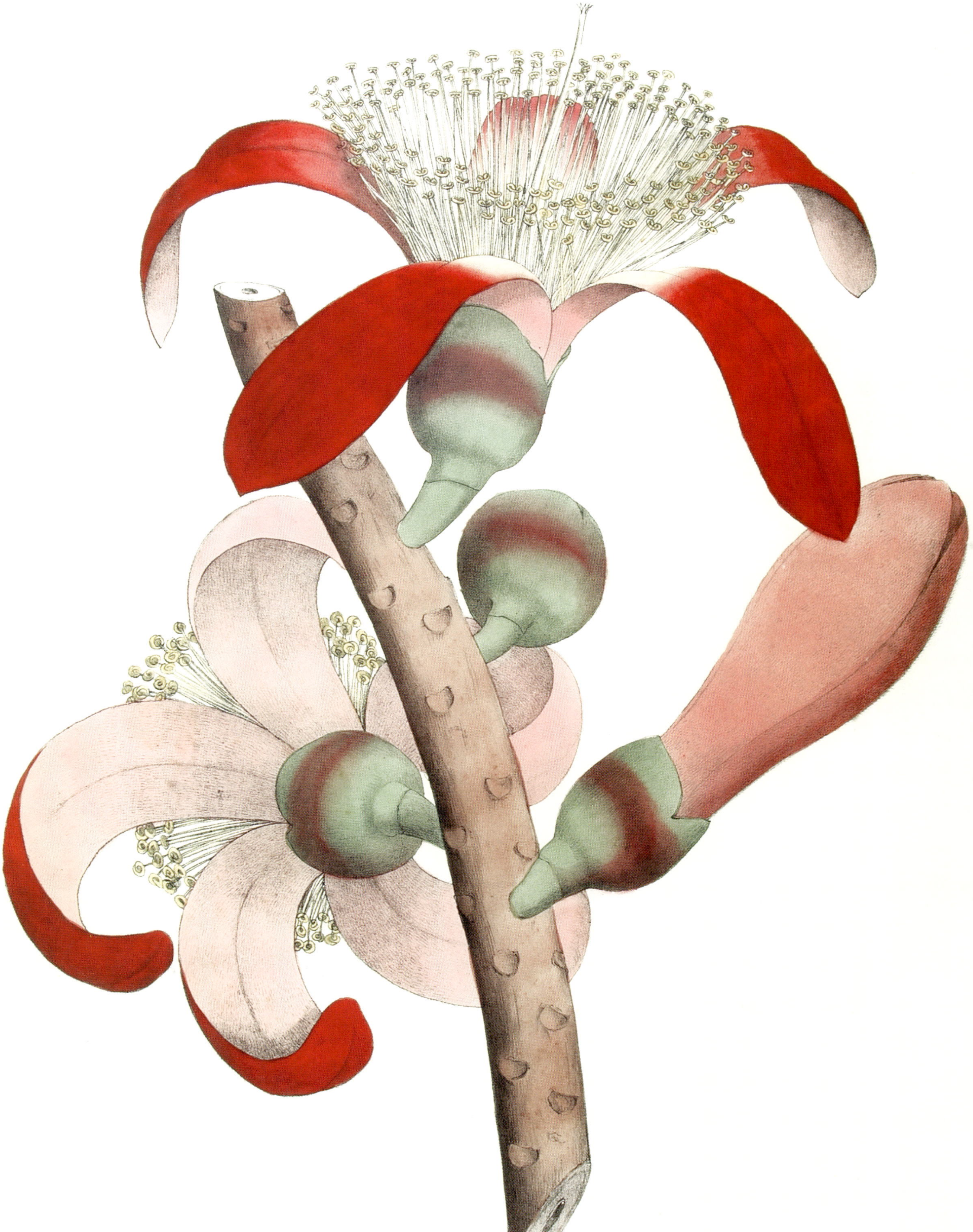

Bombax insigne

DEDICATED TO MY DEAR PARENTS
LILARAM N. CHULLANI &
MOHINI L. CHULLANI
– SATISH L. CHULLANI

Bombax insigne
Family: Bombacaceae
Synonym: *Bombax insignis*

This is a large deciduous tree found in the Western Ghats from north Karnataka southwards, and in the Andamans, Bangladesh, Myanmar, and the Malay peninsula. It is very similar and often mistaken for the Red Silk Cotton or Semal. It is called Kulasalmali in Sanskrit, Dumboil in Assamese, Semal in Bengali, Kallilavu in Malayalam, Khumantera in Manipuri, Devsavar in Marathi, and Vellaikungiliyam in Tamil. The trunk is unarmed, though the branchlets are sometimes prickly. The leaves are glabrous, with 7–9 leaflets arranged on a common petiole; the petioles are longer than the leaves. The leaflets are obovate with pointed tip. The flowers appear in February–March and are very showy (*insigne* meaning striking or spectacular), the scarlet, pink, cream, or white petals being silky within. The stamens are long and numerous; staminal bundles consist of 50 or more unequal filaments. The woody, oblong capsule is about 30 cm long, filled with silky cotton and round seeds.

From N. Wallich, *Plantae Asiaticae Rariores*, 1832.

The Valley of Deyra Doon

Captain Godfrey Charles Mundy

On the 12th of April, 1828 Colonel Dawkins and I left Hurdwar, and commenced our march through the valley of Deyra Doon. Our route was most beautiful, and reminded me much of some of the milder and least wild regions of the Alps. The road, which is made with great art, winds down through a woody declivity, sometimes closely hemmed in by abrupt rocky banks, and at others traversing a luxuriantly wooded plain. We found our tents pitched in a thick forest near the small chokee of Karsrah.

On the summit of the hill above our encampment is a small bungalow, built for the accommodation of persons travelling without tents; and similar buildings are established at nearly every stage through the Doon. From this point we enjoyed a magnificent prospect. Below us lay the beautiful valley of Deyra, luxuriant with many-tinted forests, and refreshed by the rippling little rivers, which, with as many arms as Briareus, run in a meandering course through the whole length of the vale, from the Ganges to the Jumna. The view is bounded on all sides by mountains. In the north and west, those of Gurhwal and Surmoor rise, series after series, till they are terminated by the snow-capped peaks of Himalayah. In the south, the prospect is abruptly closed by the range of woody hills which form the boundary of the Deyra Doon, and cut it off from the plains of Hindostan.

We had provided ourselves with a punkah and tattees of kus-kus at Hurdwar; and by their means, together with the umbrageous canopy overhead, reduced the temperature of our tent to 78°.

During breakfast, a hurkarah arrived with a note from a gentleman who passed by this same route yesterday, to inform us that, as he was fishing in a mountain-stream near at hand, a tiger came to drink at it within pistol-shot of him, and retired without seeing him. We determined to look for him in the evening, though our chance of success is small in a country so full of ravines, by which the tigers can sneak away to the woods when they hear the crashing approach of the elephants.

At 4 P.M., having caught a wild-looking shikkaree, we established him upon an elephant, and, descending the woody hill, entered the thick grass jungle bordering the beautiful little river which runs through the valley. The hunter promised us good deer-shooting, but denied all knowledge of tigers. We had not, however, been out more than half an hour, when, in a small patch of high silky grass, a very fine one jumped up close under my elephant's trunk, just as I was in the act of taking aim at a peacock, one of a large covey which rose before me. I immediately transferred my aim to the royal

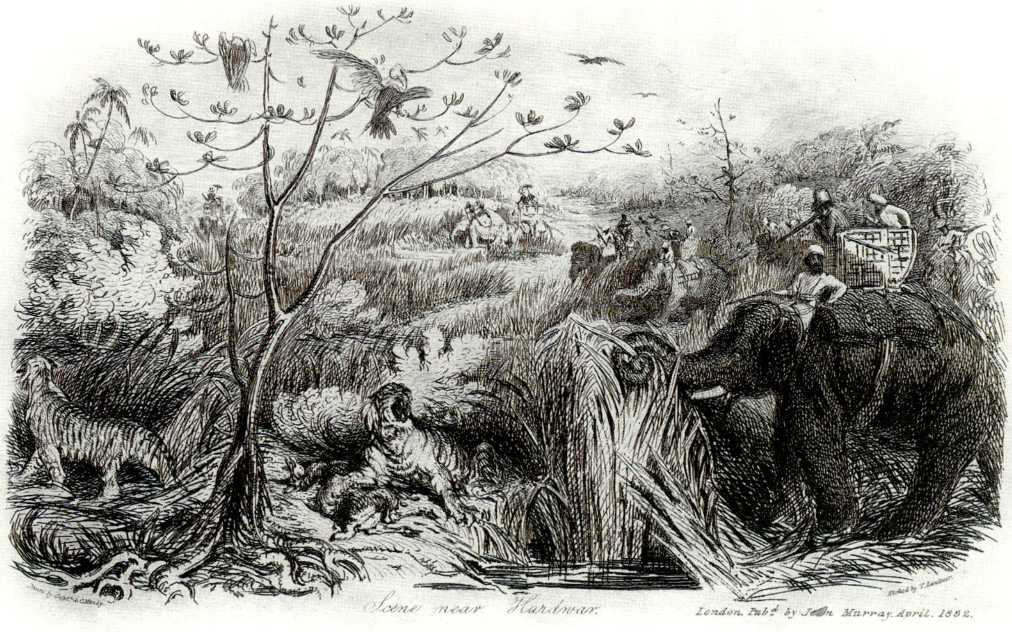

Scene near Hardwar.

quadruped, and in the hurry of the moment gave him the advantage of a load of small shot, instead of the bullet in the other barrel. He uttered a loud roar, rushed through the grass, plunged into a ravine, and I lost sight of him. We beat all the surrounding jungle without finding him, when I suggested trying a small rushy island, which was only divided from the main land by a shallow branch of the river, and was directly opposite the point where the ravine debouched upon the stream. We had no sooner entered the islet, than D. came upon the tiger, and fired two shots at him. The beast being however, upon the retreating system, escaped from the covert, and I saw him cross over to another islet, thickly set with brushwood; it was not more than fifty yards square, and as he entered it I saw by his action that he had been wounded. We, nevertheless, beat in vain for him; and though the elephants showed all the usual signs of the vicinity of a tiger, we never saw him again. My elephant imagined more than once that he had stepped upon the object of our search, and kicked a huge stump of a burnt tree into a thousand pieces in his mistaken fury.

On the road back to the camp, I shot a hog-deer and a young spotted deer, the latter of which made very good venison; the other furnishing a plentiful feast for the servants.

Burans Tree-Rhododendron,
Rhododendron arboreum
Family: Ericaceae

The most widely distributed tree-rhododendron in the Himalayan region, found from Uttarakhand to Sikkim, and in Pakistan, Nepal, Bhutan, and Tibet at a wide range of altitudes (1,500–3,600 m), and in many different forms. The bark is reddish-brown, peeling off in small flakes. The leaves, crowded at the ends of the branches, are 10–20 cm, oblong to lanceolate, acute at both ends, with grooved mid-vein and lateral veins deeply impressed, glossy-green above, underside covered with white, cinnamon, or rusty-brown hairs. The flowers are large (10–13 cm across), very showy, commonly crimson, pink, or white, the last usually found at higher altitudes. The flowers appear in large compact clusters at the ends of the branches. The capsule is 3 cm long, ribbed, and ten-chambered.

This is the national flower of Nepal. The flowers are sweet-sour in taste, are eaten and made into preserves. They are believed to be effective in the treatment of dysentery and are also applied as a poultice in headaches, besides being offered at temples in the hills. The tender leaves are cooked as a vegetable.

From N. Wallich, *Plantae Asiaticae Rariores*, 1832.

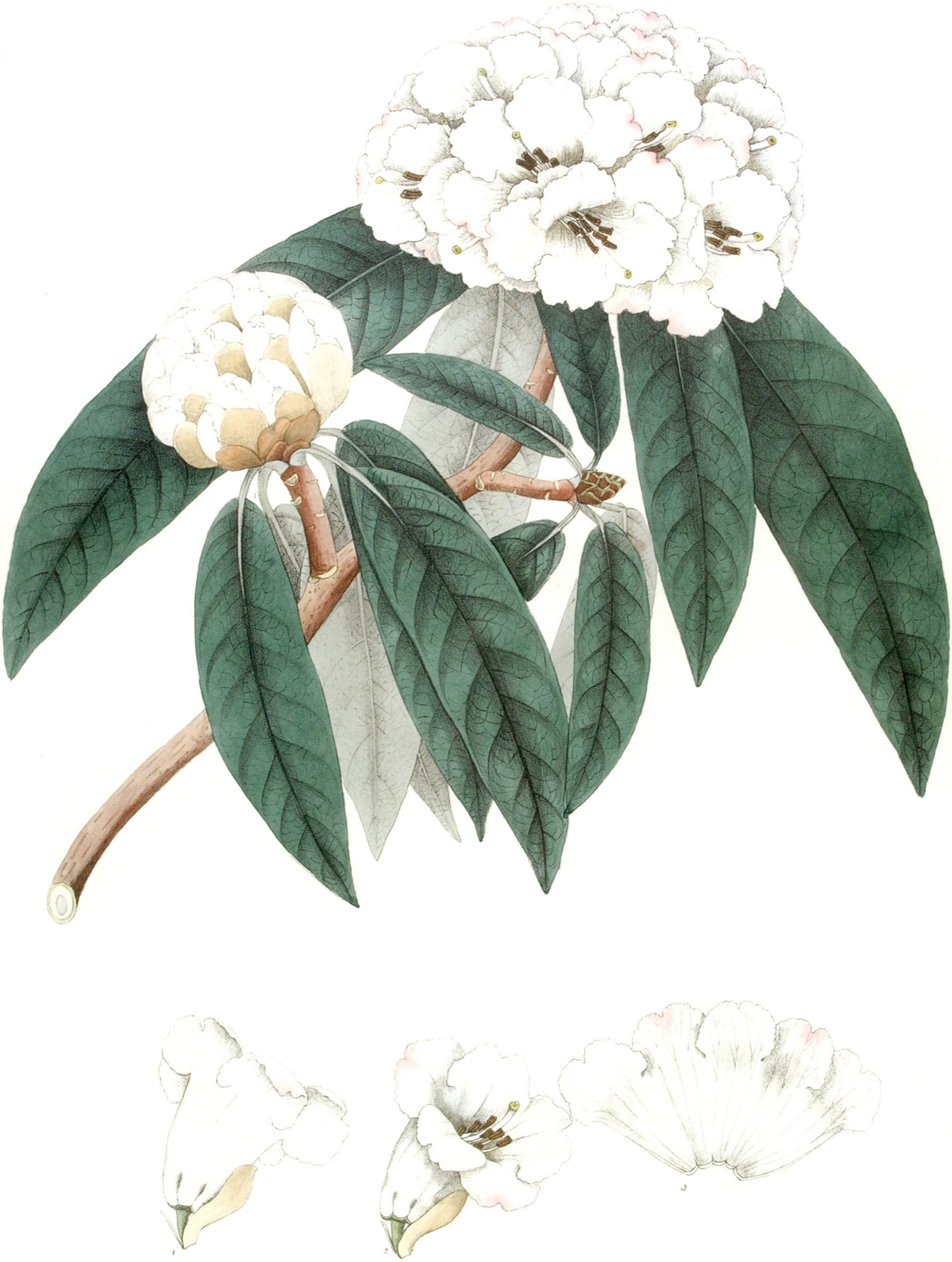

Rhododendron arboreum

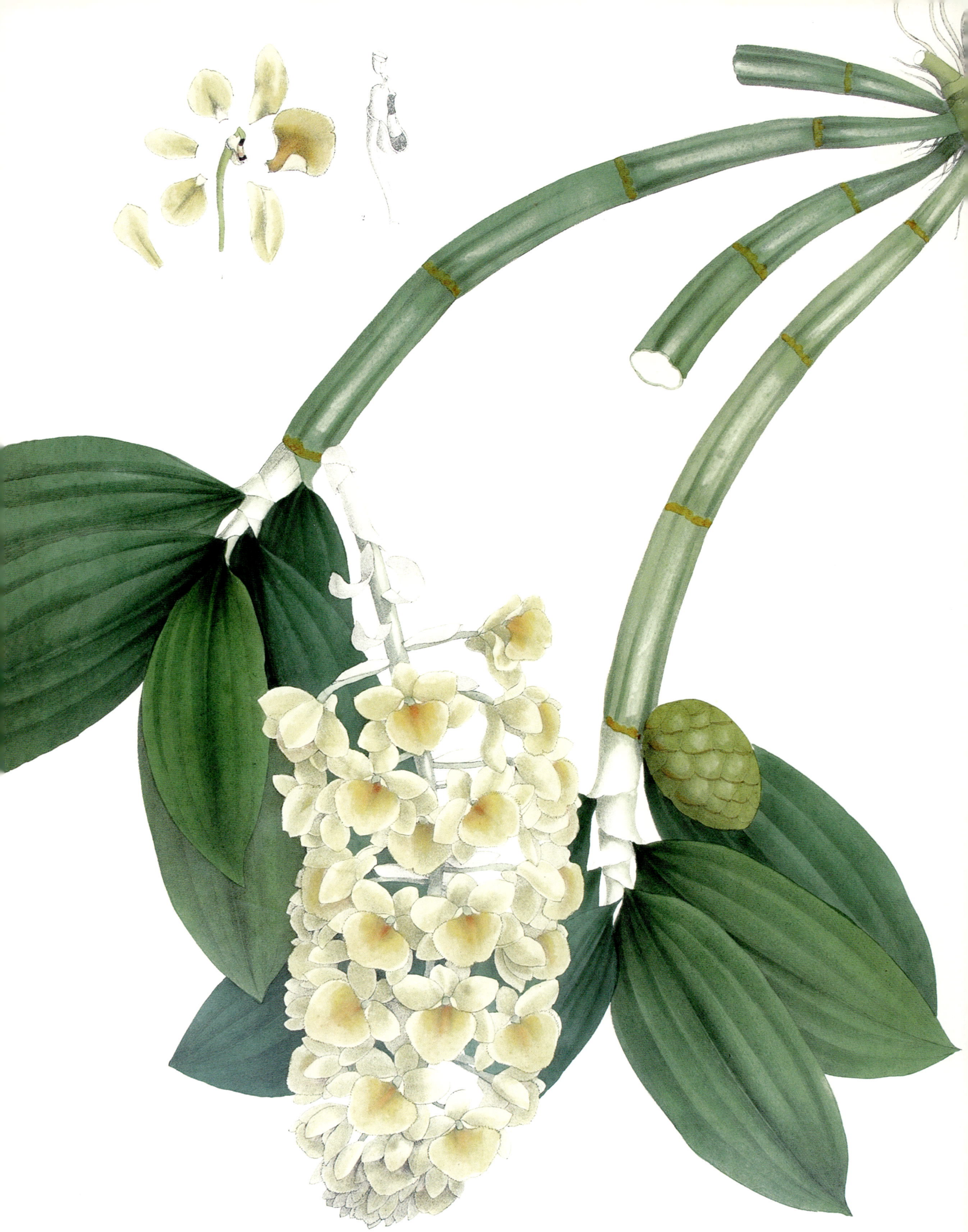

Dendrobium densiflorum

COURTESY
RAIKA & NAVROZE GODREJ

Densely Flowered Dendrobium, Pineapple Orchid
Dendrobium densiflorum
Family: Orchidaceae

Orchids form 9 per cent of our flora. It is estimated that about 1,300 species of orchids are found in India, with the Himalaya as their main home and scattered in the Eastern and Western Ghats. The largest epiphytic genus is *Dendrobium*, established by the Swedish botanist and taxonomist, Dr. Olof Swartz in 1799, which today contains about 1,200 species. The name is from the Greek *dendron* (tree) and *bios* (life). It means "one who lives on trees" or essentially an epiphyte. *Dendrobium densiflorum* is found in the Himalaya from central Nepal to Bhutan and Myanmar. The pseudobulbs are 4-angled, club-shaped, green, 30–45 cm long, increasing in thickness above. Leathery, oblong-elliptical 10–15-cm-long leaves are borne towards the apex of the pseudobulbs. The inflorescence arises in April–May from the upper nodes of the pseudobulbs, in densely arranged, many flowered, pendulous clusters, cylindrical like a pineapple, 20 cm long and 7.5 cm in diameter. The flowers are 5 cm across, yellow and funnel-shaped with orange lip.

From N. Wallich, *Plantae Asiaticae Rariores*, 1832.

The next morning – whilst our tents advanced eight miles to the village of Luckawala – we started on our elephants to look for our lost tiger of yesterday. We were again balked in our expectations, although we traced him, by the assistance of the Mohouts, who were uncommonly keen, for nearly a mile on the sand. Returned to Karsrah, quitted our elephants, and drove to the next encamping ground in a gig; the road being now nearly level.

April 14th. Marched eleven miles to the town of Deyra, whence the name of the valley. The first half of the day's journey led through a thick forest of very lofty trees, many of them strangers to the plains of India.

The underwood was formed of richly-flowering plants, among which the corinda sent forth its well-nigh sickening fragrance; and parasites of the most gigantic proportions twined round the trunks and branches of the larger trees, resembling in their grotesque writhings the folds of huge boa-constrictors. This similitude was so strong in some instances as to be almost startling.

In some places the trees on either side of the road met over head, forming a long and dark arcade of impervious foliage; the lofty arches being not uncommonly knit together at the apex by the luxuriant creepers. In one of these natural cloisters we overtook a party of peasants, who were singing in a wild and monotonous cadence; and at some distance their voices, echoed and reverberated by the vaulted forest, sounded not unlike the chanting in a cathedral. In the immediate neighbourhood of the town of Deyra, the jungle has been cleared away, and the wheat crops are remarkably fine.

On arriving at our tents we received a polite invitation from Mr. Shore (the son of the present Lord Teignmouth), the political agent for this province, to pass the day in his house. He is a remarkably tall and handsome man, has adopted the mussulman costume, and wears a long beard. He is also distinguished as the scourge of all the brigands and wild beasts which infest his province.

The verandah and rooms of this active persecutor of the wild animals of the forest are adorned with the grim skulls and stuffed skins of tigers, bears, boars, and monkeys, and other characteristic drawing-room furniture; and while we were at breakfast, two black bears with silver collars round their necks, strolled quietly into the room, and took possession of the posts of honour on the right and left of our host. These proved to be domestic pets, and ate their breakfast out of wooden bowls placed on either side of his chair. One of them was unruly; and after devouring his own meal, attempted to forage some honey from the table, upon which he received a very good thrashing with a dog-whip from his master.

Mr. Shore described to us his general method of killing tigers, which militates a good deal against my ideas of good sport. However, it is a good plan in a place where elephants are scarce, and where the thickness of the forest almost renders these animals useless. When intelligence arrives of a bullock having been killed in the jungles by a tiger (which animal generally defers his meal until evening), Mr. S. orders the carcase to be dragged under some convenient tree, among the branches of which he lays a michaun, or platform of bamboo. Here he takes his station with his guns and ammunition, and patiently awaits the arrival of his victim, who does not fail to return about sunset to his prey, and is shot before he has time to begin his meal. The accompanying plate will afford the reader a pretty correct representation of the above-mentioned sport, and of the scenery through which we are now travelling. Mr. Shore describes the approach of the tiger, on these occasions, as equally crouching and cautious as though he were in pursuit of living prey. He also gives an amusing account of the several animals which succeed each other – during the, no doubt entertaining, vigil of the sportsman – in possession of the carcase; until the jackals, the latest occupants, begin to cower and sneak away, which is a hint to the shooter to prepare for the arrival of the arch-bully, the tiger himself who is sure then to be creeping towards his feast.

At a quarter before 4 P.M. we went out to witness an eclipse of the sun, which was beautifully and distinctly visible through the medium of a looking-glass and a pair of green spectacles. After the eclipse, we had a distinct view of the new-formed invalid station of Llandowr, and Missouree, where Major Young and some others have cool summer residences. They are situated on a lofty ridge of the mountains directly north of Deyra.

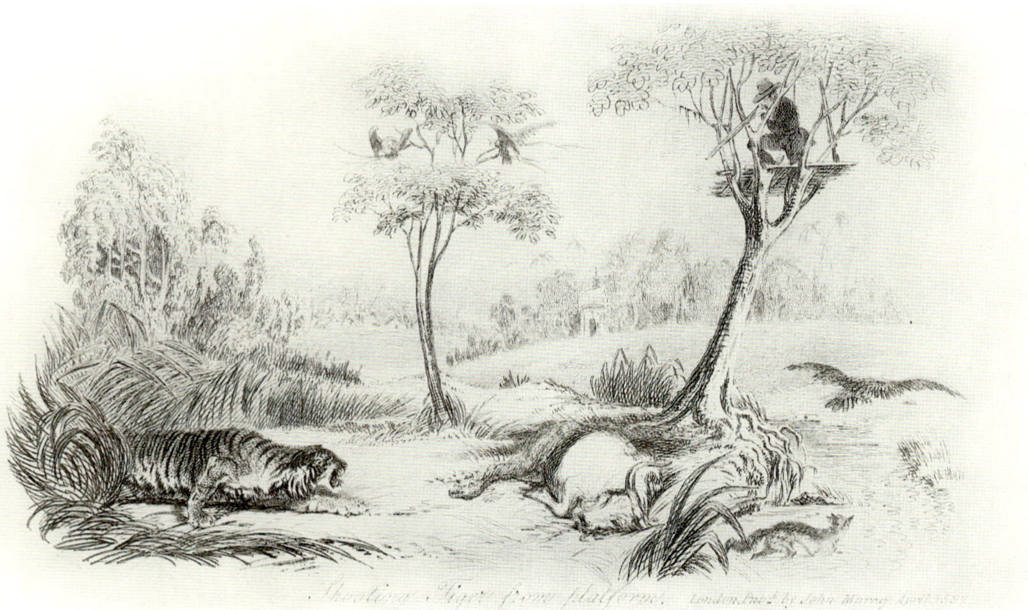

Shooting Tiger from platform.

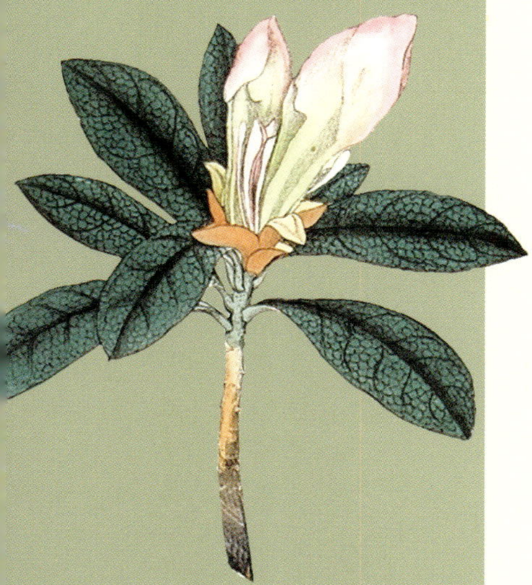

Deyra might really be a "Happy Valley". Though it is cooler than the plains, the English inhabitants retire to the mountain stations, so conveniently at hand during the hotter months; and they have a feasible and direct communication with the "cities of the plain" by the pass of Kerie, due south of Deyra. However, to counterbalance these and other utopian advantages, I am told that the surrounding forests are, during the annual rains, so loaded with malaria, that it is fatal even to travel through them.

April 15th. – Therm. in our tents at Deyra, 82°. – At 4 A.M. this morning, my friend and myself started on an expedition to the two mountain-stations above mentioned, Llandowr and Missouree. Mr. Shore was kind enough to send two capital ghoonts (mountain ponies) for us to Rajpore, a village at the foot of the mountains, seven miles from Deyra. We galloped on our own horses to this place, where we found the rough little brutes, with two guides, awaiting us.

We immediately mounted upon the well-padded saddles and commenced the ascent, being duly cautioned by the men, to lay the bridle on the necks of our ponies, and allow them to rest when they pleased. We accordingly yielded the usual prerogative of a horseman to influence the motions and will of his steed, and proceeded.

The distance from the base of the hill to Llandowr, the most elevated of the two stations, is eight miles; and the road sometimes winds round the edge of the rocks, sometimes zig-zags up the face of the hill, plunges into the dark depths of a ravine, or creeps over the giddy summit of a naked crag. It is, in its present unfinished state, in most parts not more than three feet wide, not unfrequently diminishing to one foot, and even six inches!

The journey is certainly a nervous one for beginners, for though we have both rambled through the Alps, we have been rather accustomed to trust to our own feet than to ride in mountainous expeditions. The obstinate little ghoonts, too, always select the very outermost edge of the track, so that the rider, glancing his eye over the shoulder of his beast, sees nothing between him and the unfathomable abyss below; unless it be detached atoms of stone or earth, crumbling from the passing tread of the animal, and rolling over the precipice.

We were an hour and a half in reaching a little halfway hamlet, situated, or rather perched in a small nook of cultivation; and an hour and three-quarters in accomplishing the last half of the upward journey.

To us, just emerged from the plain, it appeared disagreeably cold. The therm. stood at 54°, whilst at Deyra it ranged at 80°. Llandowr is seven thousand four hundred feet above the level of the sea.

After enjoying the most splendid view of the snowy range, the beautiful Doon, the mountains beyond it, and in the dim distance beyond them the wide-spreading plains of Hindostan, we remounted our ghoonts, and set off for Missouree, which is somewhat

Red Panda
Ailurus fulgens

The Red Panda is a tree-dwelling mammal found in the forests of the Himalaya. Characterized by its red fur, the Red Panda looks very different from its larger relative the Giant Panda. One subspecies is found only in China, while the other subspecies is found in India, and also in parts of southern China, Bhutan, Laos, Myanmar, and Nepal. It is estimated that there are less than 2,500 adult Red Pandas left in the world.

The Red Panda weighs about 3–6 kg and grows to a length of 50–60 cm with a long tail of 20–50 cm. With white-tipped ears, white markings on its face, long whiskers, and a bushy striped tail, it resembles a raccoon. These pandas prefer bamboo forests at an altitude of 1,500–4,000 m. They are also seen in coniferous, oak, and rhododendron forests. Red Pandas mainly eat bamboo leaves and bamboo shoots, but also fruits, flowers, nuts, seeds, insects, small birds, eggs, and rodents. They are excellent tree-climbers and their paws are equipped for grasping the stems of bamboo plants and branches of the trees while climbing and feeding. The rest of the time they sleep on the branches or in the hollows of trees. The Red Panda is a solitary animal. The cubs live with their mother till they are 3 years old.

From a painting in the BNHS Library collection.

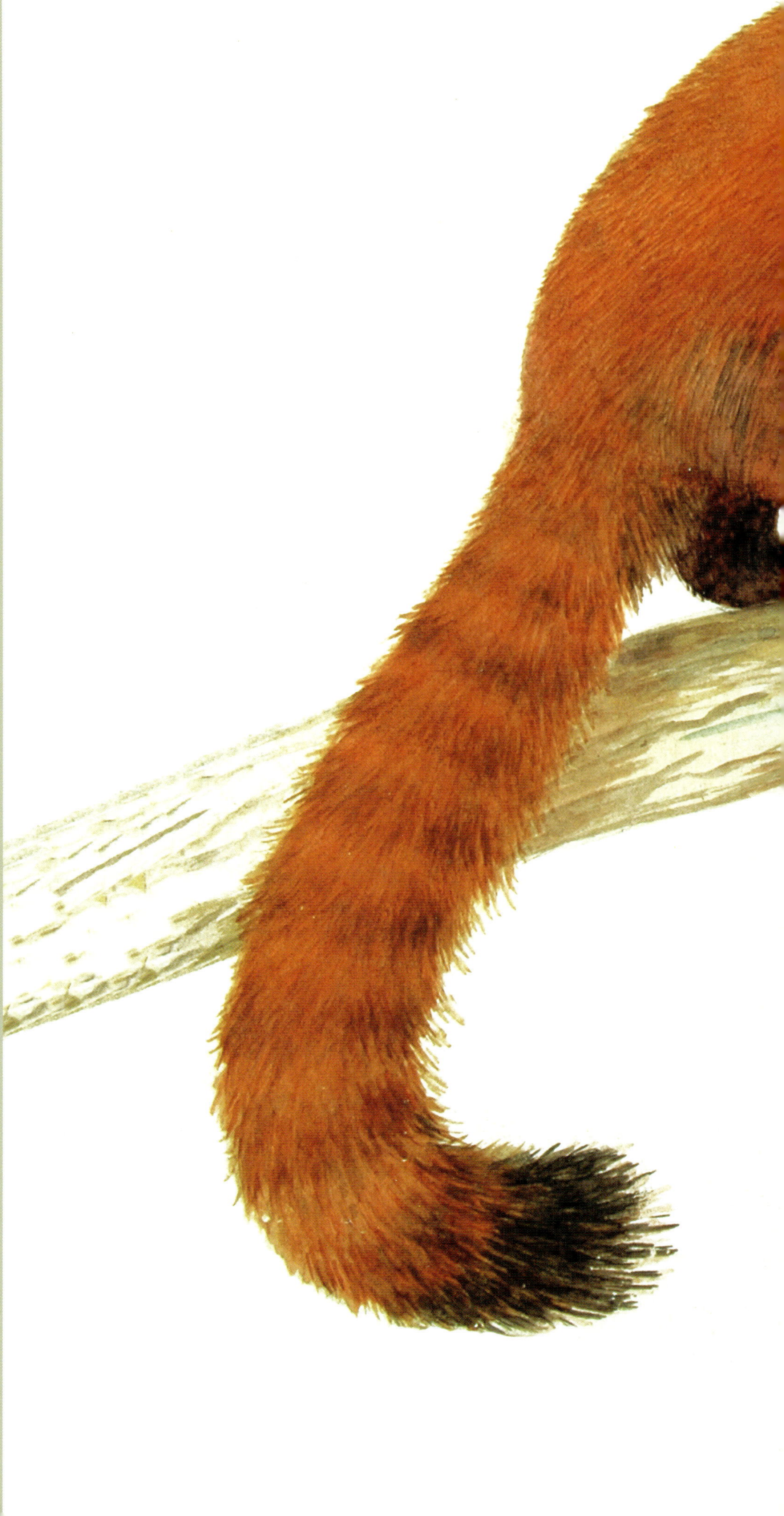

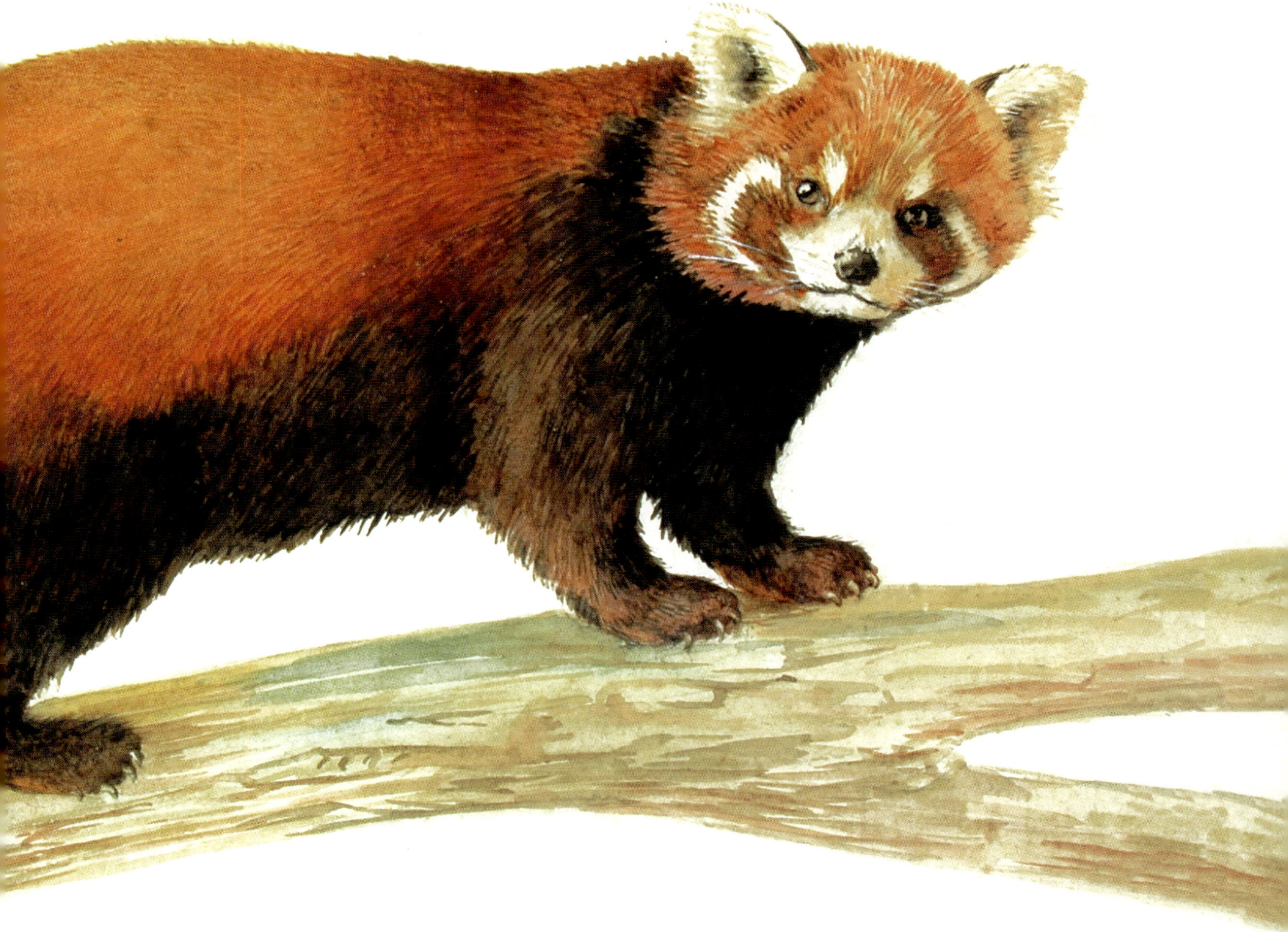

Courtesy HSBC
The Hongkong and Shanghai
Banking Corporation Limited

lower than, and three miles distant from, Llandowr. Among the various and beautiful trees and shrubs of these mountainous regions, I was delighted to recognize many old English friends. The oak and the rhododendron are the largest timber trees; and of the latter, which in Europe and America is a mere shrub, the beams of the Llandowr houses are formed. At this point they are covered with a luxuriant crimson flower, and their stem, as well as those of the oak, are thickly clothed with a long and hoary moss. During our descent I also discovered the cherry, pear, barberry, and raspberry, which are unknown in the plains.

Missouree is situated on a table hill, and is less wooded than Llandowr; but it has greatly the advantage in point of space. We called upon Major Young, who resides here, and he obligingly furnished us with directions for finding tigers in our progress through the Doon towards Simla.

The descent we found infinitely more fatiguing than the ascent, but our nerves grew callous in proportion to our fatigue; though we were obliged to dismount in a few bad places. At Rajpore we found our gig, and drove into Deyra just in time to dress for dinner. It was a good day's work. We rode twenty-six miles, nineteen of mountain equitation; and drove seven miles.

On our arrival at our tents we found five elephants sent to us from the Head-quarters commissariat, as a reinforcement for our tiger-campaign. Having already four of these animals, we shall possess a good string of beaters for the marshes of Synspore. This evening, as if on purpose to give employment for our sporting forces, a despatch arrived from one of a fishing party of gentlemen at Synspore, informing us that a Captain Kent had that morning been attacked and severely injured by a tiger. He was brought into Deyra, and his wounds examined, which though severe were not dangerous. The tiger's claws had cruelly lacerated his shoulders and breast, and one tooth had penetrated his cheek. His escape was most marvellous. He was imprudently shooting on foot in the thickest of a jungle famed for tigers, and had just fired both barrels at a deer, when the tiger sprung upon him from a tuft of grass, and knocked him down. His life was probably saved by his gun. He was carrying it on his shoulder, and the tiger seized it in his mouth, instead of the captain's head. The marks of his tusks were distinctly impressed upon the barrel, and a fragment of a shivered tooth was found afterwards in the gentleman's waistcoat. It is, therefore, probable, that his abstaining from further attack was in consequence of his finding an Englishman so much tougher a mouthful than he expected; true it is that he bounded on, satisfied with one round.

The next day, bidding adieu to Deyra, we resumed our march. We found our canvas tabernacles erected on a hilly forest-ridge, near the hamlet of Peleon. This ridge is cut up by the violence of the periodical rains into numerous deep and tangled ravines, which debouch upon the valley: the vale itself is intersected by numerous streams and

Blue Locust and Faggot Caterpillar in Gujarat

Drawing by James Forbes, 1779. Engraving by William Hooker.

James Forbes had occasion to witness a flight of locusts "extending over a mile in length, and half as much in breadth … like a black cloud, [they] cast an awful gloom like that of an eclipse over the garden, and caused a noise like rushing of a torrent, they took nearly a hour passing over our little territory." Of the Faggot Caterpillar he wrote, "The Baubul tree (acacia) afforded a curious specimen of insect sagacity in the caterpillars' nests, suspended by thousands from the branches. The little animal conscious of its approaching change … instinctively provides itself a strong mansion during that metamorphosis."

From James Forbes, *Oriental Memoirs*, Vol. III, 1812–13.

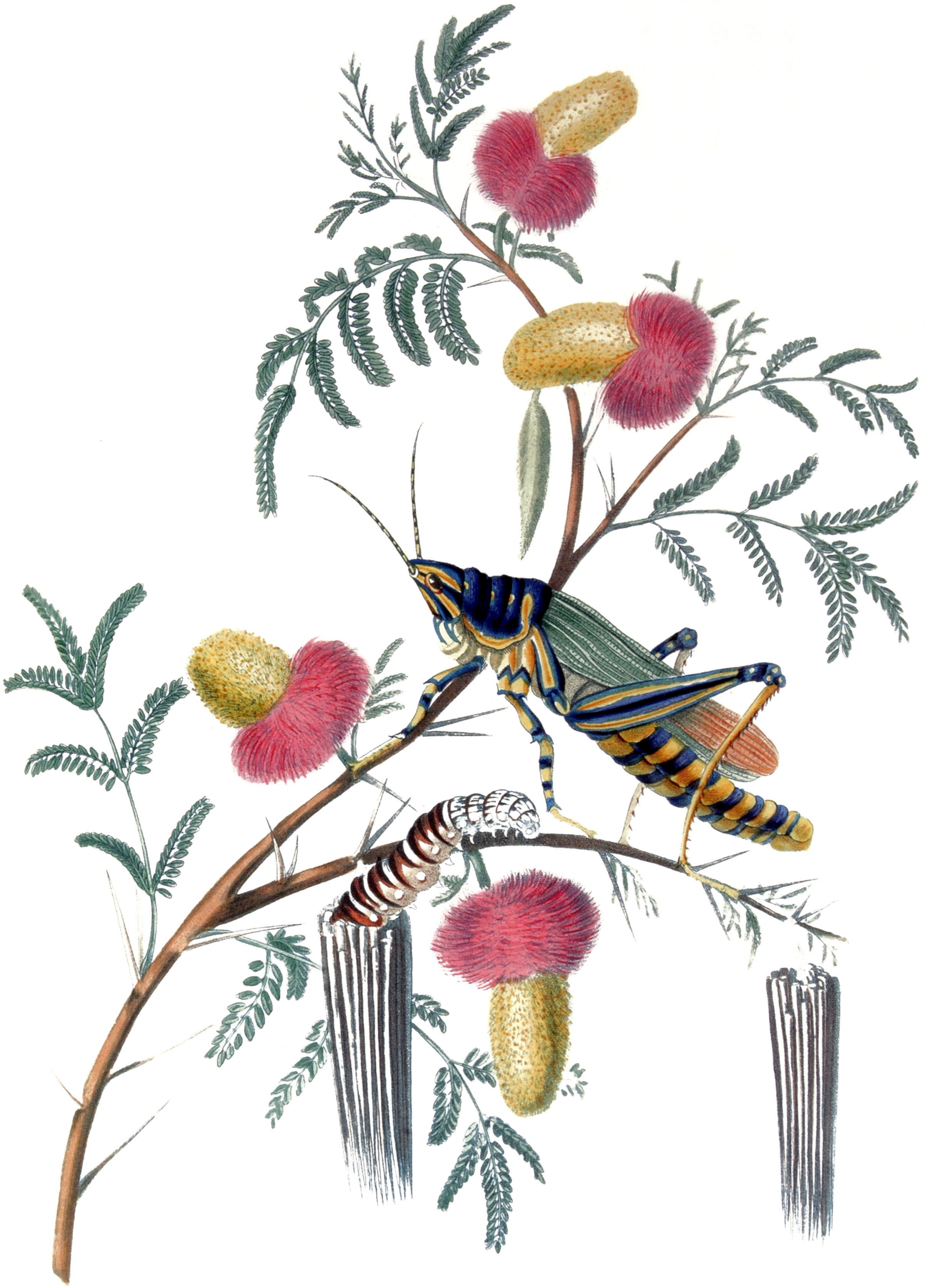

Locust & Caterpillar

COURTESY
KANSAI NEROLAC PAINTS LIMITED

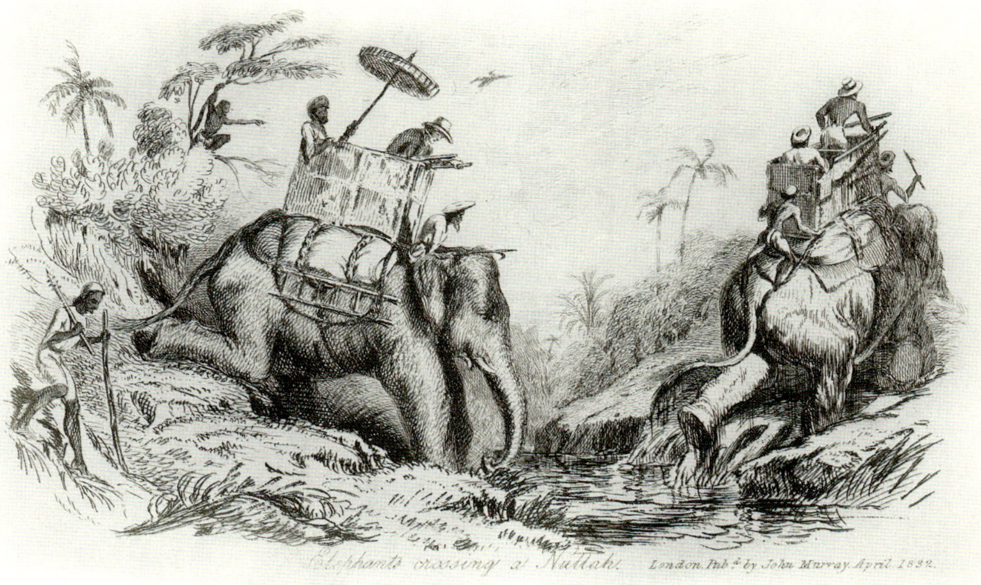

Elephants crossing a Nullah.

nullahs; the centre is grassy and swampy, and free from trees, which, however, grow thickly and wildly on the higher ground, and in the ravines. Being now in the midst of the most favourite haunts of wild beasts, our servants received directions to secure themselves and our camels, ponies, and sheep within the circle formed by the tents and the elephants, and to keep large fires blazing all night.

At 2 P.M. we took our little meute of elephants, and, descending the valley, commenced a battu for a tiger. We hunted in vain for two hours along the river side, and my companion, taking a detachment of beaters, wandered up a grassy ravine, to try his fortune in the forest jungle. We had not been long separated, before I heard one of his elephants trumpeting and snorting most furiously; and, guessing the cause, hurried to the scene of action. I found that my friend had come suddenly upon a fine tiger, which had charged him when he was most awkwardly situated among large trees, and on the steep brow of the ravine, and had then disappeared among the brushwood. Proceeding cautiously up the hollow, we soon found the mangled carcase of a heifer concealed in the dry channel of a little nullah; and, further on, a perfect Golgotha of bones. We were just congratulating ourselves on having so correctly hit upon the spot, when, looking diligently among the bushes thickly clothing the flanks of the ravine, I distinguished the well-known black stripe and yellow skin through a small aperture in a thicket, within ten paces of me, and on a ledge of ground higher than my houdah. In this contingency consists the chief danger of ravine shooting; the animal, by his advantage of position, being enabled to spring even into the houdah, a feat which, on fair ground, he could never accomplish.

Stopping my elephant, I took, as I thought, a good aim, and pulled both triggers; when, instead of the monster which I expected, out rushed a little cub tiger of about three months and charged me so courageously, that my elephant took to her heels, and was in full career towards the deep dry nullah above mentioned, when I made a successful *Parthian* shot with my favourite Joe Manton, and slew my determined little pursuer.

My friend's elephant, which was equally ill-behaved on this occasion, in her anxiety to escape slid down the steep bank about forty yards, dragging her hind legs after her in the most ridiculous but *periculous* manner. The mahouts declared that there must be another cub; but, as it was growing dusk, we deferred our pursuit until the morrow.

The next morning, whilst our camp advanced to Synspore, we proceeded in due form to our appointment with the tigress and the remaining cub. Mr. Grey, a young medico from Deyra, joined us by invitation, and we mounted him for the occasion: it was his first tiger-chase. We proceeded as direct as we could to the scene of yesterday's skirmish. Grey and I were close together, and D. a little on the flank; when, about one hundred yards from the spot where the tiger was seen yesterday, G. suddenly exclaimed, "Here she comes!" and I saw the beautiful animal creeping quickly out of a richly-flowering lemon-thicket, about forty paces from us, and approaching us in a crouching attitude, with her fierce eyes fixed upon us. We fired so simultaneously, that neither heard the other's shot, and the tiger rushed back into the thicket. Fearing her escape, we hurried to the spot, and my elephant tearing up the bush, discovered her in the centre. I immediately fired, but the shot was useless, for she was stone-dead.

On dragging her out of the odoriferous lemon-bush, she proved to be a very fine tigress, her dugs full of milk, and measuring in length ten feet five inches. Coming out of the thicket, we fell in with the surviving cub, the last scion of this ill-fated family; and Dawkins, dismounting from his elephant, and arming himself with a kookaree, or mountain-dagger, attempted to take the beautiful little creature alive; but it made so furious an attack, that he was obliged to knock it down.

Having secured our game upon the elephants, we drove on to our camp at Synspore, Mr. Grey accompanying us. Our encampment occupied a beautiful site on the edge of the forest, with a green lawn in front, gently sloping down to the rushy level through which the little river meandered. For miles in advance the banks of the stream are clothed with thick grass; and the frequent patches of tall, rank rushes and reeds, with the filmy marsh-miasma floating and dancing in aguish wreaths above them, wore greater charms in the eyes of us tiger-hunters, than would have the far-famed gardens of Shalimar, or even those of royal Kensington.

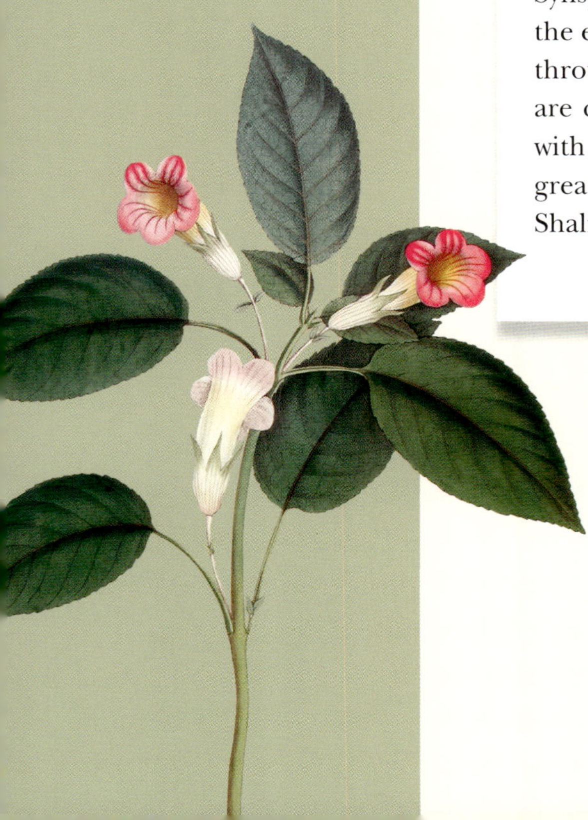

We had no sooner arrived at Synspore, than the Jemadar of the village presented himself, and, with a thousand "khodabunds!" and "maharajs!" besought our aid against a tiger, which, he said, had this season killed nine bullocks and a horse belonging to his village, and had two days ago wounded a veleitee sahib (English gentleman), doubtless Captain Kent. Making sure of our prey the next day, we determined to give our elephants a good rest (for they were beginning to look thin, and grow foot-sore), and employed this morning in casting balls, and manufacturing squibs and crackers for the swamps on the morrow.

Synspore, April 19th. Out all day, with a line of twelve elephants, in search of Captain Kent's tiger; but returned without finding even a trace of him. The villagers, who accompanied us in great numbers, were as much disappointed as ourselves, and supposed the tiger had retired to the inaccessible parts of the marshes. Into some of these we obliged the mahouts to force our elephants in spite of their sagacious resistance; and more than once we were nearly thrown out of the houdahs, by the violent exertions made by the frightened animals to extricate themselves from the sinking quag. We threw fireworks as far as we could among the reeds, and were even imprudent enough to dismount, and try to proceed further into the swamp on foot; but after getting thoroughly wet, and finding ourselves, when above our ankles in a most retentive bog, in a poor situation for resistance should the tiger attack, we retraced our way back to the elephants, and returned, well tired, to a late dinner. The evening was beautiful; one side of the tent was raised to admit the coolness of the night breeze, which wafted gently to us a delicious tribute of fragrance from the surrounding forest; and as we sat in the bright moonlight, sipping our lall-sherab (claret), and inhaling long draughts of complacency from our rose-odoured hookahs, we all agreed that this gypsy-like life in the wilderness is, for a season, enjoyable in the extreme.

Next day, my servant had a curious escape; – he requested permission to dismount from the houdah to drink at the river, and was in the act of stooping to raise the water in his hands to his lips, when I heard him utter a scream, and then a grand splash in the river. I rushed forward through the long grass, fully expecting that a tiger had seized the poor fellow, and found him standing in the stream, feeling his limbs all over, and blubbering from fright. An immense boar, roused from his lair, had charged him in the rear, and, ripping up his voluminous cummerbund, tumbled him headlong into the stream – fortunately, without the least injury. The animal continued his career without repeating his attack.

21st. – Therm. in tent 87° – Dr. G. left us, and returned to Deyra. Shot our way to Raj Ghaut, a timber depot and ferry on the Jumna. This great river, the next in rank to the holy Gunga, emerges at this point from the Himalayah mountains, and is even here, so near its source, wider than the Thames at Windsor.

Magnolia hodgsoni

Family: Magnoliaceae

Synonym: *Talauma hodgsoni*

This is a native of northeast India, Bhutan, Nepal, northern Myanmar, Thailand, southwestern China, found at altitudes from 800 to 1500 m. It is an evergreen tree, up to 15 m tall. The leaves are leathery, obovate-oblong, 20–50 cm long and 10–13 cm broad. The fragrant white flowers appear in April–May, and have nine petals up to 9 cm long. The fruit resembles a large pine cone 13–15 cm long, composed of 40–80 follicles. The wood is soft and worthless as is the case of almost all Himalayan Magnoliaceae. According to Hooker, "This is not an uncommon plant in the Sikkim forests, as at Khersiong and below Leebong, where it grows close to the road. It is always densely clothed with its large, handsome, coriaceous evergreen leaves. The flowers are very fragrant, aromatic; are exceedingly handsome, from rich plum-bloom on the purple outer petals, contrasting with the ivory whiteness of the inner ones, all the pieces of the flowers are thick, hard and fleshy."

From J.D. Hooker, *Illustrations of Himalayan Plants*, 1855

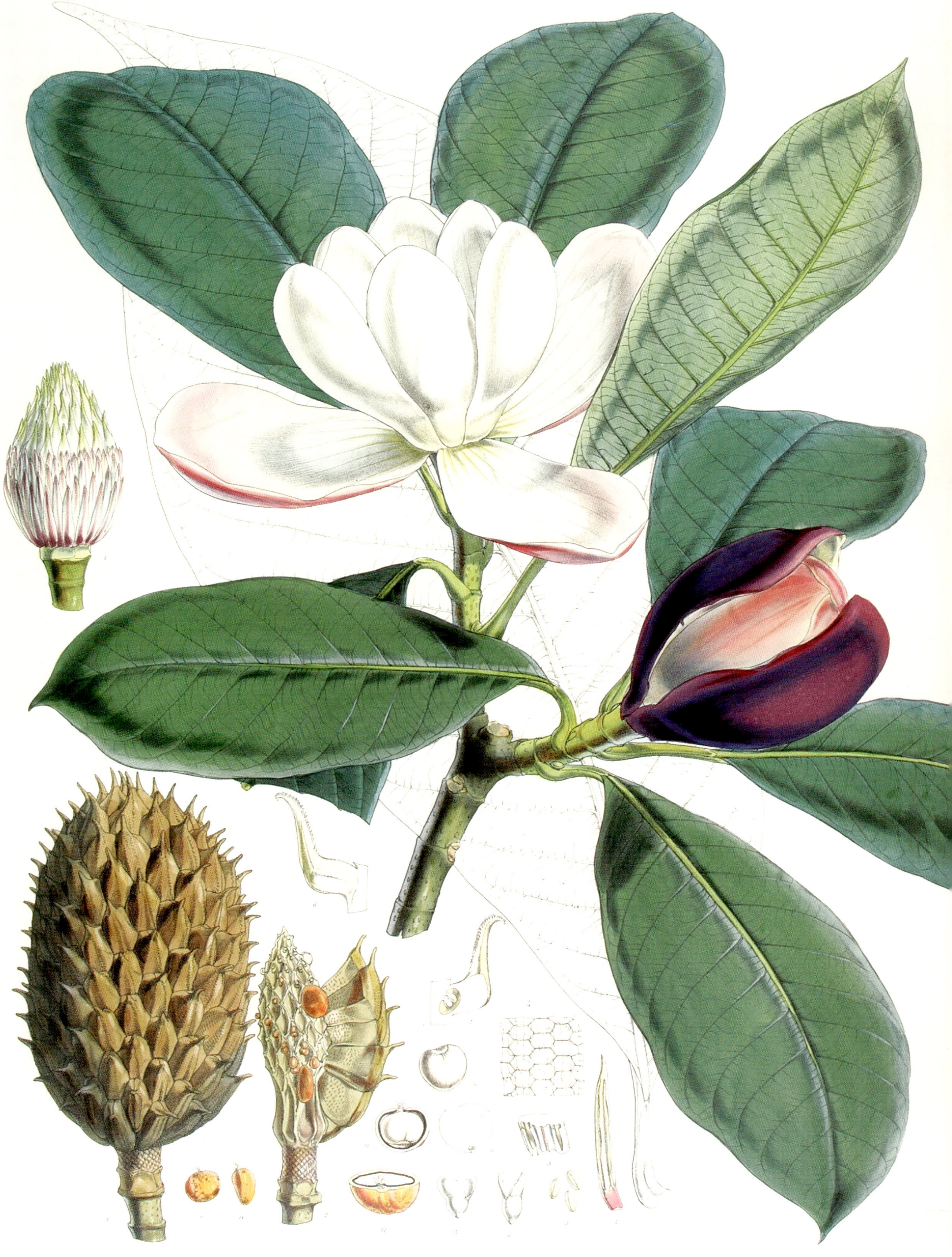

Magnolia hodgsoni

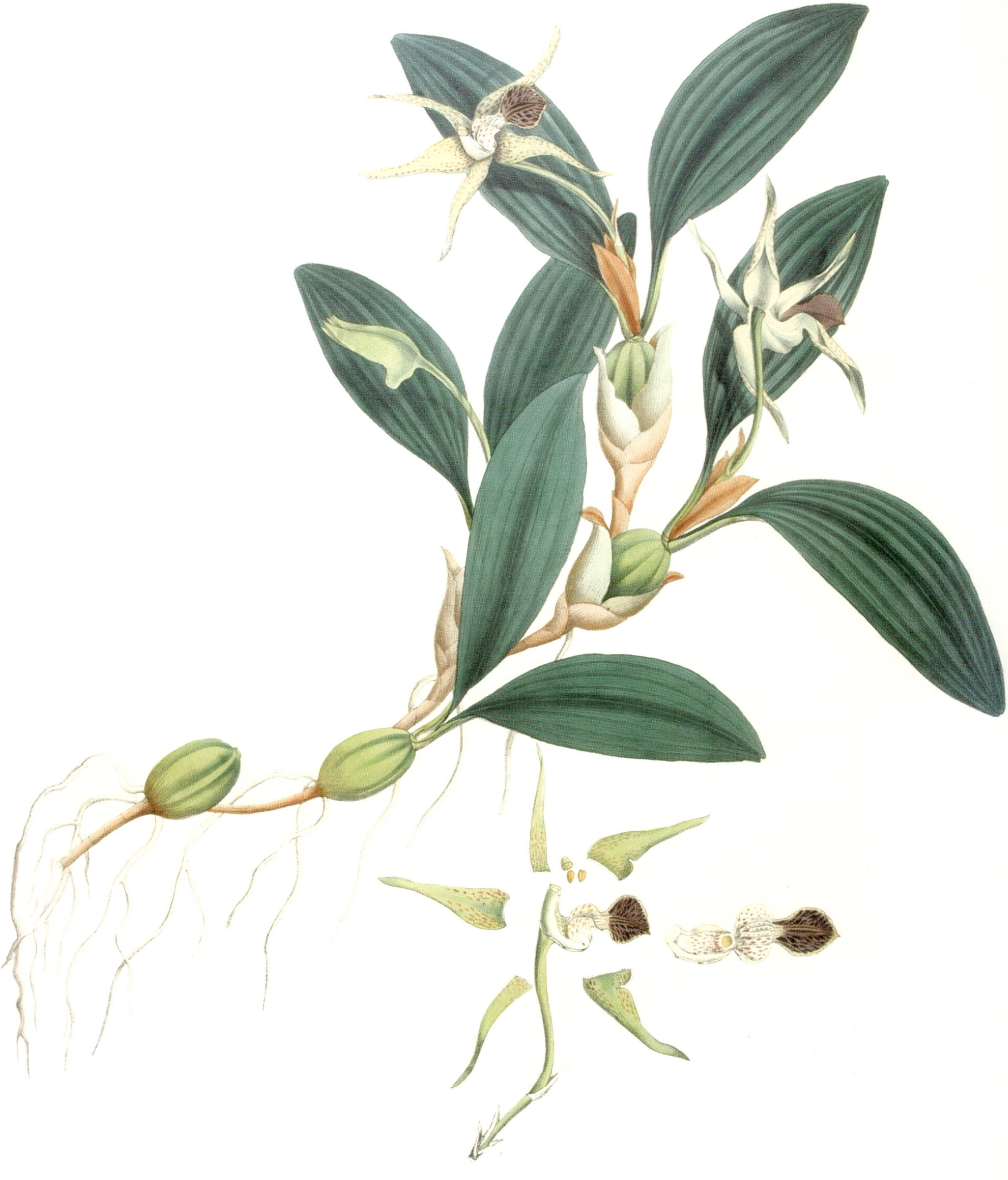

Dendrobium ampulum

IN MEMORY OF
SOHRABJI PIROJSHA GODREJ,
A NOBLE SOUL, AN ARDENT NATURALIST
AND TREE-LOVER

Black-tongued Orchid
Dendrobium ampulum
Family: Orchidaceae
Synonym: *Epigeneium ampulum*

Found in Sikkim, Assam, and Meghalaya at 1,500 m, this pendulous woody creeping epiphyte bears 4-sided oblong pseudobulbs, at intervals on a creeping rhizome. The leaves are elliptic-oblong, 10 to 15 cm long and about 3.5 to 5 cm broad. The flowers appear in October–December, are large, solitary, 7 cm in diameter, borne on the apex of the pseudo-bulb; the sepals and petals are white to dull greenish-brown, profusely spotted and blotched with darker brown. The lip of the flower is large, trilobed; the midlobe is dark purple with brown edges, so deep as to appear black.
According to Wallich, "This plant attains a very large size, and is one of the most beautiful of the genus to which it belongs. It is very rare in Nepal, and I have only found it towards the summit of one of the highest mountains. It grows parasitically on trees and rocks in moist and shady situations, and produces the large flowers towards the end of the rainy season."

From N. Wallich, *Plantae Asiaticae Rariores*, 1832.

It issues upon the Doon through a beautiful defile flanked by richly-wooded hills, and, sweeping across the valley, receiving the tributary waters of the little stream along whose banks we have been travelling, forces its way through the southern range of mountains, and bursts upon the plains; where, after running through the great cities of Delhi, Muttra, and Agra, it finally blends itself with the Ganges at Allahabad. In the vicinity of Raj Ghaut there is very good trout-fishing, both in the Jumna, and in the smaller streams.

The following day we crossed the river in a rude boat – the elephants wading and swimming over – and entered the Kearda Doon, a continuation of the Deyra valley, to which it bears in its features a miniature resemblance. Our day's march was eight coss, or about twelve miles; the roads were execrable, and almost dislocated every joint of the poor buggy. The valley gradually narrowed as we proceeded, and we found our tents at the village of Kearda, romantically situated in a dell completely encompassed by woody heights, on the summit of the nearest of which are the ruins of what appears to have been a petty Goorkah fortress.

A little, ugly, but athletic Shikkaree, who visited our camp, informed us that the surrounding forests abounded in bears, leopards, hyaenas, wild elephants, and even tigers; but the coverts are so heavy, and cut up by ravines, that they are not accessible to mounted sportsmen. A small species of pheasant and the jungle-poultry are also found in the hills – the latter were crowing all day on the heights above us.

We took a sporting stroll in the afternoon, and brought home a good bag of partridges and hares: we also roused a very large boar, but he escaped. During the chase I found myself in a fine grove of mulberry-trees, and regaled myself with the fruit, which was very refreshing on the hottest day I almost ever experienced. The berry is sweet and juicy, but much smaller than it is in Europe. Returning towards camp, in the dusk of the evening, I saw a large snake, (the only one I ever met with when out shooting,) which, as it rolled its scaly length over the brow of a ravine, seemed to be about the thickness of a man's arm. I had but a momentary glimpse of it, and fired, the ball splintering the piece of burnt timber under which it was gliding.

23rd. Continued our march and found our advanced tents scarcely pitched. The servants, and the Goorkah Sepoys who formed our guard, told us that they were much delayed during their night march by the wild elephants, who made several attacks upon their tame brethren carrying our tents, and were only repulsed by the firing of the Sepoys. These huge inhabitants of the woods are rarely seen except in the night, when they commonly descend to the valleys.

That night nothing alarming occurred; except, perhaps, that we heard the wild elephants trumpeting and crashing in the distant forests as they commenced their descent to the valley to drink at the nullahs.

From Captain Mundy, *Pen and Pencil Sketches, being the Journal of a Tour in India*, Vol. I, 1832. (Dr. Ashok S. Kothari Collection.)

In memory of the late Shri Mukund Desai
from Naina M. Desai and Sonal & Ratnesh Desai

When the Tiger Was King

T.C. Jerdon

The Tiger (*Felis tigris*)
Descr.– Bright fawn colour, more or less tinged with rufous, and with dark stripes.

"The peculiarly striped skin of the tiger," says Blyth, "at once distinguishes it from every other feline animal, and equally so does the intensity of the bright rufous ground hue, so exquisitely set off with white about the head." Again, "unless the lion, no other cat approaches it in the massive proportions of the fore paw, as compared with the hind. Some of both sexes are made more heavily than others, with a greater development of the fold of skin along the belly, which adds to their apparent bulk. The stripes too vary much in different individuals, and occasionally are almost throughout double."

The tiger is found throughout all India, from Cape Comorin to the Himalayas, ascending the hills occasionally to an elevation of 6,000 or 7,000 feet. It is found in all the forests and jungles throughout the peninsula, occasionally visiting the more open and cultivated parts of the country, and harbouring in thickets, long grass, and especially in brushwood on river banks, and on churs covered with tamarisk. In the hot weather, indeed, these are its favourite resorts in many parts of Central India, and from there it sallies forth towards the villages in search of food. In lower Bengal the heavy grass jungles and swamps are his usual lair. Tigers are perhaps more abundant in lower Bengal than in most other parts of the country, and are said to be both larger and more savage than those from other parts of India. Those of Central India, however, are perhaps as large, and quite, if not more ferocious than their Bengal brethren.

The average size of a full grown male tiger is from 9 to 9½ feet in length, but I fancy there is very little doubt that, *occasionally*, tigers are killed 10 feet in length, and perhaps a few inches over that; but the stories of tigers 11 feet and 12 feet in length, so often heard and repeated, certainly require confirmation, and I have not myself seen an authentic account of a tiger that measured more than 10 feet and 2 or 3 inches. Major Sherwill who was for some years in Dinagepore district, told me that the largest he had seen killed was 9 feet 8 inches. The skin is very often measured either when fresh taken from the carcass, or after it has been stretched out to dry; and Mr. W. Elliot records an instance of a lion measured by himself at 9 feet 4 inches. This was noted by one of the party as 11 feet and by another as 12 feet, the first measurement being taken from the skin when taken off; the other from the skin stretched out by pegs for drying.

In the 19th century, when this account was written, tigers were plentiful all over India. Dense forests existed, where the tiger ruled. Over the decades, the felling of trees for cultivation, construction of roads and railways through forested areas, hunting, and lately poaching have greatly affected the conditions of jungle life. Today the alarming decrease in the number of tigers is a matter of national concern.

"Mundlah" is Mandla in today's Madhya Pradesh; it was then part of Madras Presidency.

Kharsu Oak
Quercus lamellosa
Family: Fagaceae
Synonym: *Cyclobalanopsis lamellosa*

A medium-sized to large evergreen tree, a native of the Himalaya, growing up to 40 m tall with a trunk up to 1.5 m in diameter. It grows in Assam, Sikkim, eastern Nepal, Bhutan, southwest China, and Thailand, at altitudes of 1,300–2,500 m. The leaves are large, dark green, spirally arranged, oval-elliptic, 16–45 cm long and 6–15 cm broad, with a sharply serrated margin, and with many pairs of strong lateral veins. The flowers are catkins and appear in April–May. The female flowers mature into broad acorns, 2–3 cm long and 3–4 cm broad set in concentrically ringed cups that almost enclose the nuts. Hooker remarks, "The present is one of the commonest trees around Darjiling and is certainly by far the noblest species of Oak known, whether from the size of the foliage or acorns, the texture and colour or imposing appearance of the tree."

From J.D. Hooker, *Illustrations of Himalayan Plants*, 1855

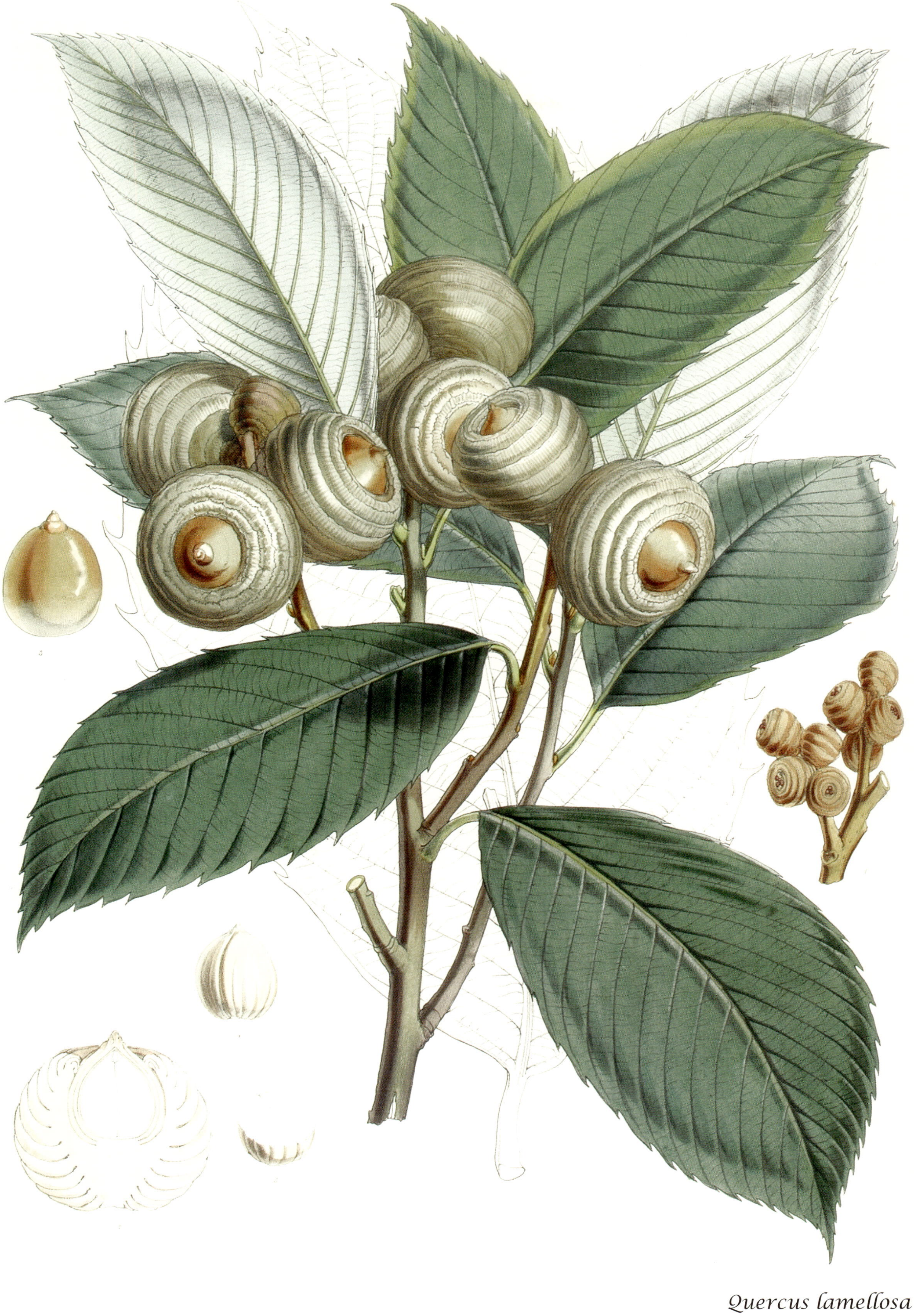

Quercus lamellosa

In memory of Bidyanand Dubey
– Pradeep Dubey, Rajeev Dubey, Ravi Dubey

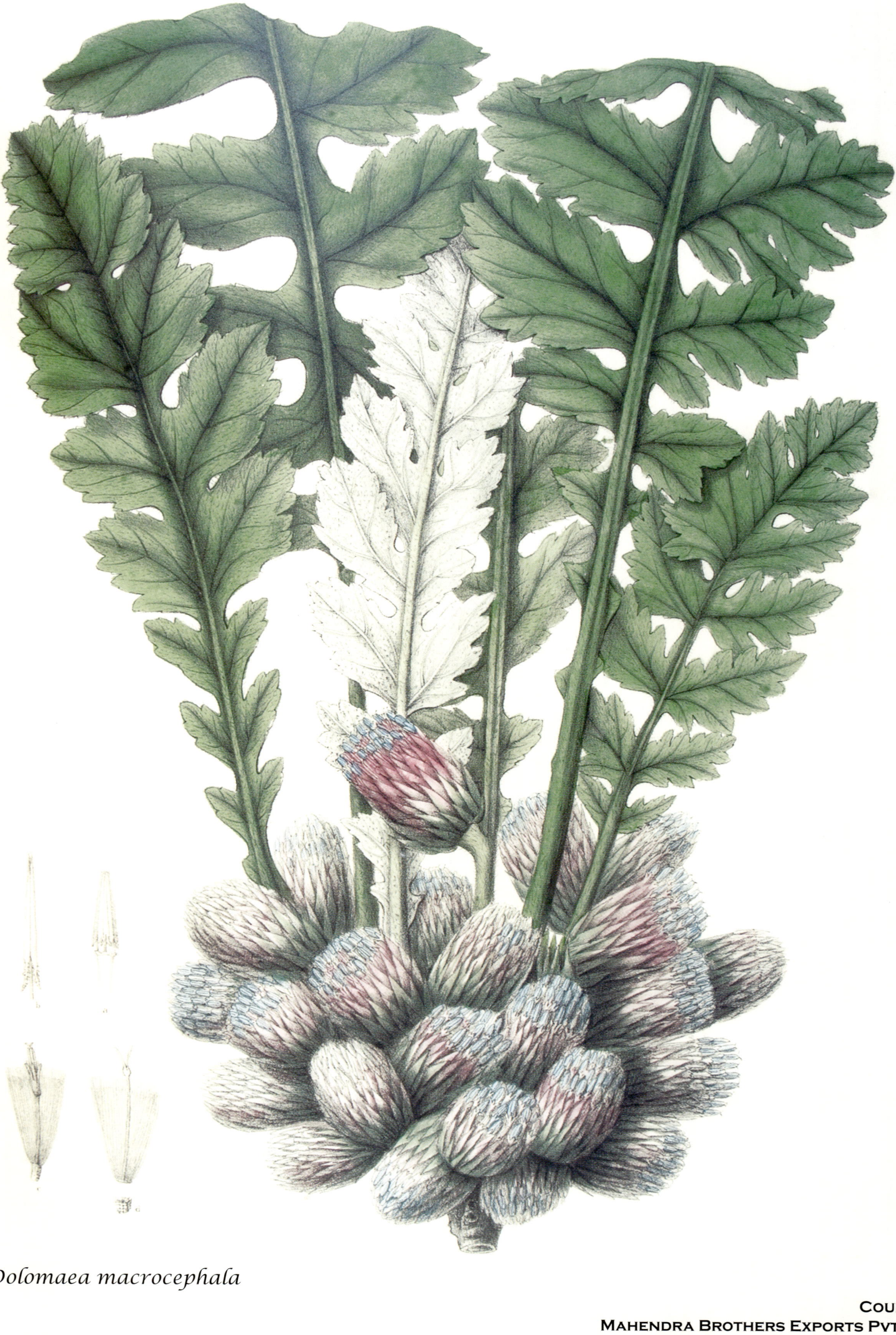

Dolomaea macrocephala

COURTESY
MAHENDRA BROTHERS EXPORTS PVT. LTD.

Dolomaea macrocephala
Family: Asteraceae

Also spelt *Dolomiea* (or *Dolomiaa*) *macrocephala* and locally called Dhupa or Gugal in Punjab, the root being known as *pokhar mul*. In his *Cyclopaedia of India* Surgeon General Edward Balfour notes the following: "A plant not uncommon from the Sutlej up to the Indus, at from [3,500 to 4,300 m], often growing on the crest of the ridges. The odorous root is locally used as incense offered at shrines and to rajas and the flowers are also placed in temples on the Sutlej." In the *Journal of the Asiatic Society of Bengal*, No. 81 (September 1837) Lieut. Hutton's account of his tour to the Borenda Pass is quoted: "Here I found some beautiful flowers growing among the moss and lichens above which they scarcely peeped, as if afraid to lift their heads into the chill and desolate regions around them. Some of them occur in Royle's book of the Himalayan flora such as *Dolomiaa macrocephala* which was abundant and in flower."

From J. Forbes Royle, *Illustrations...*, 1839.

Mr. Walter Elliot (Cat. Mammalia, South Mahr. Country) has the following remarks on the distribution and habits of this animal in Southern India. "The tiger is common over the whole district, breeding in the forest and mountain tracts, and coming into the open country when the grain is on the ground, in some places they do much mischief, and have been known to carry off the inhabitants out of the villages whilst sleeping in their verandahs during the night. The female has from two to four young, and does not breed at any particular season. Their chief prey is cattle, but they also catch the wild hog, the sambar, and more rarely the spotted deer. It is naturally a cowardly animal, and always retreats from opposition, until wounded or provoked. Several instances came to notice of its being compelled to relinquish its prey by the cattle in a body driving it off. In one case an official report was made of a herd of buffaloes rushing on a tiger that had seized the herd-boy, and forcing it to drop him. Its retiring from the wild hog has been already adverted to. Though the wild hog often becomes its prey, it sometimes falls a victim to the successful resistance of the wild boar. I once found a full-grown tiger newly killed, evidently by the rip of a boar's tusk; and two similar instances were related to me by a gentleman who had witnessed them – one of a tiger, the other of a panther. It is generally believed a tiger always kills his own food, and will not eat carrion. I met with one instance of a tigress and two full grown cubs devouring a bullock that had died of disease. Another instance was related in a letter from a celebrated sportsman in Khandeish, who having killed a tigress, on his return to his tents sent a pad elephant to bring it home. The messenger returned, reporting that on his arrival he found her alive. They went out next morning to the spot, and discovered that she had been dragged into a ravine by another tiger and half the carcass devoured. They found him close by, and killed him also. The Bheels in Khandeish say that in the monsoon, when food is scarce, the tiger feeds on frogs, and an instance occurred some years ago in that province of one being killed in a state of extreme emaciation from a porcupine's quill that had passed through his gullet, and prevented his swallowing, and which had probably been planted there in his attempt to make one of these animals his prey. Many superstitious ideas prevail among the natives regarding the tiger. They imagine that an additional lobe is added to his liver every year; that his claws arranged together so as to form a circle, and hung round a child's neck preserves it from the effect of the evil eye; that the whiskers constitute a deadly poison, which for this reason are carefully burnt off the instant the animal is killed. Several of the lower casts eat his flesh."

The late Major Sherwill, of the Revenue Survey, gave me some interesting information on the habits of the tiger, as observed by him in the Dinagepore district – the substance of which is as follows:–

Tigers are very partial to certain localities, and avoid others to all appearance quite as favorable cover. Year after year they will be found in one locality and killed, and never be seen in another close at hand apparently just as suitable. They are fond of ruins, and may often be seen lying on the top of old walls, temples, &c.; sometimes three or four together. Generally speaking the Bengal tiger is a harmless, timid animal, but when wounded he becomes ferocious and dangerous. He seldom molests man without provocation, and man-eaters are very scarce in Bengal, except in the vicinity of the Sunderbuns. A tigress has from two to four cubs at a time, which remain with her until they are able to kill for themselves. Young tigers are by far the most mischievous, occasionally killing as many as four or five cows at once, whilst an old one seldom kills more than what it requires for its food. An old tiger will kill a cow about once a week, and for this purpose will quit its place of retreat in dense jungle, proceed to the vicinity of a village and kill a bullock or cow. It will remain near the "*kil*" for two or three days, and sometimes longer, gnawing the bones before retreating to deep cover. A tigress deposits her young in good cover. Two taken by Major Sherwill were laid under a thorn bush in dense jungle. Null grass appears to be a favourite place for breeding in. The mother appears much distressed on losing her young, and for three or four nights afterwards remains at the spot roaring all night in a very excited manner.

The few remaining observations must be considered as supplementary to the previous observations of Messrs. Elliot and Sherwill. When once a tiger takes to killing man, it almost always perseveres in its endeavours to procure the same food; and, in general, it has been found that very old tigers, whose teeth are blunted or gone, and the vigour of whose strength is faded, are those that relish human food, finding it a much more easy prey than cattle. In some parts of Central India, however, it appears to be more the rule than the exception; and in the Mundlah district, east from Jubbulpore, in 1856, and previous years, on an average between two and three hundred villagers were killed annually. So dire was the destruction that Major Erskine, the Commissioner, applied to the Madras Government to furnish an officer for the special work of thinning these cannibals. In the Bustar country, south-east of Nagpore, when I traversed part of that then unexplored district, I found that in several parts the villages were deserted, entirely, as I was informed, from the ravages of tigers, although in some instances, the villages had been surrounded by a high stockade. In the Bengal Sunderbuns too many wood cutters are annually carried off.

Tiger hunting is generally done from elephants in Northern India; and a well trained shikaree elephant will stand the charge of a tiger well, occasionally even rushing to meet it, which is by no means agreeable to the sportsman in the howdah. In Southern India, where there are but few elephants kept, the tiger is often successfully slain on foot; but it is at all times a dangerous sport, and many serious and fatal accidents are well known to have occurred. Occasionally a tiger is shot by night from a platform, on a tree, either close to where the tiger has killed, but not eaten all his prey, or, with a fresh bullock picketed near. In the Wynaad one class of Hindoos assemble in large numbers; and forming a large circle, drive the tiger into a net, where it is speared. Various modes of capture are practised all over India, and strychnine has been had resort to occasionally to destroy this animal; but in spite of the numbers killed annually by sportsmen, and by native shikarees for the sake of the Government reward, in many districts its numbers appear to be only slightly diminished.

The native idea about the tiger getting an additional lobe to its liver every year has been fully taken up by English sportsmen, and in the pages of the Bengal Sporting Magazine, &c., the number of lobes in the livers of tigers whose death is there chronicled, are duly recorded. The clavicle lies loosely imbedded among the muscles near the shoulder joint, and is considered of great virtue by natives. The whiskers are, in some parts of Southern India, considered to endow the fortunate possessor with unlimited power over the opposite sex. The claws are mounted in silver and made into bracelets.

The tiger is peculiar to Asia, extending as far west as Georgia, through Persia to Bokhara, and is also found in Amurland, in the Altai region and China; thence extending south through Burmah to the Malayan peninsula, and some of the neighbouring larger islands. It is not found in Ceylon.

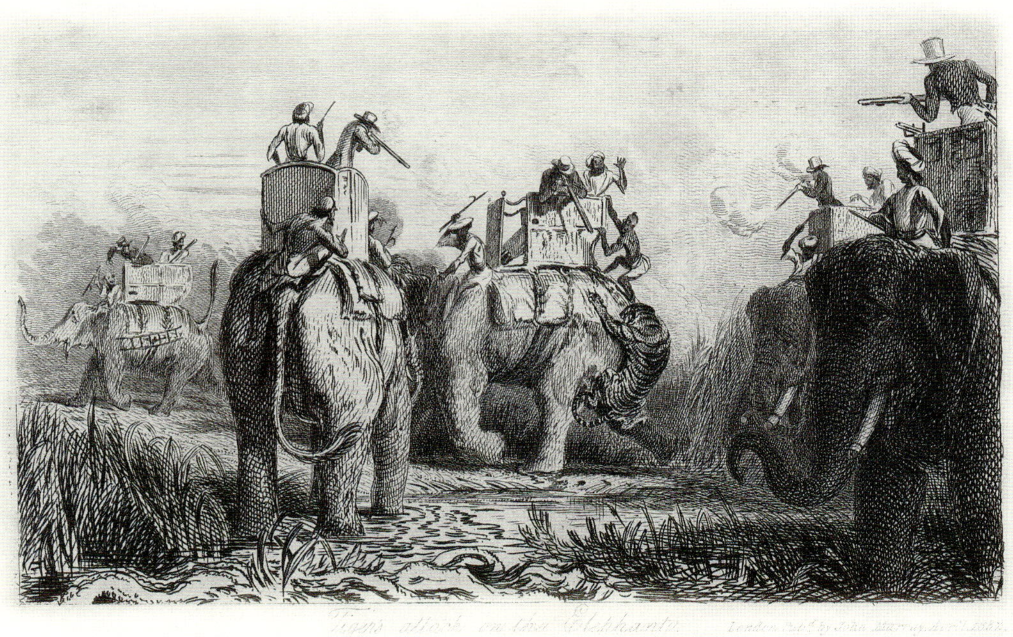

Tiger's attack on the Elephant, from Captain Mundy, *Pen and Pencil Sketches, being the Journal of A Tour in India*, Vol. I, 1832. (Dr. Ashok S. Kothari Collection.)

From T.C. Jerdon, *The Mammals of India*, 1862.

Our Tiger Shooting

"You would like to try your luck at tiger shooting? ... Now you may have a chance to-day ... for this morning an aggrieved agriculturist came howling into the cantonments, reporting the sudden apparition of a tiger, only fifteen miles off, which has abstracted his oxen and his fatlings, and evinced a desire, moreover, to elope with his wife. So Byle has got his ... battery of guns, ... Topey has lent some elephants, a troop of beaters have been enlisted, and we shall be off in the morning by daybreak. ...

"Away then we start. With cunning eye the beaters track the foot-prints in the shade. ... we strain our eyes in every direction; occasionally we have a false alarm, then the elephants turn up their trunks, and go trumpeting along, crushing down the jungle – at times into the deep ravines, into which they lower their huge bodies so carefully and so well, then clambering up again, press onward on their way. A beater gives a shout – he is on the track – he sees blood upon the trampled grass! The sun has begun to shoot down his penetrating rays upon our turbaned heads which need the still greater protection of the umbrella. Another shout! Byle, with a practical eye, judges where the game might be, points to our friend, the agriculturist, who is beside the elephant, where he might explore. A yell, accompanied by a precipitate flight on the part of the bereaved Oriental, followed by a roar from the tiger, and two sharp cracks from a double rifle, announce to the field that the game has been found."

From Capt. G.R. Atkinson,
Curry and Rice, 1859.

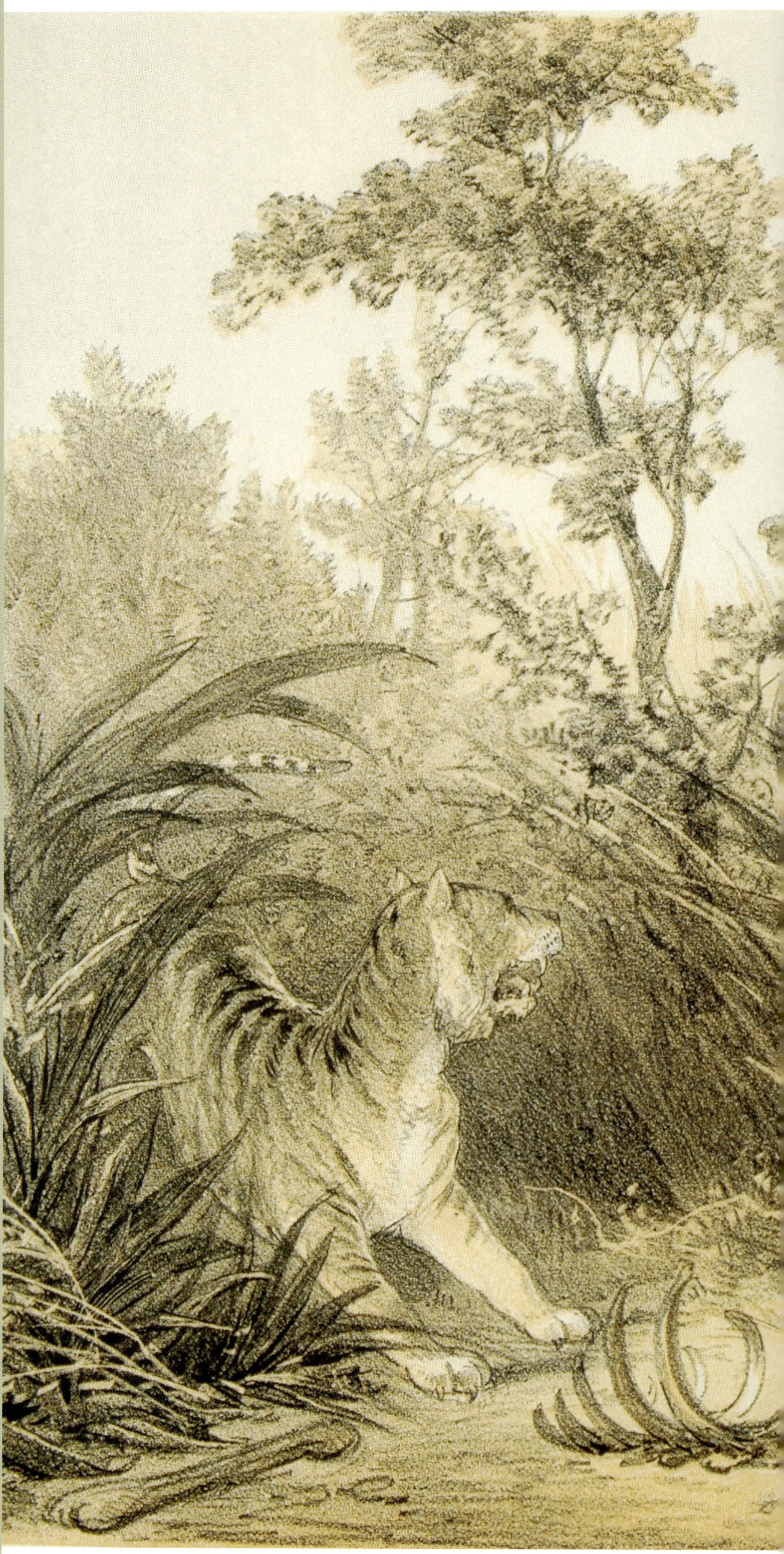

RAJSHEKHAR & SHAILA PARIKH
SAUMIL, SALONI, SAREENA, SHANAYA
JAY, MILAUNI

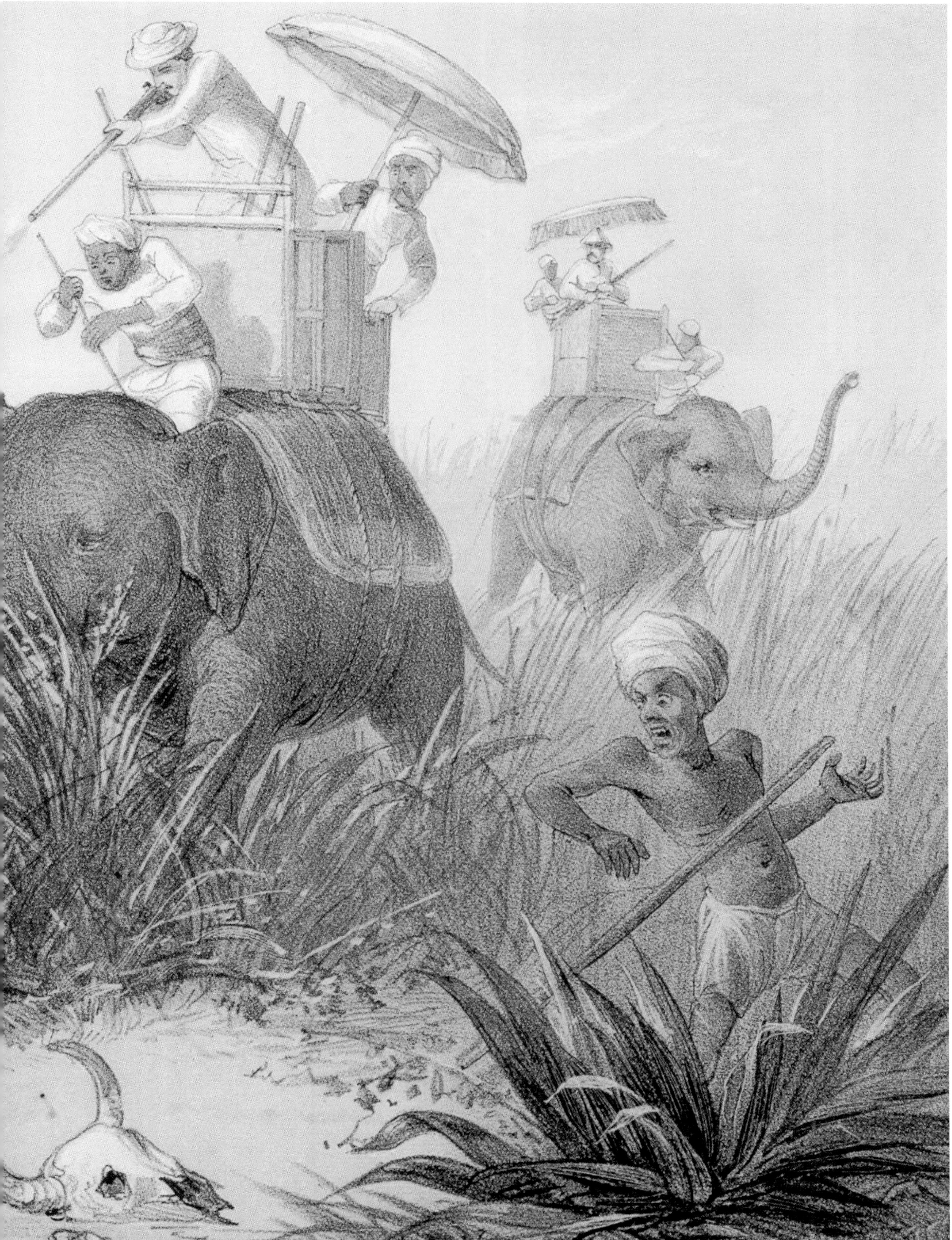

Suresh S. Kothari
Mehul, Dharini, Monil, Megan
Shardul, Rupal, Rhea, Sanya

Antelope Hunt

Captain Godfrey Charles Mundy

ON the last day of the year 1827 we marched to the village of Boingong. Just before we reached our ground, the shutersuwars (camel couriers), who always moved on our flanks in search of game, reported a herd of antelopes about half a mile out of the line of march; and the chetahs being at hand, we went in pursuit of them. The mode of conducting this sport I may as well describe. The leopards are each accommodated with a flat-topped cart, without sides, drawn by two bullocks, and each animal has two attendants. They are loosely bound by a collar and rope to the back of the vehicle, and are also held by the keeper by a strap round the loins. A leathern hood covers their eyes. The antelopes being excessively timid and wild, the best way to enjoy the sport is to sit on the cart alongside the driver; for the vehicle being built like the hackeries of the peasants, to the sight of which the deer are accustomed, it is not difficult, by skilful management, to approach within two hundred yards of the game. On this occasion we had three chetahs in the field, and we proceeded towards the spot where the herd had been seen, in a line, with an interval of about one hundred yards between each cart. On emerging from a cotton-field, we came in sight of four antelopes, and my

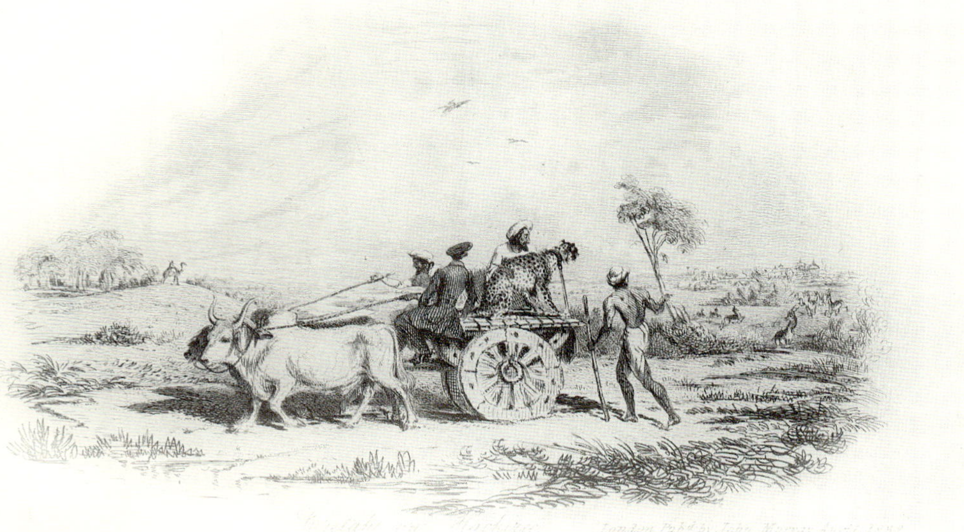

Chetah on Hackerie.

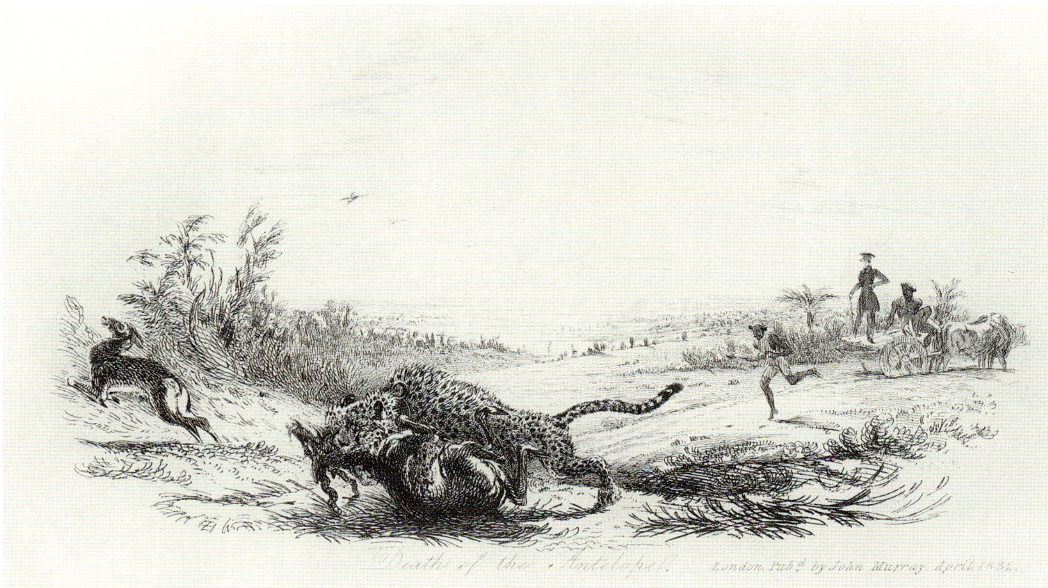

Death of the Antelope.

driver managed to get within one hundred yards of them ere they took alarm. The chetah was quickly unhooded, and loosed from his bonds; and as soon as he viewed the deer he dropped quietly off the cart, on the *opposite* side to that on which they stood, and approached them at a slow, crouching canter, masking himself by every bush and inequality of ground which lay in his way. As soon, however, as they began to show alarm, he quickened his pace, and was in the midst of the herd in a few bounds.

He singled out a doe, and ran it close for about two hundred yards, when he reached it with a blow of his paw, rolled it over, and in an instant was sucking the life-blood from its throat.

One of the other chetahs was slipped at the same time, but after making four or five desperate bounds, by which he nearly reached his prey, suddenly gave up the pursuit, and came growling sulkily back to his cart.

As soon as the deer is pulled down, a keeper runs up, hoods the chetah, cuts the victim's throat, and receiving some of the blood in a wooden ladle, thrusts it under the leopard's nose. The antelope is then dragged away, and placed in a receptacle under the hackery (Indian cart), whilst the chetah is rewarded with a leg for his pains.

The hunting leopard is a long, slight flat-sided animal, with a very neat, small head; and, unlike the rest of the feline species, its tail is flat, instead of round. In the first plate I have endeavoured to represent the sporting party manoeuvring to *gain the wind* of the game: the death of the antelope is portrayed in the second.

From Captain Mundy, *Pen and Pencil Sketches, being the Journal of a Tour in India*, Vol. I, 1832. (Dr. Ashok S. Kothari Collection.)

The Man-eater of Botta Singarum

H.A.L., the Old Shekarrey

I WAS hunting in the Deccan, in the neighbourhood of Mulkapore, when I heard that a man-eating tiger, which I had been after for some days, had been seen skulking near the outskirts of the village of Botta Singarum. I had on a former occasion tracked this cunning brute to one of his lairs, where the remains of several of his victims were discovered, and had twice beaten all his usual haunts in the jungle; but up to this time had never been able to get a shot at him. Sending my gang of trackers on before I mounted my horse, and guided by the villager who brought the news, I made my way to the place where the marauder had been seen the evening before, where I found unmistakable signs that the information I had received was true, as his fresh pugs were plainly visible.

I sent my horse back to the village, and, accompanied by the gang, followed his track through a narrow ravine densely wooded. Here the trail became exceedingly difficult to follow, as the brute had evidently been walking about backwards and forwards in the bed and along the banks of a dry nullah, and we could not distinguish his last trail. I caused the band to separate, and for half an hour or so we were wandering about as if in a maze, for the cunning brute had been describing circles, and often, by following the trail, we arrived at the place we started from.

Whilst we were all at a loss, suddenly I heard a low "Coo" twice repeated, and I knew that Googooloo, who was seldom at fault, was now on warm scent, and from his call I was as certain that the game was afoot as any master of hounds would have been, while breaking cover, to hear his favourite dog give tongue. The gang closed up, and, guided by the sound, we made our way through thick bush to where Googooloo was standing by a pool of water in the bed of the nullah.

Here were unmistakable marks of his having quenched his thirst quite lately, for when we came up the water was still flowing into the deeply-imprinted pugs of his fore feet, which were close to the edge of the pool, and I noticed that the water had still the appearance of having been disturbed and troubled. After having drunk, the brute made his way to some very thick jungle, much overgrown with creepers, through which we could not follow without the aid of our axes. Thus, stalking with any hope of success was out of the question, so I held a solemn consultation with Kistimah, Chineah, Googooloo, and the dhoby, as to the best means of proceeding.

I felt convinced that the brute was still lurking somewhere near at hand in the jungle, for, besides the very recent trail we were on, I fancied I heard the yelling of a

This incident, on the death of a man-eater in Botta Singarum (a village near Malkapur in the Telangana region), is taken from Henry Astbury Leveson's famous book, *Sport in Many Lands*. His other well known books are *The Hunting Grounds of the Old World* (1860) and *The Forest and the Field* (1867). Henry Astbury Leveson (1828–75) was considered among the four most famous big game hunters. The others were Sir Samuel White Baker, Rovaleyn George Gordon Cumming, and William Cotton Oswell. Their exploits are described in *Four Fathers of Big Game Hunting* (1901).

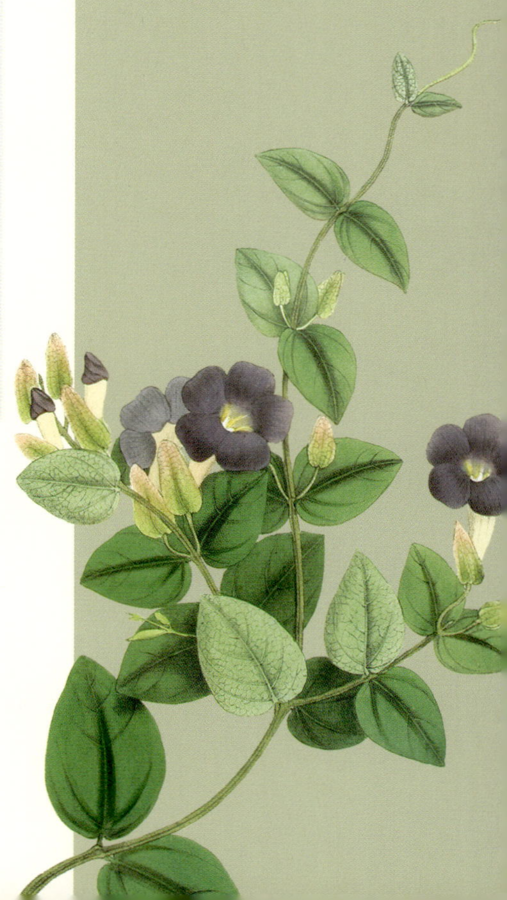

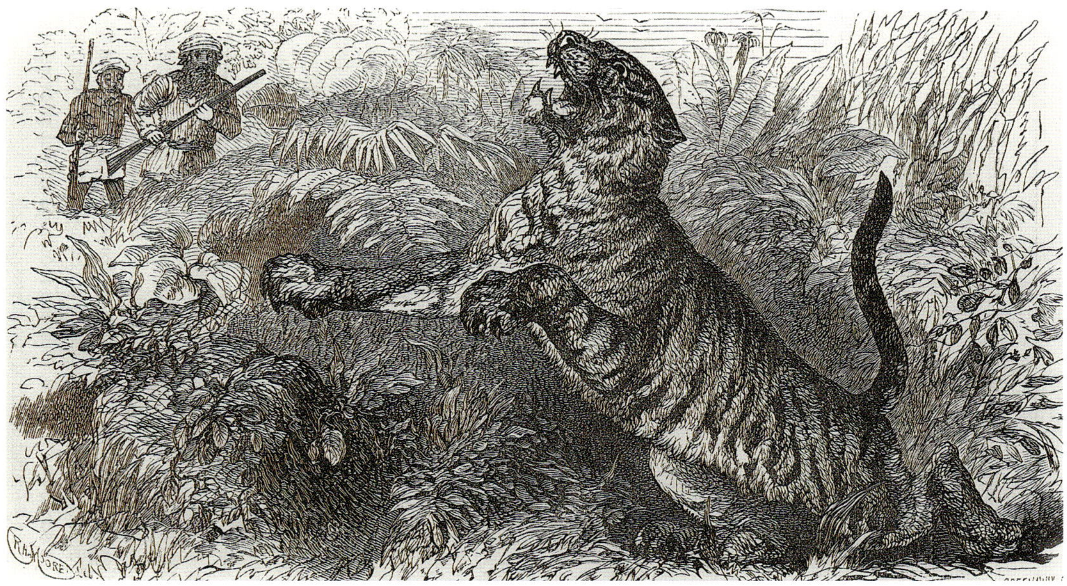

Tracking up a wounded tiger.

swarm of monkeys, which I attributed to their having been frightened by his appearance; besides, this was just the kind of place that a tiger would be likely to remain in during the heat of the day, as it afforded cool shade from the sun, and water. All the gang were of my opinion, and Kistimah observed that, on two different occasions, after a post-runner had been carried off, he had remarked that the trail of the tiger led from this part of the jungle to a bend in the road, where he had been known frequently to lie in wait for his prey. "These maneaters," added he, "are great devils, and very cunning, and I should not at all wonder if even now he was watching us from some dark thicket." As he said this I carefully examined the caps of my rifle, and I observed some of the gang close up with a strange shudder, for this brute had inspired them all with a wholesome fear, and prevented their straggling. Two or three spoke almost in whispers, as if they were afraid of his really being sufficiently near to hear them conspiring for his destruction.

At length Kistimah said that he had been thinking of a plan which, though dangerous in the execution, might be attended with success. It was for me to go, with a man dressed as a runner, down the main road at sunset, being the time the tiger generally carried off his victims, and to run the chance of getting a shot. At this proposition sundry interjectional expressions, such as "Abah!" "Arrez!" "Toba!" "Toba!" escaped from the lips of the bystanders, and, from sundry shaking of heads and other unmistakable signs, I could see that it had not found much favour in their eyes. Chineah, the dhoby, and one or two of the gang, however, approved of the plan, and Kistimah offered to accompany me as the post-runner. This, however, I objected to, for I thought that I should have a better chance of meeting the tiger if I went alone than in company; besides, I preferred having only myself to look after. The plan of action once settled, I returned to the village and obtained from the patel the bamboo on which the tappal-runners sling the mail-bags over their shoulders. To the end of this is an iron ring with a number of small pieces of metal attached, making a jingling noise as the man runs, which gives warning of the coming of the post to any crowd that might be obstructing the path, allowing them time to get out of his way. Having broken off the ring, I fastened it to my belt, so as to allow it to jingle as I walked; and, arming myself with a short double rifle by Westley Richards, a brace of pistols, and a huge shekar knife, I made Kistimah lead the way down the road towards the place where the man-eater was said to lurk.

About a mile from the village I made the gang and the villagers who accompanied me halt, and went on with Kistimah, Chineah, and Googooloo to reconnoitre the ground. The road was intersected by a narrow valley or ravine, along the bottom of which was a dry, sandy watercourse, the banks of which were overgrown with high rank grass and reeds, intermixed with low scrubby thorn-bushes. To the left was a low rocky hill, in some place bare and in others covered with thick jungle, with wild date or custard-apple clumps here and there. Kistimah pointed me out a clump of rather thick jungle to the right of the road, where, he said, the tiger often lurked whilst on the look-out for his prey, and here we saw two or three old trails. He also showed me a rock, from behind which the brute had sprung upon a post-runner some weeks before; but we saw no signs of his having been there lately. It was, however, quite what an Indian sportsman would term a "tigerish-looking spot", for bold, scarped rocks, and naked, fantastic peaks rose in every direction from amongst the dense foliage of the surrounding jungle, whilst here and there noble forest trees lowered like giant patriarchs above the lower verdure of every shade and colour.

Not a breath of air was stirring, nor a leaf moving; and as the sun was still high up, without a cloud visible to intercept his rays, the heat was most oppressive, and respiration even was becoming difficult on account of a peculiar closeness arising from the decayed vegetation underfoot, and the overpowering perfume of the blossoms of certain jungle plants.

Having reconnoitred the ground round about, I felt rather overcome with lassitude, and returned to the rest of the gang, whom I found sleeping in a clump of deep jungle a little off the roadside. Here I lay down to rest, protected from the piercing rays of the sun by the shade of a natural bower formed by two trees, which were bent down with the weight of an immense mass of parasitical plants in addition to their own foliage. I must have slept several hours, for when I awoke I found the sun sinking low

Orange-bellied Leafbird, Orange-bellied Chloropsis, *Chloropsis hardwickii*

This bird is resident in the Himalaya from Himachal Pradesh to Arunachal Pradesh, and Bangladesh. It is seen singly or in pairs, often with other birds on flowering trees, making its presence felt by a wide variety of harsh and whistling notes. While both sexes are leaf-green above, with orange-ochre belly and vent, and blue moustachial stripes, the male has dark blue wings and tail, the black on the throat extending to the breast. The female has a green throat. The species breeds in May–August, the small cup-shaped nest, made of fine roots, fibres, and lichen, suspended between twigs at the end of thick branches.

From *JBNHS*, Vol. 8, No. 1.

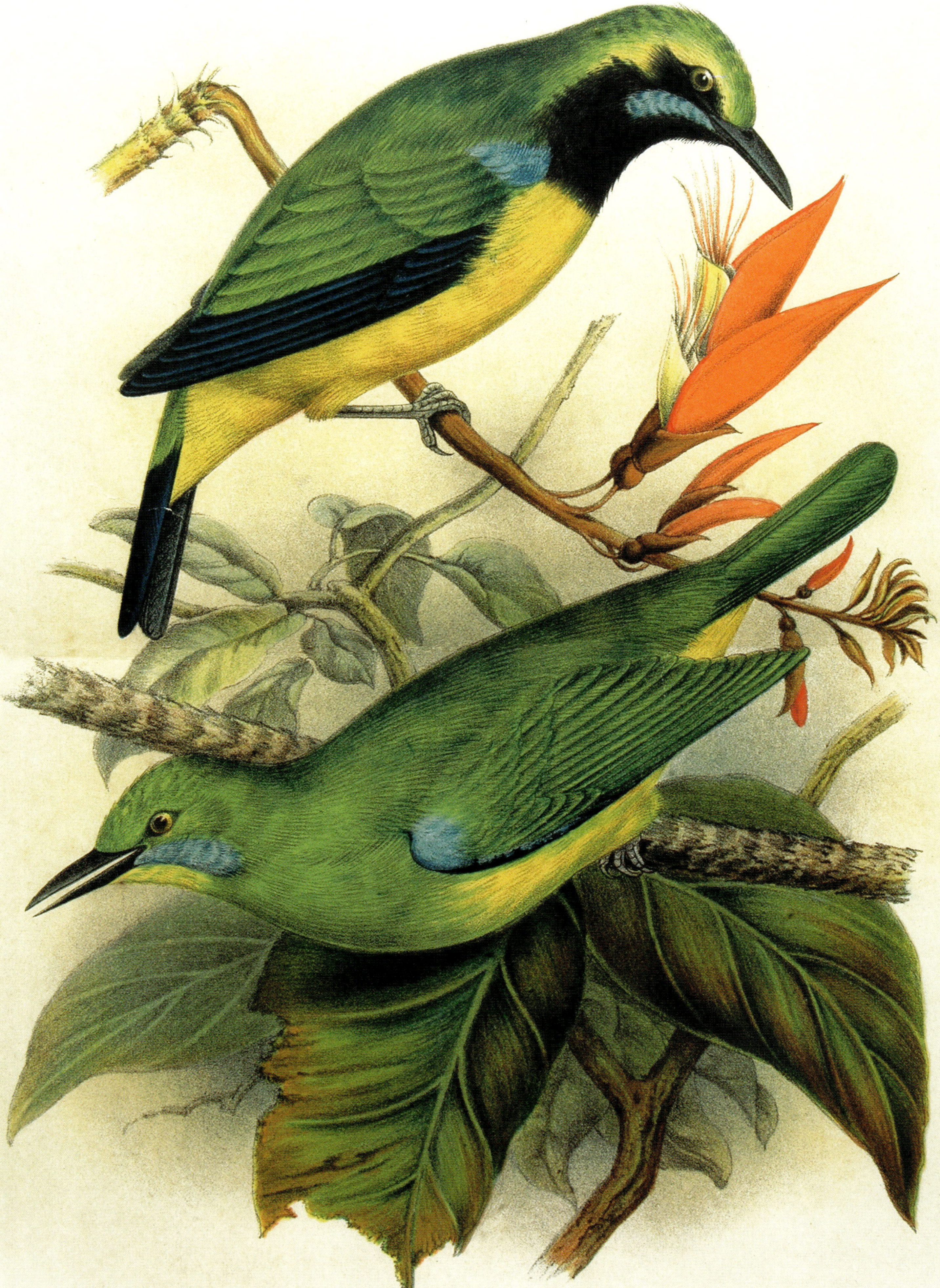

Chloropsis hardwickii

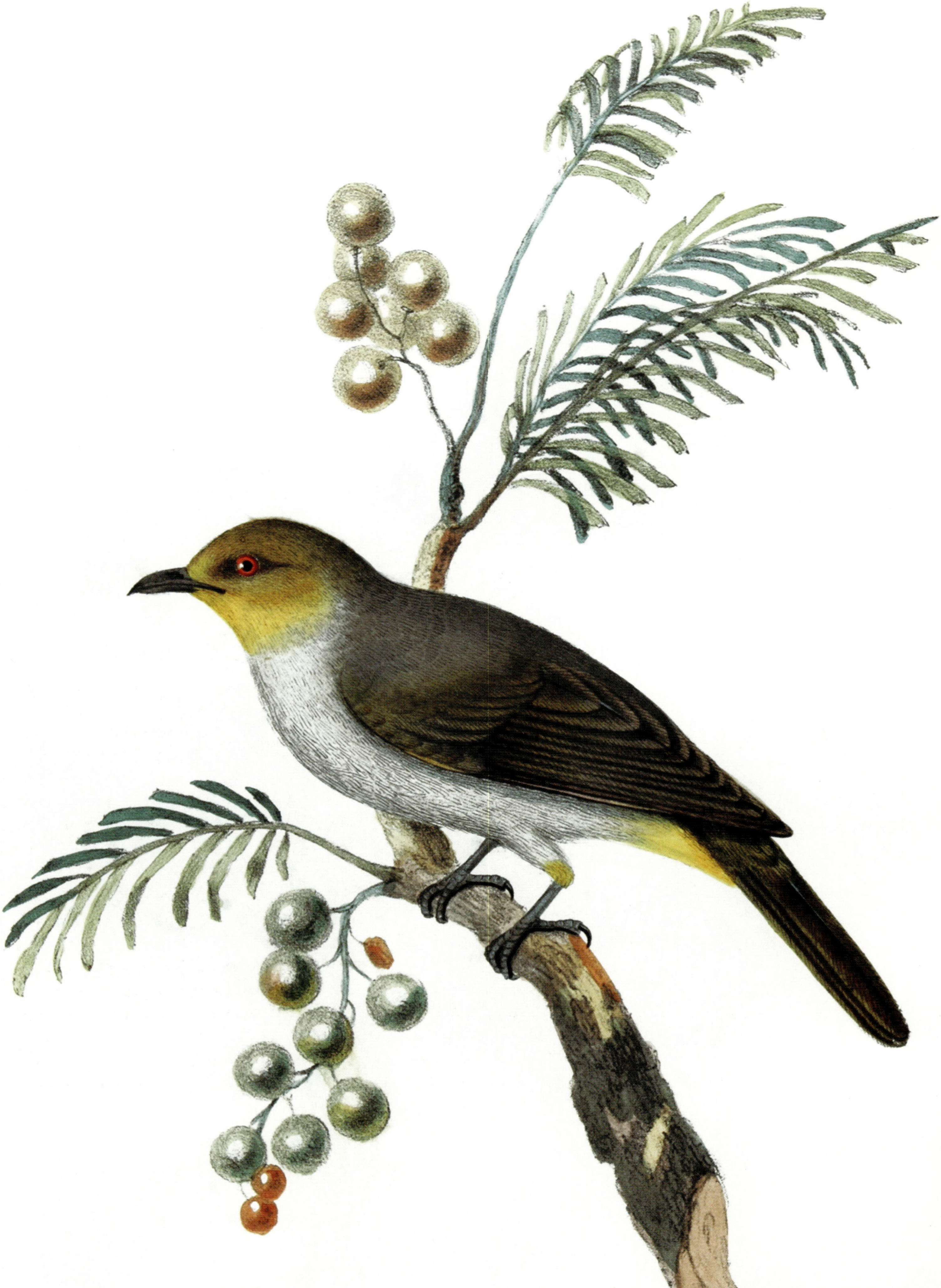

Pycnonotus xantholaemus

COURTESY
BHANSALI & CO.

Yellow-throated Bulbul
Pycnonotus xantholaemus

This uncrested grey and yellow-olive bulbul has yellow on its crown, chin, throat, thighs, and undertail-coverts. The wings and tail are brown; the tail yellow-tipped. It is a shy and restless bird, moving in pairs, often in flocks, seen in scrub jungle with trees and boulders, searching for berries and insects. It is resident in the area between the Krishna and Kaveri rivers – eastern Karnataka, northern Tamil Nadu, and southern Andhra Pradesh at 600–1,300 m. The call is high-pitched but soft. It nests from May to July, building a shallow cup of twigs bound and lined with cobwebs and fibres, creepers, etc. It lays 2 or 3 eggs, white blotched with purple and brick red.

From T.C. Jerdon, *Illustrations of Indian Ornithology*, 1847.

in the horizon; however, I got up considerably refreshed by my nap, and giving myself a shake, prepared for the task I had undertaken. I carefully examined my arms, and having ascertained that nothing had been seen by any of my gang, some of whom had kept a look-out, I told my people to listen for the sound of my gun, which, if they heard, they might come up, otherwise they were to remain quiet where they were until my return. I ordered Chineah, Kistimah, Googooloo, and the dhoby, to accompany me down the road with spare guns in case I might want them; and when I arrived at a spot which commanded a view of the ravine which was supposed to be the haunt of the man-eater, I sent them to climb different trees.

Kistimah begged hard to be allowed to accompany me, as he said this tiger never attacked a man in front, but always from behind; but I would not permit him, as I thought that two people would perhaps scare the animal, and his footsteps might prevent me from hearing any sound intimating his approach.

The sun had almost set as I proceeded slowly down the road, and, although I was perfectly cool and as steady as possible, I felt cold drops of perspiration start from my forehead as I approached the spot where so many victims had been sacrificed. I passed the rock, keeping well on the look-out, listening carefully for the slightest sound, and I remember feeling considerably annoyed by the chirping made by a couple of little bulbuls, that were fighting in a bush close to the roadside. Partridges were calling loudly all around, and as I passed the watercourse I saw a jackal skulking along its bed. I stopped, shook my jingling affair, and listened several times as I went along, but to no purpose.

Whilst ascending the opposite side of the ravine I heard a slight noise like the crackling of a dry leaf. I paused, and turning to the left, fronted the spot from whence I thought the noise proceeded. I distinctly saw a movement or waving in the high grass, as if something was making its way towards me: then I heard a loud purring sound, and saw something twitching backwards and forwards behind a clump of low bush and long grass, about eight or ten paces from me, and a little in the rear. It was a ticklish moment, but I felt prepared. I stepped back a couple of paces in order to get a better view, which action probably saved my life, for immediately the brute sprang into the middle of the road, alighting about 6 feet from the place where I was standing. I fired a hurried shot ere he could gather himself up for another spring, and when the smoke cleared away I saw him rolling over and over in the dusty road, writhing in his death agony, for my shot had entered the neck and gone downwards into his chest. I stepped on one side and gave him my second barrel behind the ear, when dark blood rushed from his nostrils, a slight tremor passed over all his limbs, and all was still. The man-eater was dead, and his victims avenged.

My gang, attracted by the sound of my shots, came rushing up almost breathless, and long and loud were the rejoicings when the tiger was recognized by Kistimah as the cunning man-eater who had been the scourge of the country for months. He was covered with mange, and had but little hair left on his skin, which was of a reddish brown colour, and not worth taking.

I have killed many tigers both before and since, but I never met with such a determined enemy to mankind, for he was supposed to have carried off more than a hundred individuals. He fully exemplified an old Indian saying. "When a tiger has once tasted human blood he will never follow other game, men proving an easier prey." On the spot where the tiger was killed a large mausoleum now stands, caused by the passers-by each throwing a stone until a large heap is formed. Since that day many a traveller who has passed that way has been entertained by the old pensioned sepoy, who is in charge of the travellers' bungalow, with an account of the terrible man-eater of Botta Singarum.

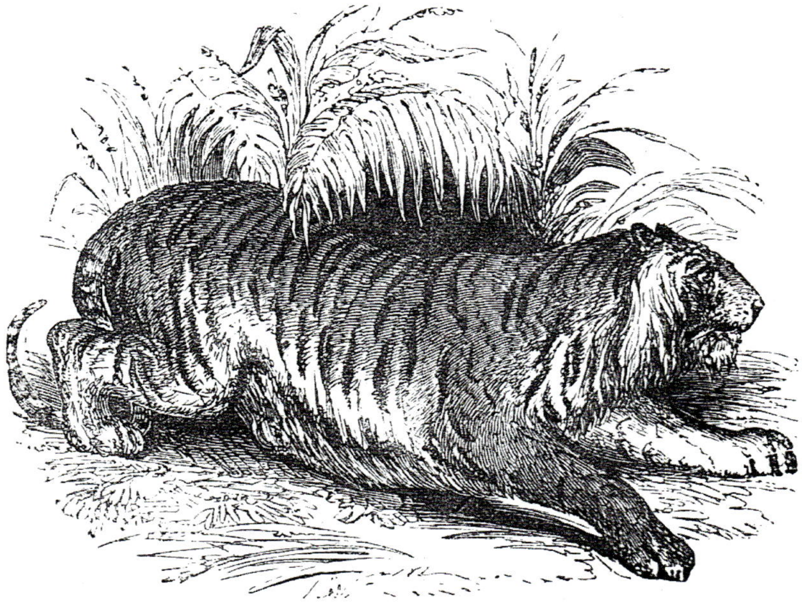

From H.A.L., the Old Shekarrey, *Sport in Many Lands*, 1877. (Dr. Ashok S. Kothari Collection.)

Sterculia roxburghiana

Family: Malvaceae

Synonym: *S. lanceifolia*

A small tree with ashy bark, found in Darjeeling, Sikkim, Assam (Lakhimpur, Sibsagar), the Garo Hills in Meghalaya, also in Nepal, Bangladesh, and China up to an altitude of 1,800 m. The leaves are oblong or oblong-lanceolate, suddenly acuminate, with base rounded, 7.5–20 cm long, 2.5–8.7 cm broad. Numerous brick-red flowers appear in erect 5–10-cm-long racemes during May–July. The flowers are axillary, about 1.2 cm across. The fruits are pink, lance-shaped, beaked, spreading pods 5–10 cm long with 4 to 8 black seeds. The roasted seeds are eaten.

From N. Wallich, *Plantae Asiaticae Rariores*, 1832.

IN MEMORY OF NATHUBHAI RAMABHAI PADHIAR OF DEESA, PIONEER AMONG POTATO CULTIVATORS IN INDIA
– PURUSHOTTAMBHAI N. PADHIAR & SHANTABEN P. PADHIAR
CHETAN P. PADHIAR, KUNAL P. PADHIAR, & CHARULATABEN SHAH

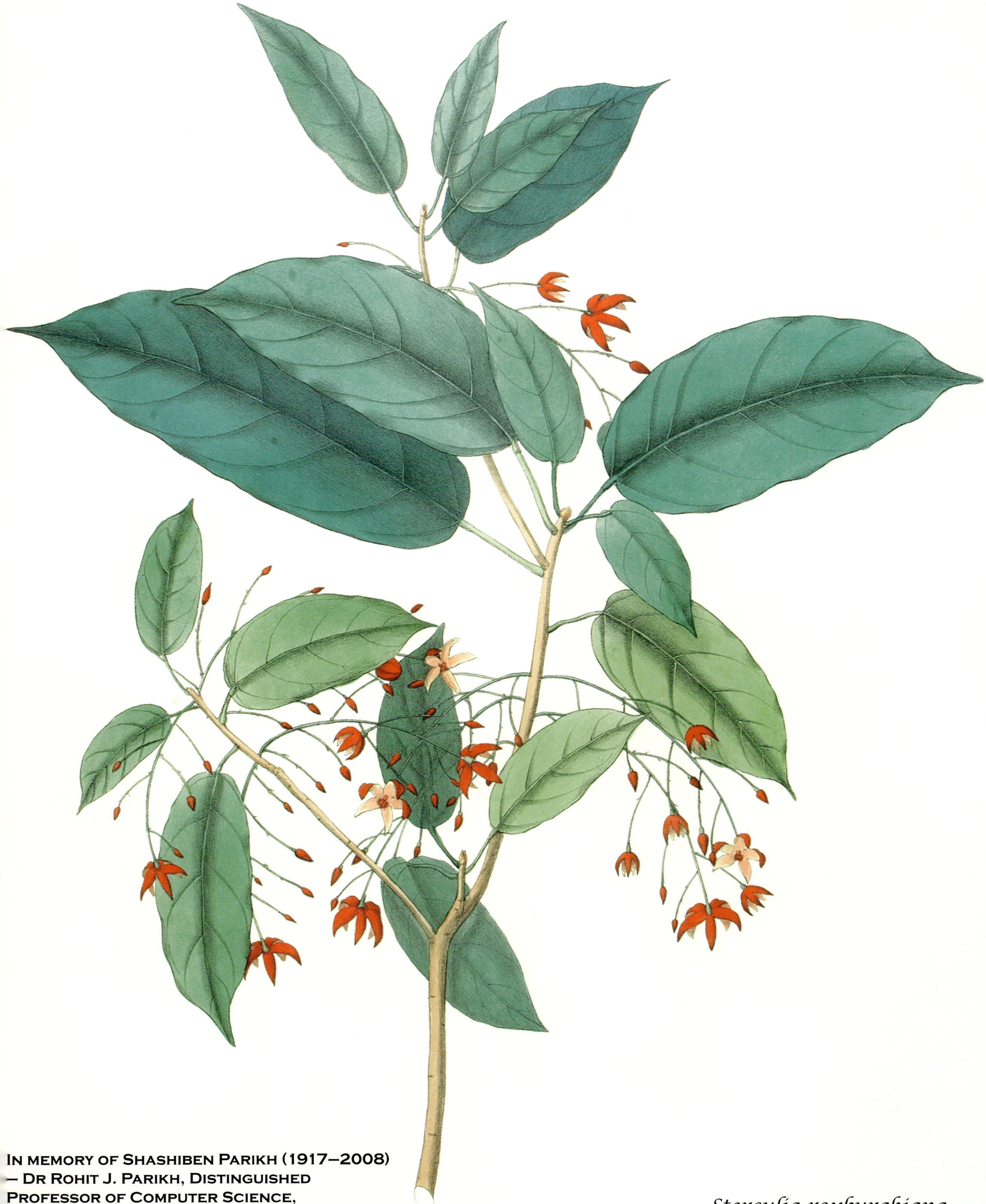

Sterculia roxburghiana

In memory of Shashiben Parikh (1917–2008)
– Dr Rohit J. Parikh, Distinguished
Professor of Computer Science,
Mathematics, and Philosophy at the
University of New York,
and Carol, Vikram, and Uma Parikh

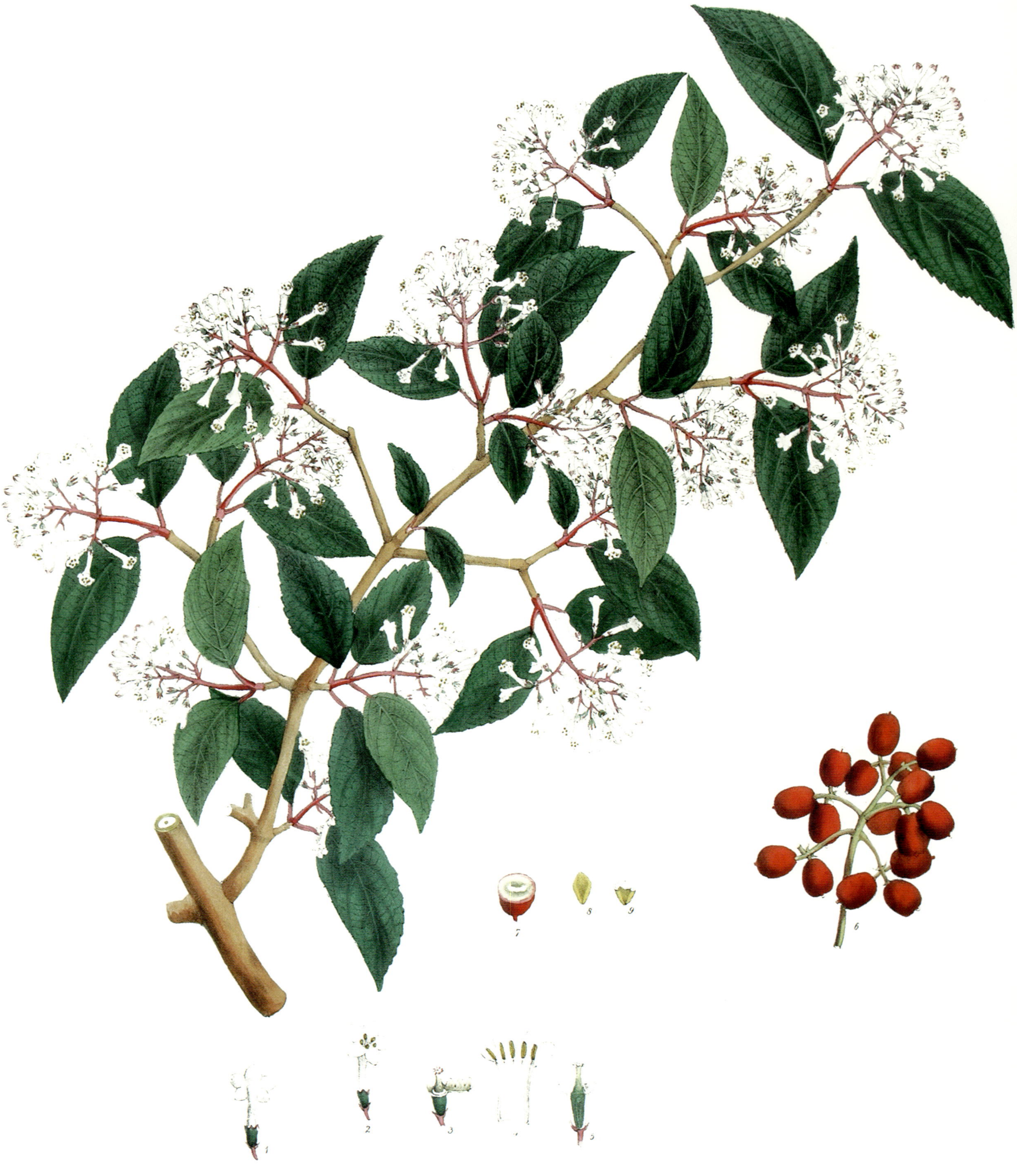

Viburnum erubescens

COURTESY
M. SURESH & CO. PVT. LTD.

Viburnum erubescens
Family: Caprifoliaceae

This is a large bushy, deciduous shrub or small tree growing up to 3 m tall. It is found in the temperate Himalaya up to 2,000 m, from Uttarakhand to Darjeeling and Sikkim, the Khasi hills, Assam, Arunachal Pradesh, Nepal, and Bhutan. The tree is distinguished in April–May by its lax, drooping, long-stalked, branched clusters of white, cream, or pink tubular flowers borne with the leaves at the ends of the short branchlets. The flowers have slender corolla tubes and round spreading lobes; the anthers are dark purple. The drupe is ellipsoid, slightly compressed, and red. Wallich states: "This elegant large shrub is common on the mountains of Nipal. The flowers are sweet scented like honey. The leaves when rubbed emit a very peculiar sumptuous smell."

From N. Wallich, *Plantae Asiaticae Rariores*, 1832.

Man-eating Tigers of Nagpore, Bansda, Toongar, and Jaunsar

Reginald Gilbert

I HAVE selected this title, not because I have had particular experience on the subject, or because I am an expert, but because I wish to place on the records of our Society a few facts relating to man-eaters which can be considered as reliable, several of them being cases of man-eaters killed by my friend, Mr. W.B. Mulock, Bombay Civil Service, of our Society, now at home on furlough, and who has most successfully devoted a great deal of his time to the destruction of man-eaters; another being the case which is known as the Nagpore man-eater, another the "Jaunsar" man-eater, well known in the N.-W. Provinces, and lastly, the case of an alleged man-eater, which I killed this year in Bansda. You must not expect me to give you any thrilling account of some personal adventure where I risked my life to rid the district of a brute long the terror of the inhabitants, because I may say at once that the only man-eater I have killed, exposed me to no more danger than I should incur in any ordinary day's shooting after small game. Indeed, it would scarcely be in accordance with the object of our Society to read a paper relating to personal adventure of this kind. I only wish to touch on various points which I think may chiefly be of interest to our members from a Natural History point of view, and in the hope that other members may be able to supply us with information on this very interesting subject which they can personally vouch as correct.

Now the general impression prevailing about man-eaters is, that the man-eater is an old brute, more often decrepit than otherwise, perhaps lamed from some former wound, with his teeth broken and his skin always mangy, unable from his infirmities to kill game, his natural food, but obliged to conceal himself near a village path and then to pounce on some solitary human being and devour him, never attacking when there are more than two or three human beings together and always displaying very great cunning, so that his destruction becomes almost impossible. It is difficult to read books of Indian sport without coming to that conclusion. No reliance, however, can be placed, I fear, on books of sport, with one or two bright exceptions, one of which is Mr. Sanderson's book. Books of sport are written to be read by the masses, and the first idea of the author is to romance and to write something of an exciting kind to please his readers, and not a strictly accurate account of what the author has himself witnessed or known. These books of sport have therefore little value from a Natural History point of view. I only mention this, because my own opinion is that the general impression about

Reginald Gilbert, a solicitor in Bombay and member of the Bombay Natural History Society, read out his experiences with man-eating tigers at the Society's meeting held on 4th September 1889. The Nagpore man-eater took a heavy toll of the people working on the Bengal-Nagpore Railway. Her depredations can be compared with those of the notorious Man-eaters of Tsavo who preyed upon Indian labourers laying rails for the Mombasa-Uganda Railway. Encounters with man-eaters in Bansda (Vansda in south Gujarat), Toongar (Tungareshwar near Mumbai), and Jaunsar are also described. Shri Digveerendrasinhji >

man-eaters is altogether wrong, except as to his display of cunning. I am aware that one swallow does not make a summer, and that many of my hearers may rightly think my opinion on this subject is of little value. Man-eaters are happily so few and far between that the most experienced shikari can in a lifetime only come across a very few. I start then by saying that I believe man-eaters are not different in any way from the ordinary game or bullock-eating tiger, and that age, deformity, injury or otherwise, have nothing whatever to do with the question. Why a tiger turns man-eater I can offer no opinion, and why a tiger never kills a goat, but nevertheless kills such small fry as peacocks, porcupine, or monkeys I also cannot reply to. The universal fear that all animals have to man is no doubt the reason why the tiger seldom happily does turn man-eater. There are some large districts in India infested with tigers where a man-eater is never heard of, whilst there are other smaller districts, one of which I intend hereafter to refer to, where man-eaters are constantly appearing. I wrote to a friend of mine, a Forest Officer in the Berars, who is a very successful and keen tiger slayer, and who, I thought, could give me some very important information. He however tells me that he has not known a single case of a man-eating tiger, although however he has known of a man-eating panther in his districts in the Berars.

THE NAGPORE MAN-EATER

The first man-eater I wish to introduce you to is the tiger we have often read of in our local newspapers as the Nagpore man-eater. As regards this one, I have obtained my information from Messrs. George Anderson and George Moule, Engineers on the Bengal-Nagpur Railway, who have been out on several occasions after the beast, and have reliable means of obtaining accurate information. This man-eater is a tigress, and has the following peculiarities of character, *viz.,* her love of feasting on the employees of the Bengal-Nagpur Railway, of frequenting only a small tract of country about nine square miles in area, and her great cunning and audacity. She has been killing for three successive years; as far as my informants know, she has killed from twenty-eight to thirty-eight human beings, but these must be taken as much less than the correct actual number, because my informants only return what the railway officials hear of and confirm, and these returns are mostly of people connected with the railway. In 1889, up to June, she has killed seven people besides wounding others. The district she works in is as I have said about nine square miles only, and is near the Darekasa Railway Station. She appears to live in a rocky and precipitous spur, through which a tunnel has been cut. This spur carries heavy bamboo and other jungle. Several springs of water rise from out of the spur. In many places at the foot of the scarps there are delightfully cool places for her to lie up in, where the ground is always moist. There is also a cave in a detached mass of the spur, which shows many signs of being used by the tigress and the family. A

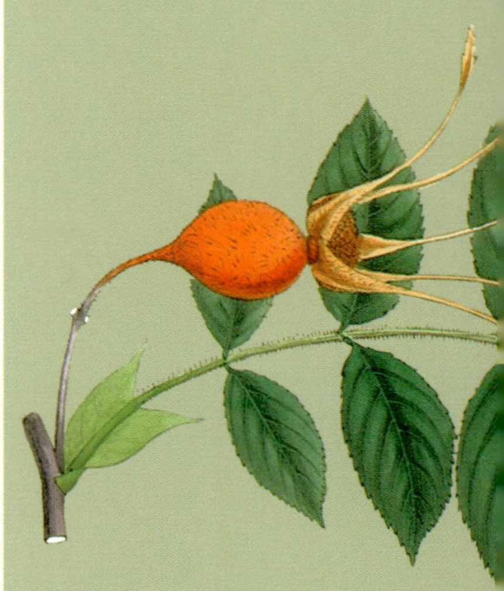

>I. Solanki, the present Maharaja of Vansda and one of the seniormost among the Society's members, regrets that today there is not a single tiger in the forests of Vansda and Dangs, or in the entire state of Gujarat.

The author also narrates how a wounded man-eater killed Mr. G.L. Gibson, another member of the Society.

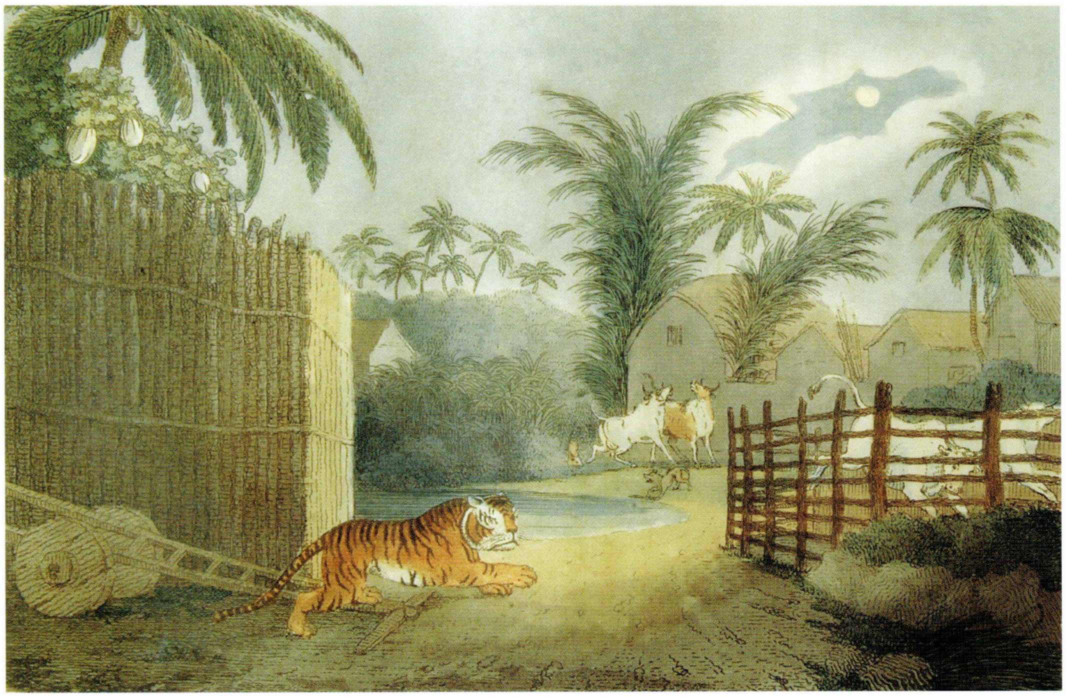

Tiger entering a village, from Thomas Williamson, *Oriental Field Sports,* Vol. I, 2nd edition 1819.

big stone just outside the entrance is scored deep and long with many scratches of their claws. The jungle around the cave is very thick, and the cave is very awkward to get at. "The whole area hunted by the tigress," writes Mr. Anderson, "is hard to determine, but for weeks together it is believed she has hunted within this area of nine square miles or even less." A great number of sportsmen (in fact too many) have been after her without success. She will not return to a kill; if she cannot carry off a carcase to a safe place, she will abandon it altogether. About the middle of January last she began to frequent the railway, being seen at all hours in broad daylight. On 24th February 1889, at 2-20 p.m., she jumped from the top of the slope of a cutting about twelve feet on to the line, where a gang of permanent-way men was at work, snatched up one of them and vanished up the opposite slope in a second. She carried the body to a pool of water about 300 yards off and there ate it. On the 25th February a boat was organised, and three tigers were found at home, in the cave, of which two were shot by Mr. Cleveland and Captain Silver, Adjutant of the B.-N. Railway Volunteers, both of which tigers were young ones, not fully grown, the cubs of the old sinner. On the 29th February, she killed a boy near the same place and carried his body a long way. Mr. Anderson has seen the pugs of a young cub with her, apparently one of her last litter. The cubs that were killed are probably

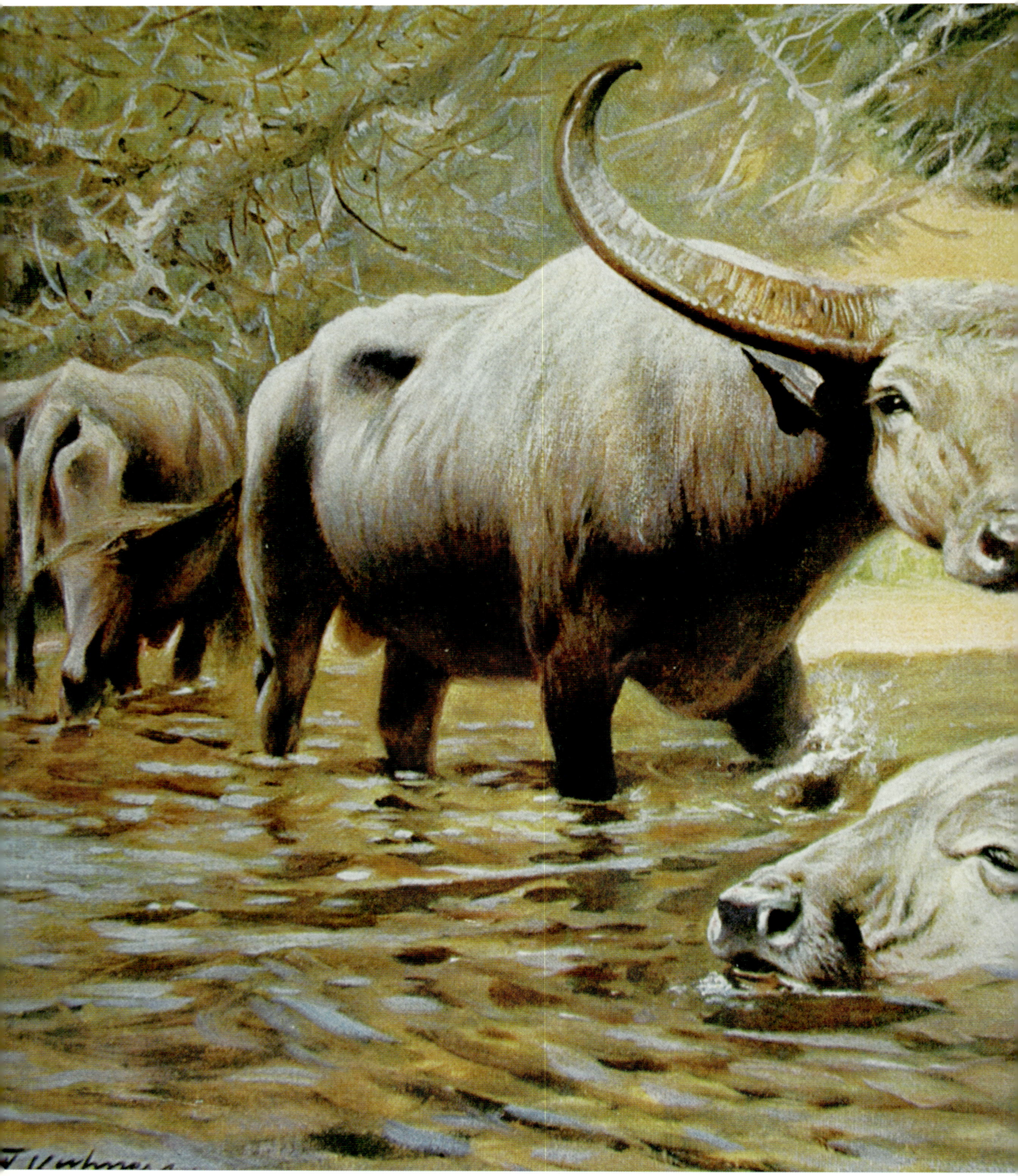

Courtesy HSBC
The Hongkong and Shanghai
Banking Corporation Limited

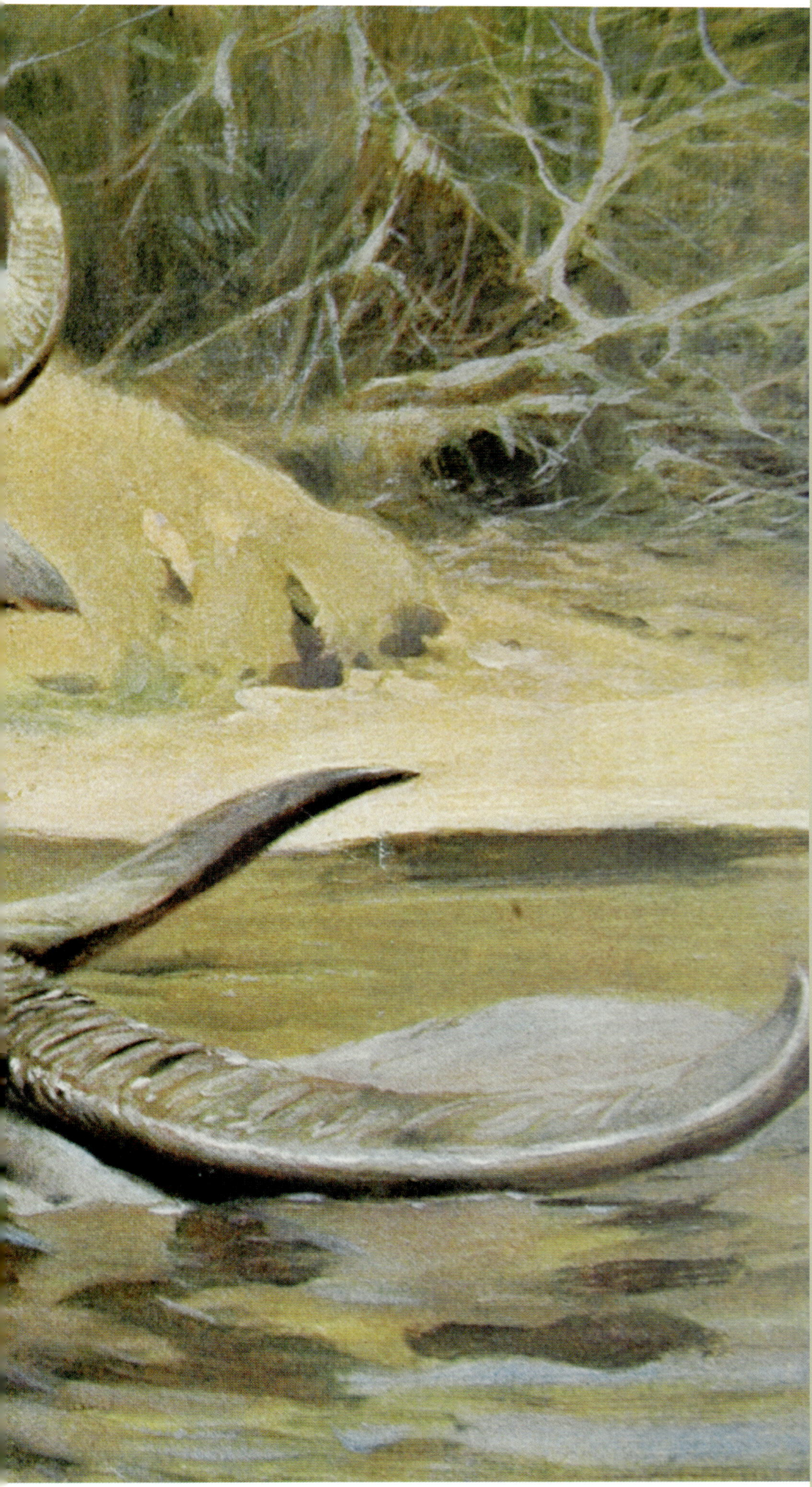

Indian Wild Buffalo, Asiatic Wild Buffalo, Wild Water Buffalo
Bubalis arnee

This large ungulate survives today in India, Nepal, Bhutan, Myanmar, and Thailand, while it is extinct in Pakistan, Bangladesh, Laos, and Vietnam. Its numbers are steadily declining, the estimated population worldwide being less than 4,000. In India, 90 per cent are found in Assam, in Manas and Kaziranga National Parks; Udanti Wildlife Sanctuary in Chhattisgarh has 60, while Indravati National Park and Pamed Wildlife Sanctuary have 49 and 9 respectively. In the Bastar region, "Parath" or community hunting takes place when these animals are trapped and speared. In many other places in central India where they were found in the 19th and early 20th centuries, as mentioned by J. Forsyth and A.A. Dunbar-Brander, they are extinct today. Intensive hunting, habitat loss, diseases and parasites transmitted by domestic stock, and interbreeding with domestic buffalos have affected their numbers in the wild. Once they were found in abundance in Orissa and in the Nepal Terai. Seals found in excavated Harappan sites depict Wild Buffalos, confirming that during ancient times they existed on vast tracts of the country. These Wild Buffalos are much larger than domestic buffalos and have more hair and larger curving horns. They are slaty-black in colour. An adult male weighs up to 1,200 kg while the female is 800 kg. They can stand as tall as 2 m, with 3 m body length. The newborn calf is light-coloured, almost yellow. They live in herds in grasslands where they can find small ponds with water and mud, and are not found in dense forests or in the mountains. They are very fierce and have no fear of humans because of their sheer size and sharp, massive horns. They come out to graze during the mornings and early evenings. During the day they come to ponds where they may be seen taking mud baths, or sleep amidst the tall grass.

Painting by William Kuhnert, from Lydekker et al., *Harmondsworth Natural History*, 1910.

of another former litter. On the 4th March, the tigress attacked a woodcutter near the railway, but was driven off pluckily by his companion, who attacked her with an axe. All April she appears to have kept to the same ground, and in the middle of May she killed another man near the railway. It appears there have been previous man-eaters in this neighbourhood, so that this tigress may have been educated in this vice by a wicked ancestor or companion of hers. In the year 1883, this particular spur was infested with tigers, and seems to have long been a regular house of call for tigers. In May last Mr. Moule, whilst seated on a *machán* at night, had a shot at this tigress, but he was unable to see her properly, and she got off scatheless. It is hoped she will be bagged before long.

Man-eaters of Bansda

I will next take you to the neighbourhood of Sardardevi, Bansda State, where the Dangs Baroda territory and Bansda territory lie contiguous. Near here a tigress in 1884 and 1885 killed a very large number of people, and in 1888 and 1889, in the same neighbourhood, another man-eater sprang up, which became the terror of that district, and killed a large number of people. Mr. Mulock killed the former in May 1885, and I killed that which is supposed to be the latter on 2nd April 1889, so that after an interval of only four years the inhabitants of this district were so unfortunate as to have a second man-eater spring up to take the place of the first. Mr. Mulock has written a very graphic account of how he killed this tigress and two three-quarter grown cubs with her, which he printed, and a copy is in our library. This district is hilly, well watered, and covered with jungle. There is scarcely any cultivation, but villages inhabited by various jungle tribes are scattered about here and there. There are plenty of chetul, sambhur, pig, &c., about, and also plenty of cattle grazing in the jungle, so that the man-eater has little excuse for taking to his evil ways. Mr. Mulock mortally wounded this tigress on the 4th May 1885, and it was only on the 13th May that he actually got it. It was wounded in the jaw, and was unable to eat from the 4th to 13th May. Maggots got into the wound, so that when she was killed "she was exceedingly finely drawn from hunger and emaciation". Mr. Mulock describes her as having a beautifully marked skin. There is no appearance of mange in her skin, nor does Mr. Mulock describe her as appearing injured, so as to prevent her from finding her food in a legitimate feline manner. Mr. Mulock found the same difficulty as I did in getting exact information of the number of people killed, but the evidence he collected showed that a large number of people were eaten by this tigress, and there can be no doubt, as he says, this was the culprit, as, writing six months afterwards, he states not a single person had been killed by tigers, and so far as I can discover not till 1888 did another human being fall a prey to a tiger in this district.

I received an invitation from Mr. A.W. Crawley-Boevey, who was then Collector of Surat, to shoot with him in April 1889 in the Native State of Bansda. I was informed of

1 *Primula calderana*
Family: Primulaceae
Synonym: *P. obtusifolia*

This beautiful wild flower is found in the Himalaya from western Nepal to southeastern Tibet; in Himachal Pradesh, Jammu & Kashmir, and Sikkim, at altitudes of 3,600–4,800 m. The flowers appear in June–July. They are usually dark purple with golden centre, borne in clusters, erect or drooping. The petals are broad and overlapping, with corolla tube 1–2 cm long, twice as long as the calyx. The plant has a strong smell. The leaves are long (10–15 cm) and obtuse (hence *obtusifolia*), narrow at the base, with toothed margins.

2 *Primula macrophylla*
Synonyms: *P. purpurea, P. nivalis* var. *macrophylla*

A fairly common alpine plant found in Kashmir, Sikkim, Bhutan, Afghanistan, and southeastern Tibet, gregarious along snow banks, moist meadows, and streams, from 3,300 to 5,400 m. The leaves are erect, narrow, lanceolate, entire or toothed, with a powdery white layer beneath. The flowers are purple, violet, or lilac and appear in June–August in dense clusters of 5–25. The capsule is cylindrical.

From J. Forbes Royle, *Illustrations...*, 1839.

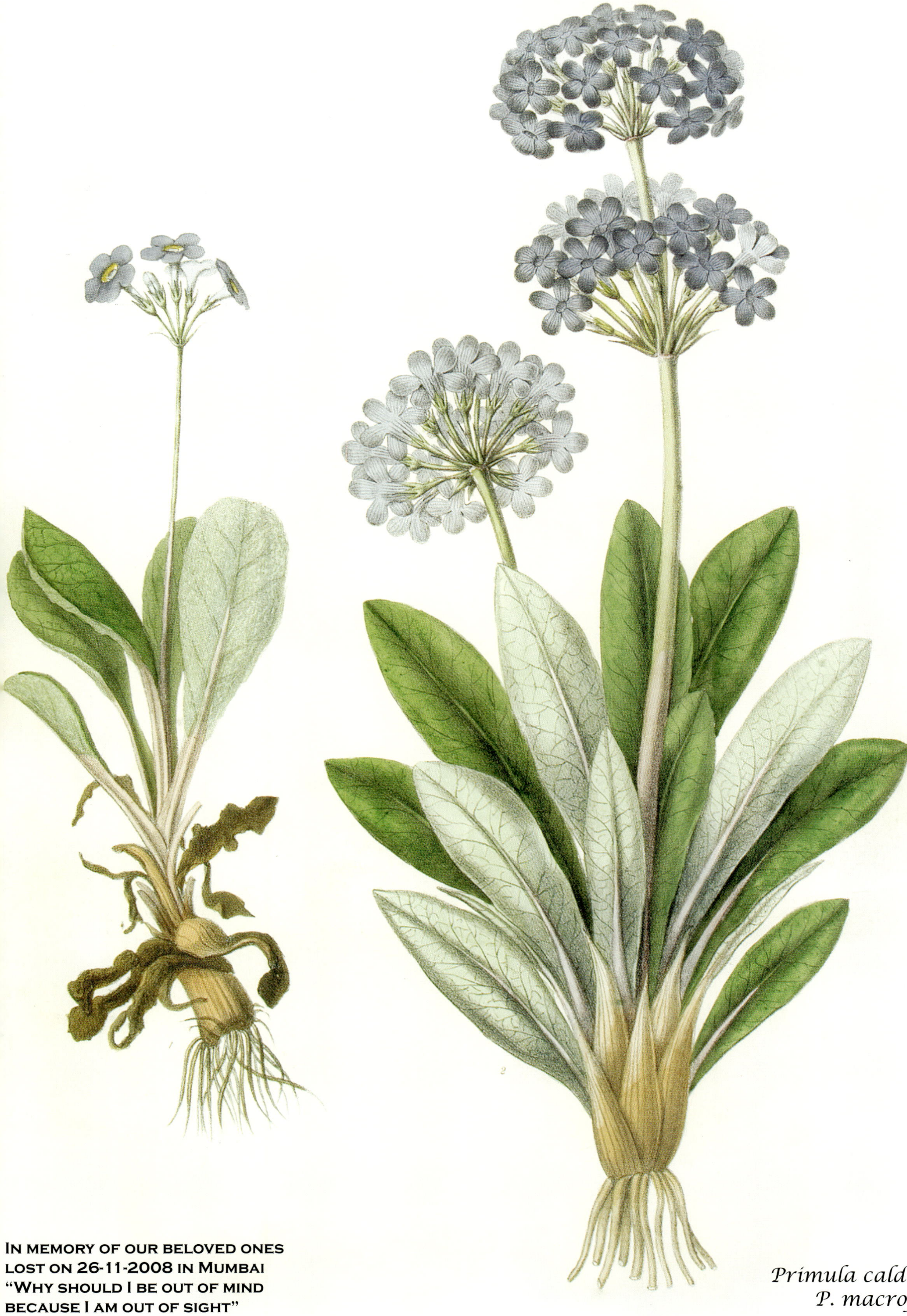

Primula calderana,
P. macrophylla

IN MEMORY OF OUR BELOVED ONES
LOST ON 26-11-2008 IN MUMBAI
"WHY SHOULD I BE OUT OF MIND
BECAUSE I AM OUT OF SIGHT"
(SCOTT HOLLAND)
– SARLA & SEVANTI PAREKH

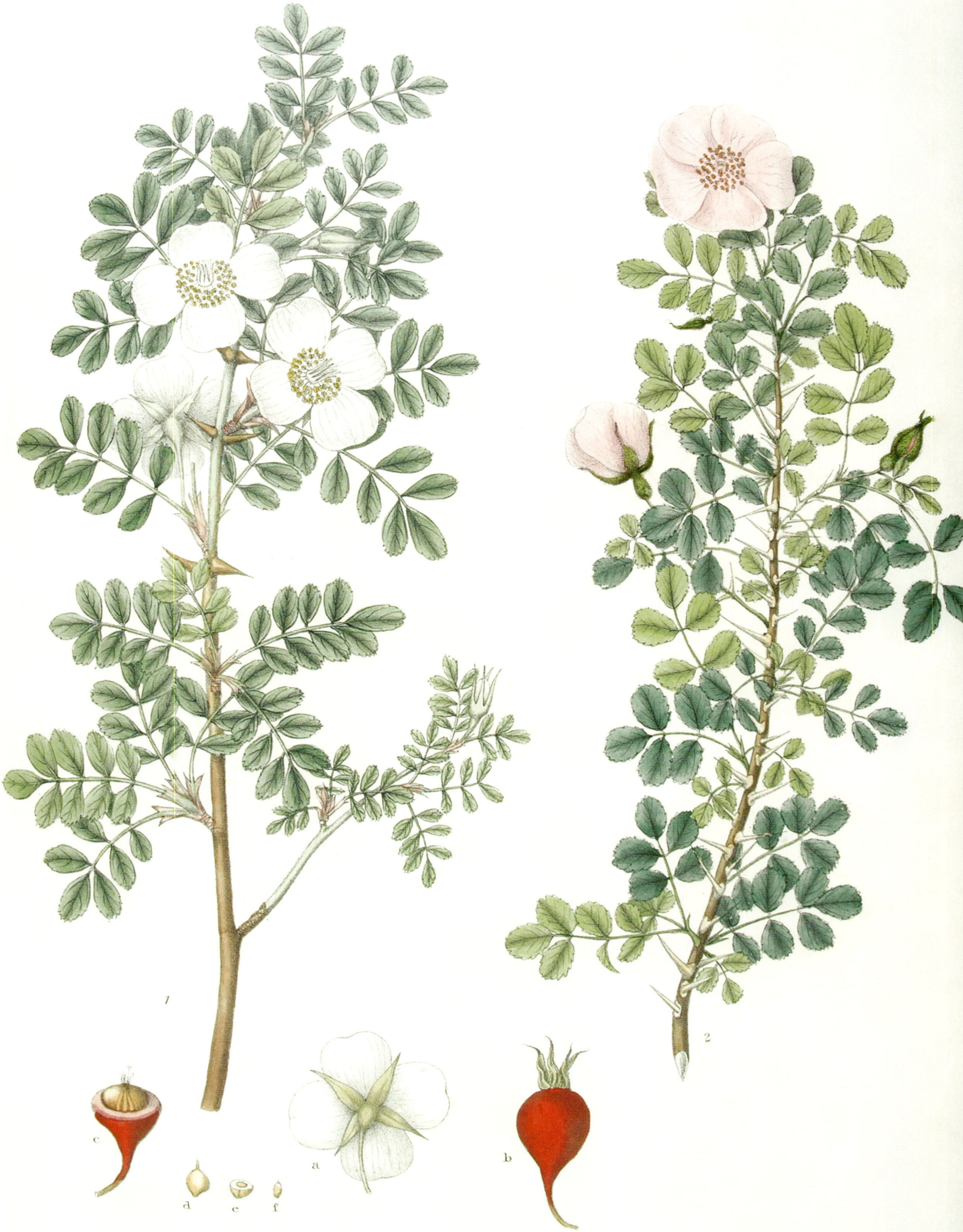

Rosa sericea, R. webbiana

1 Winged Thorn Tree, Wingthorn Rose
Rosa sericea
Family: Rosaceae

This is a thorny deciduous shrub found on alpine slopes at 2,000–4,400 m in Tehri Garhwal, Har ki Dun, and the Valley of Flowers in Uttarakhand, in Sikkim, and in Bhutan, Nepal, Myanmar, Tibet, and southwestern China. It grows to about 2 m tall, with greyish-brown bark peeling off in papery flakes. The thorns are large and flattish, brown in colour. The leaves are 4–8 cm long, composed of 7–11 leaflets with serrated margins. The plant blooms in May–July. The flowers are pure white or with a pinkish shade, 2.5–3 cm in diameter, solitary and somewhat drooping, with 4 petals and 4 sepals that are elliptic with a long apex, glabrous above and silky along the nerves beneath, persisting in the fruit. The scarlet, globular to pear-shaped fruit is eaten, raw or cooked.

2 *Rosa webbiana*

This is a deciduous shrub with slender branches having straight prickles and leaves with small rounded leaflets and dense clusters of pink flowers in June–August. The flowers are 2.5–7 cm across with 5 petals, the sepals being shorter. Each leaf has 5–9 coarsely toothed leaflets. The stems and leaves often turn pinkish. The fruit is ovoid to flask-shaped; it is edible with high vitamin C content. The shrub is common in the Valley of Flowers and other parts of Uttarakhand, and is also found in Nepal, Pakistan, Afghanistan, Mongolia, and China.

From J. Forbes Royle, *Illustrations...*, 1839.

the ravages of a man-eater in these same Sardardevi jungles, and Mr. Boevey suggested we should go after it if the *kubber* was reliable. I arrived at Bansda on 30th March, when Mr. Boevey at once informed me that the tiger had killed a woman near Sardardevi during the Holi holidays, about fifteen days back, and we agreed it would be well to go over there some ten miles distant, and spend a few days after the man-eater or some bears said to be in that neighbourhood. The Rajah of Bansda showed us every kindness and facility, had a camp pitched for us at Sardardevi, and placed a number of sowars, sepoys, shikaris, &c., at our disposal. He is an enlightened Rajah, but not so enlightened as is the English ruler of a neighbouring province, who with one hand doles out a reward to those who kill a tiger, whilst with the other he from time to time pens rules placing needlessly heavy restrictions on those who spend time and money in killing tigers in his province! No rules or restrictions were placed in our way and no permit was required; a hearty welcome was given us by the Rajah and his Dewan, Mr. Jhaverbhai Nathoobhai. Whilst riding out to Sardardevi in the early morning, a sowar met us bearing a message from the shikari Hubib, who had for years been Mr. Mulock's shikari, telling us that the man-eater had killed three bullocks belonging to some Brinjarees on the previous afternoon and two bullocks the day before. We hurried on to Sardardevi, where we met Jemadar Abdulla, head of the Dharampore State Police, and a noted shikari who had come to assist us. He informed us that there were two tigers, one they had marked down, the smaller one having eaten and gone off some distance. We found the beaters ready, and at once went off to the jungle, where we found everything ready, passing on the road the Brinjaree camp from whence came the bullocks which had been killed. Mr. Boevey was placed up a tree near one of the kills which had not been eaten at all. Soon after the beat commenced, the tiger roared. The beat came on almost up to us when the tiger broke back with a roar through the beaters without damaging any one of them. In fact, Hubib told me he turned out of the way of a beater standing on his direct path, instead of knocking him down and giving him a pat, as often does occur when an unwounded tiger breaks back. I attributed his breaking back to his being driven, over his kill. I have noticed tigers always break back when an attempt is made to drive them over the kill. The second time the beat commenced in the same way, and the tiger was driven into some high grass near Mr. Boevey's tree. He came out with a bound under Mr. Boevey's "mahla". Mr. Boevey saw that he was going straight towards my tree in an open space, and generously sacrificed his shot so as to enable me to get an easy one. I killed him without any trouble. He was not a large tiger, nor a mangy one, nor did he appear to be different in any way to an ordinary tiger. He was nine feet long. All the beaters, shikaries, &c., declared that he was the identical man-eater, but could bring me no other evidence than their oft-repeated assertion and the fact that many people had been killed lately in the jungles in the neighbourhood. There was still the tigress to be accounted for; which had been pugged to a distant jungle that morning, and this tigress probably was a man-eater too. In the night I heard the tigress roar several times within a mile from my tent. The next morning and the morning after that we found her pugs at the water where she had drunk close to our camp. These were carried into some likely jungle which we beat on both days without seeing anything of her. On the third day she drank at the same water, passed close to our camp along the same path, and the shikaries declared her to be lying down on the side of a hill. No one had seen her, but they pointed out some vultures sitting in a tree up the hill, and said that she had killed a pig or a chetul, and they stated most positively she was lying down in a certain spot pointed out to us. In this they were correct. She had killed a pig, and she was lying down in the place indicated. It is difficult to account for the certainty and accuracy of these jungle men in placing the exact whereabouts of a tiger which no one has seen. How we ought to proceed was a question of woodcraft, and our Bheels held a council of war. After considerable discussion they took us up hill, saying they would beat up hill to the guns, but after taking us part of the way they concluded it was too hot for her to go up the hill and it would be better to drive her through some shady jungle at the bottom. We offered no opinion, but simply left it to them to decide. Our knowledge of woodcraft was as nothing compared to theirs. This tigress also roared when the beat commenced, and came up to some high grass opposite to Mr. Boevey. When the beaters came quite up she went at a rush past Mr. Boevey, who fired twice with a Magnum express. She rapidly made off, and as we could find no blood we had another beat. Whilst walking on ahead we heard a bekri deer bark and saw a peacock fly, which made us sure she was there. To make a long story short, she did not come up to the guns, but started off directly the beat began, passed out at the side under a tree upon which sat a sepoy who saw she had a bullet in her stomach. We found lots of blood, and pugged her up a bit, and then stopped as the sun was sinking, and the shikaries said we should find her dead next day, and if we went on some one would get mauled. Each of us being a *paterfamilias,* under solemn promises to our wives to do nothing rash, we fell in with this view. Next day we found she had drunk at the same place and pugged her into a jungle a mile off. Drops of blood were found on her path. However, although we beat for her all day we could not find her, and after that all trace of her disappeared. All said she was dead somewhere, but we could not find her, and there is no doubt that with an Express bullet in her stomach she must have died. Next day, which was a cruelly hot one, I went out at noon alone to look for her, and also to get information about the depredations of the man-eater. It is impossible to place much reliance on the various statements made to me by the various jungle men I spoke to, because it was very evident they grossly exaggerated. I was told the tiger had killed various people, from forty to 500 in number, and that last rains he had been particularly vicious in killing people engaged in cultivating their small

Dillenia scabrella
Family: Dilleniaceae

Wallich says: "This tree was introduced from Chittagong by John Roxburgh in 1807. It is deciduous and slow-growing. At the time of flowering it is entirely destitute of leaves, and is covered by countless numbers of elegant and fragrant yellow blossoms. The flowers last for a few weeks and are soon succeeded by a plentiful crop of small round orange coloured fruits. The fleshy leaves of the calyx have a pleasantly acid taste, and are much used by the inhabitants of Chittagong." The tree is also common in the Western Ghats also where it is known as Kulgul (Kannada) and Karmal (Marathi). The straight, large trunks provide wood for boats in Canara. In Assam it is found at the foot of the Mikir hills and in Sibsagar and Nowgong.

From N. Wallich, *Plantae Asiaticae Rariores,* 1832.

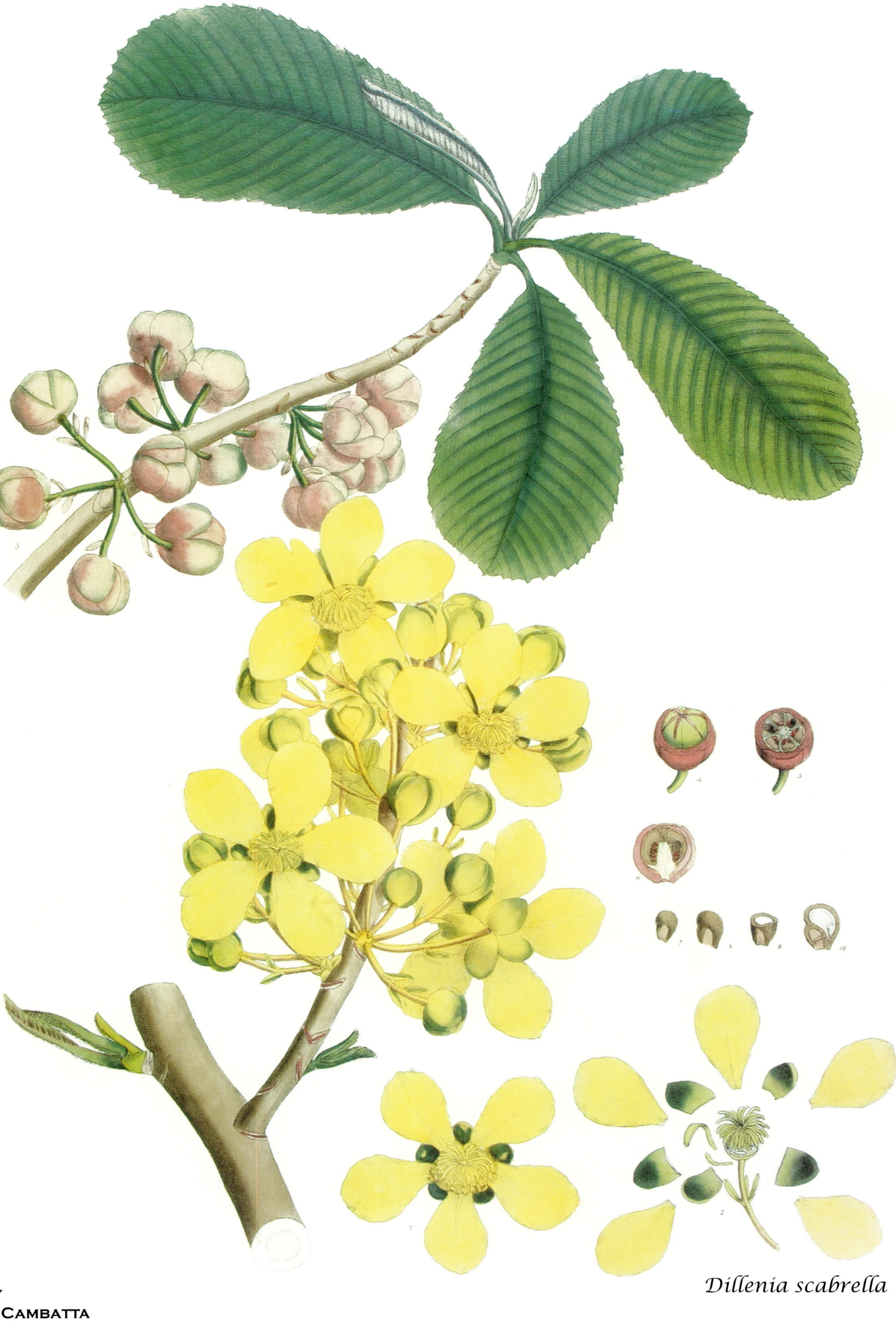

Dillenia scabrella

Courtesy
Firdos S. Cambatta
The Serenity Trust

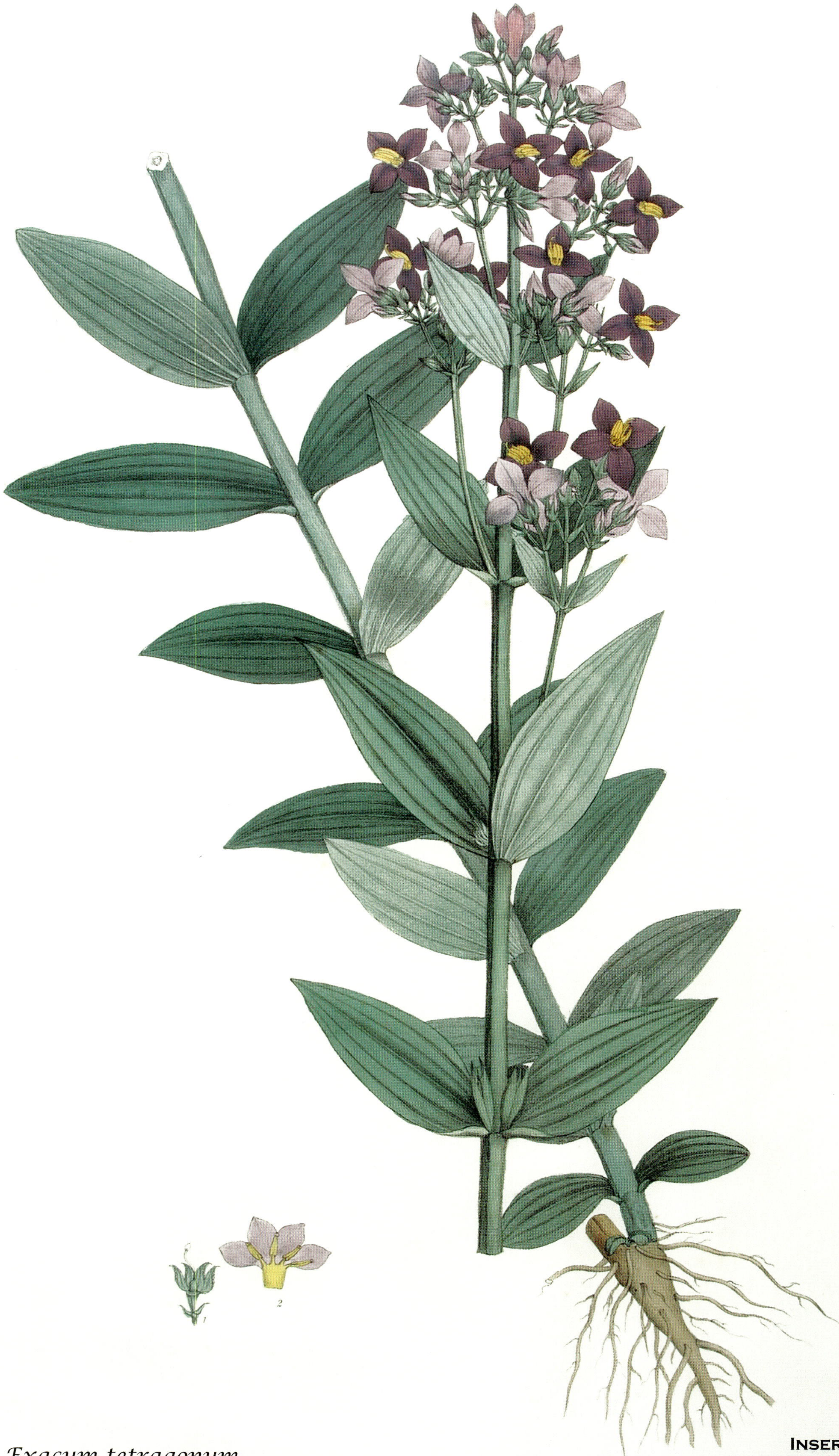

Exacum tetragonum

INSERTED IN FOND MEMORY OF
AND AS A MARK OF GRATITUDE TO
THE LATE PROFESSOR P.V. BOLE
– DHUN PANTHAKI

Purple Chiretta
Exacum tetragonum
Family: Gentianaceae

A very elegant plant with a 4-angled erect stem, 30–120 cm high, found on open slopes and damp places at altitudes of 1,000–1,800 m, from Uttarakhand to Sikkim, Bhutan, southern China, and Southeast Asia. The leaves are 3- to 5-veined. The large flowers appear in August–October; they have 4 purple-blue petals and golden anthers, and are borne in terminal spike-like clusters on the branching stems. The capsule is globular, with persistent calyx and corolla.

From N. Wallich, *Plantae Asiaticae Rariores*, 1832.

strips of land in the jungle, and even taking people out of bullock carts carrying timber from the forests. I went to see two or three places where people were actually seized. One of these was close to a jungle road to the Dangs, over which a large timber traffic passes. I have a photo of one of the jungle roads over which the beast used to pass, and also of a temple to the tiger god which the Bheels erect all over these districts. On the next day Mr. Boevey had to break up camp, and leave for Surat hurriedly, as fast as we could march, in consequence of the Surat fires. I tried unsuccessfully to reason with him, that it was more important to bag man-eating tigers than inspect a fire which would be put out long before he could reach it. However my arguing did not convince Mr. Boevey, and we had to put an end to a very pleasant trip and march rapidly for the line of rails. The only real accurate information I was able to get was from Mr. Jhaverbhai Nathoobhai, the Dewan of Bansda, who obtained for me the names of persons killed in the Bansda State; only the villages, dates of death, age, sex, &c., which was embodied in an official document I have. This was made up from the police records of the inquests on the persons killed, and from the reports of the cattle killed in the various villages reported by patels. This only applies to the villages in the Bansda State and not to the adjoining villages in the Dangs, Baroda, and British territory. During 1888, out of thirteen jungle villages in the State he killed in six villages altogether eight people, of which seven were males, the dates of killing 28th January, 31st July, 22nd October, 2nd November, 4th November, 12th November, 15th December, and 18th December. Of these all were adults except one boy, named Ganda Kalia, aged 13 years. In the same villages during the same period thirty-five cattle were returned as killed. I am sorry I am not able to get any record of those killed in the Dangs or Baroda territory during the same period. I have written to enquire at Bansda if there have been any more cases reported of persons killed by tigers, and I am informed by the Dewan there have been no more cases, so that this strongly supports the view that the tiger I killed was the man-eater.

Man-eaters of Toongar and Jaunsar

Some eight years ago or so, Mr. Mulock killed an undoubted man-eater near Toongar, some thirty-five miles from where we are now sitting. I say an undoubted man-eater, because Mr. Mulock was close by when the man was carried off, and went after it at once, and found the half-devoured corpse. Whilst waiting by the corpse the tiger or tigress (I forget which) returned, and Mr. Mulock killed it and made a *post mortem* examination of it, finding parts of the deceased man inside the tiger, so I think we may safely assume this was a man-eater. Mr. Mulock published an account of it at the time, and gave me a copy which I have unfortunately mislaid, but I think I may trust my memory so far as to say there was nothing peculiar about this tiger in any way, and that his skin was not mangy.

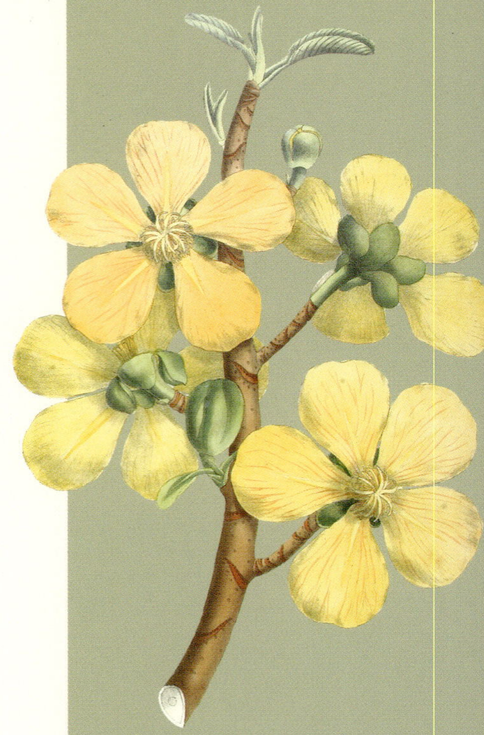

I have no doubt you read in the papers a short time ago of a man-eater being killed, called the Jaunsar man-eater. An account was published, in which it appeared that Mr. B.B. Osmaston, of the Forest Department, was out near Chakrata after her with a companion, when Mr. Osmaston shot her whilst she was worrying his friend. The papers further published an account by an anonymous correspondent of the career of this tigress, and an extraordinary story of the tigress allowing her cubs to play with a man she had taken out of a house in the Himalayas. This story I have not been able to verify, so that I am not able to give you particulars. I wrote, however, to Mr. Osmaston for certain particulars about this tigress, and he has very kindly given me some interesting information about her, which I will give to you almost in his very words.

Man-eaters have been numerous in the Jaunsar district for many years. The district which this tigress frequented was a very large one. She used to make long journeys in a very short time. Having killed a man in one place, she would appear the next night at a place twenty miles off. She frequented a somewhat high belt of the Himalayas, mostly from 5,000 to 10,000 feet high. The spot where Mr. Osmaston killed her at is about 8,000 feet above the sea. No old wounds were visible, but she was very old. Report makes her out to be a man-killer of at least ten years' standing. Her skin was not mangy but a very fair one. The cold climate might, of course, give her a good skin. Her teeth were exceptionally bad. Of her four canines none were sound, and the two upper ones were worn and broken down to about half their original length. They also had two slight cavities in their centres, which were found by probing to extend to a depth of three-quarters of an inch. About seven or eight porcupine quills, mostly broken off to about three inches in length, were found in the tigress's body. Two were actually embedded in her tongue. There is not very much game suitable for tigers in the district. Buffaloes and men are easily obtainable, and excellent opportunities afforded of seizing them on the march on the mountain sides.

This tigress appears undoubtedly to have been the man-eater, because she actually attacked Mr. Osmaston's companion without provocation, except in so far as was due to the fact that these two gentlemen went up in broad daylight to a buffalo killed by her. The papers stated Government offered a reward of Rs. 500 for her.

Other Adventures

I once had another adventure with an alleged man-eater. I have, however, no evidence to offer that it was a man-eater, except the fact that all the shikaries and villagers of the neighbourhood declared that this was the particular miscreant that had eaten a number of people in the neighbourhood, and that every villager for miles round, in number nearly 500, turned out to assist as beaters. This was in the State of Rewa, a few years back one of the best tiger countries in India, where tigers are always ready for visitors of distinction in a State preserve specially kept for tigers and other large game. Through the kindness of Major Barr, the then Political Agent, I was asked to form one of the usual hot weather party who made an annual campaign against the tigers. We shot in royal style: we had State elephants, shikaries, sowars and all the assistance we required. All we had to do was to go to the places fixed for us by the head shikari when we got *kubber,* and if we did wound a beast, we had only to get on two of the best elephants in India and follow up. Following up a wounded tiger on an elephant I must say is a royal form of sport. It is a grand thing to see a wounded and irate tiger absolutely at your mercy and without incurring any risk to yourself. You have all the fun and none of the danger one experiences in going after a beast on foot. I was once so excited in watching a wounded tiger charging from a long distance at my elephant, that I invoked the anger of the mahout for not firing soon enough. I forgot for the moment that if any one was in danger it was him and not me in the *howdah.* To return to the man-eater. Soon after the beat commenced I heard the tiger coming up towards my tree, and it remained within about sixty yards of me for some time, giving low growls. I could not, however, manage to see him. However, a few minutes afterwards, I saw him lying prone on the open side of a hill opposite. I would not fire at first, thinking he might go to another gun; but after

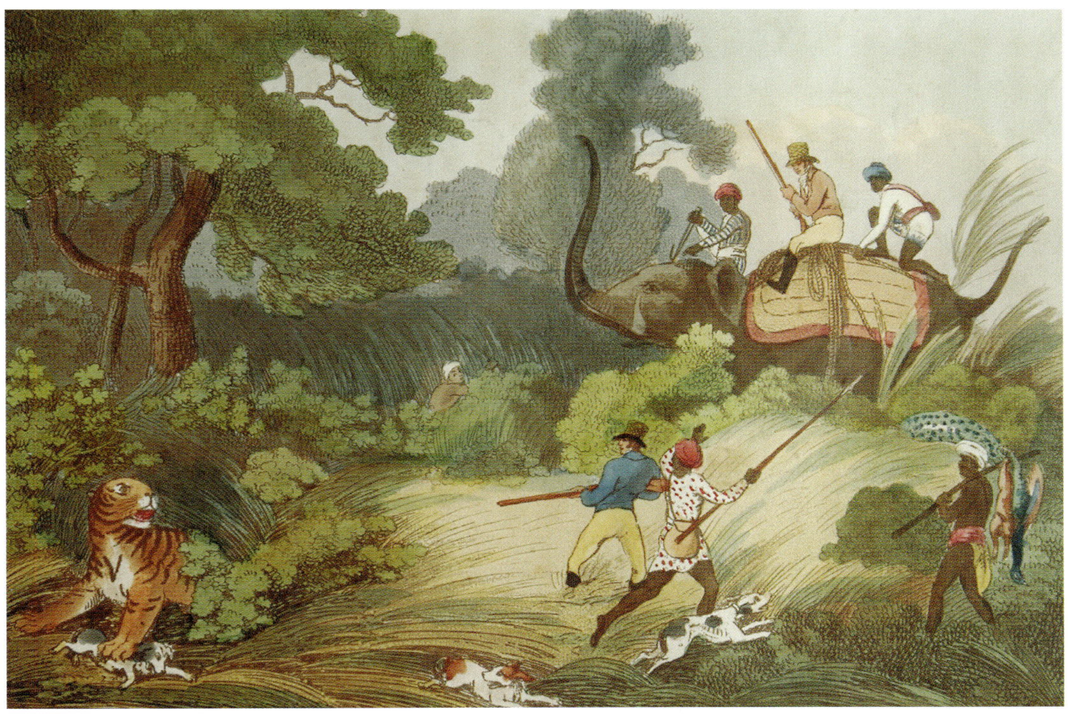

The tiger at bay, from Thomas Williamson, *Oriental Field Sports,* Vol. I, 2nd edition 1819.

some time I very wrongly determined to fire at him, believing he was bound to go out at the side of the beat and not in sight of the other guns. I afterwards learnt there were stops everywhere, and the tiger was bound to come close to one of the guns. I computed he was 200 yards off, and sighted my rifle accordingly. I missed him. We afterwards all agreed he was only 100 yards off, which gave me a good and valid excuse for the miss. The tiger then bolted into a patch of jungle, and the beaters all got up trees terribly frightened. Two of our party got on elephants, and he passed within twenty yards of them, giving each of them a right and left easy shot. All the shots missed him, and the tiger went away in sight of every one up another hill. The beaters said he was a regular "shaitan", and no bullet could hurt him. I suppose this was said to let us down easy, but the State head shikari, Moti Singh, was terribly downcast about it, and I was horridly depressed in spirits also. However, two days afterwards we killed two tigers in one beat, and we got our good spirits back.

A common theory appears prevalent that a wounded tiger often turns man-eater, and lately in the columns of the *Pioneer,* I think, reference has been made to certain cases of wounded tigers having turned man-eaters in the Central Provinces. Wounded tigers often turn man-*killers,* but I have not obtained any evidence of their turning man-*eaters.* The difference is very great, except, perhaps, to the victim. A wounded tiger no doubt, until its wounds are healed, attacks every person who comes near to it. It does this not for the purpose of obtaining food, but because it is smarting under a painful wound, and it believes that the person approaching it is going to inflict another wound. Many instances can be quoted of wounded tigers killing persons approaching them after they have been wounded, and I need only mention the case of my friend the late Mr. G.L. Gibson, a member of our Society, who died here from wounds inflicted by a wounded tiger he was seeking for, and whilst he was examining the body of a native boy which he found killed by this wounded tiger. This is the chief danger of leaving a wounded tiger, as one knows that the first person who unfortunately comes near the place where the tiger may be lying down will undoubtedly be killed, and many sportsmen therefore very properly prefer to run considerable risk in killing a tiger they have wounded, rather than allow it to live and kill the first innocent person who may be so unfortunate as to come near it. Mr. Mulock writes me as follows, viz., "My theory is that if one member of a tiger family takes to the man quarry they all lose their fear of the biped and kill him when hungry. I have found this in one or two instances." I observe also that Mr. Saunderson in his book scouts the idea of man-eaters being mangy, and wonders how this idea became prevalent.

To sum up then, I have no particular theories, with one exception, to put before you as to man-eaters. The one theory I can advance is that the man-eater inherits this vice from its parents, or that the parent having previously learnt this vice from a parent

Nepal Lily
Lilium nepalense
Family: Liliaceae

This plant is a native of the southern slopes of the Himalaya from Uttarakhand through Nepal to Bhutan. The stem grows from 60 cm to about 1 m height and bears lanceolate leaves about 8 cm long. The mildly fragrant flowers appear in June–July are few or solitary, pendant and greenish-yellow with a mahogany-red zone within. The protruding stamens bear reddish anthers. The bulbs are cooked and eaten like potatoes

From N. Wallich, *Plantae Asiaticae Rariores,* 1832

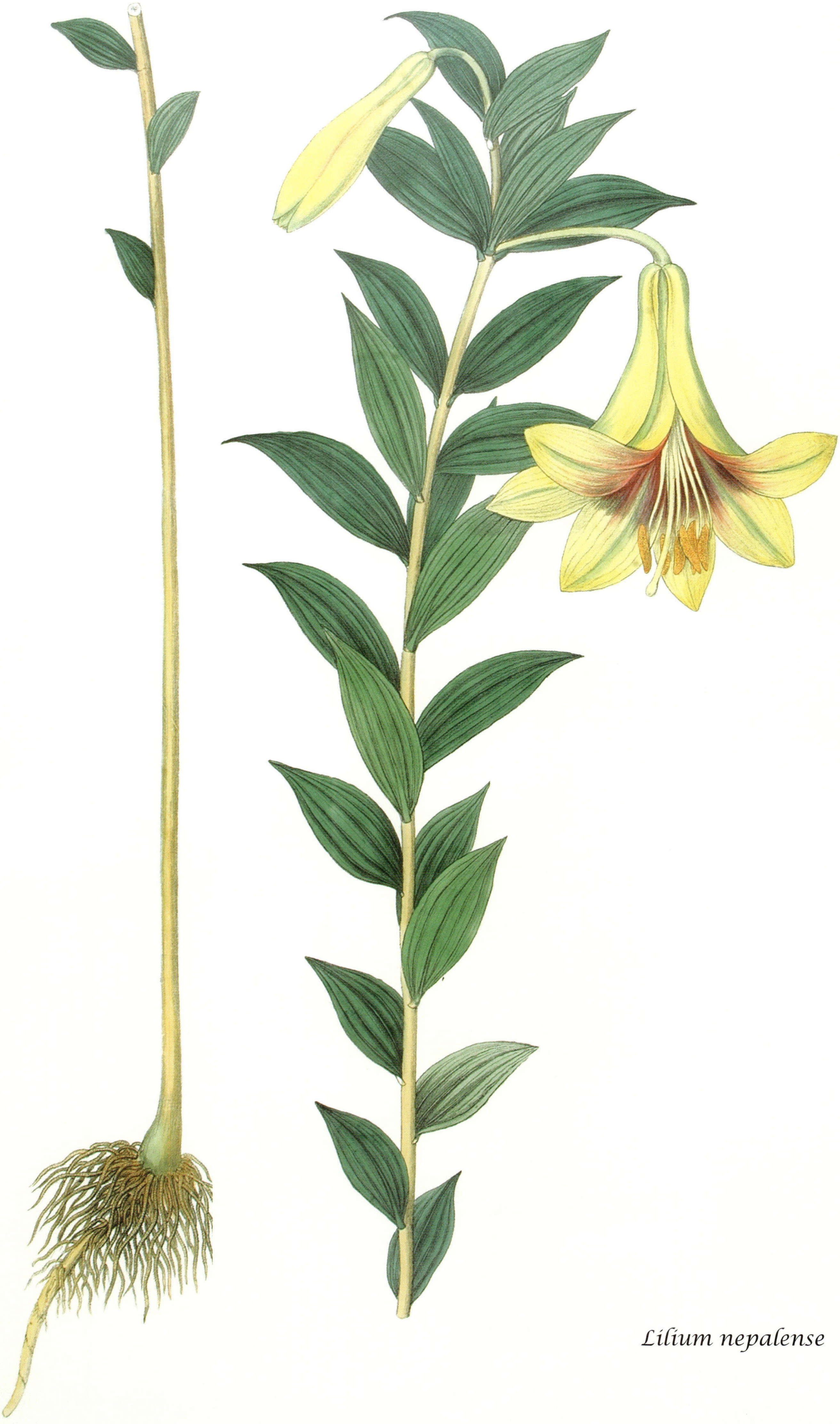

INSERTED IN
FOND MEMORY OF
MY COLLEAGUE,
THE LATE
DR NARIMAN A. IRANI
– DHUN PANTHAKI

Lilium nepalense

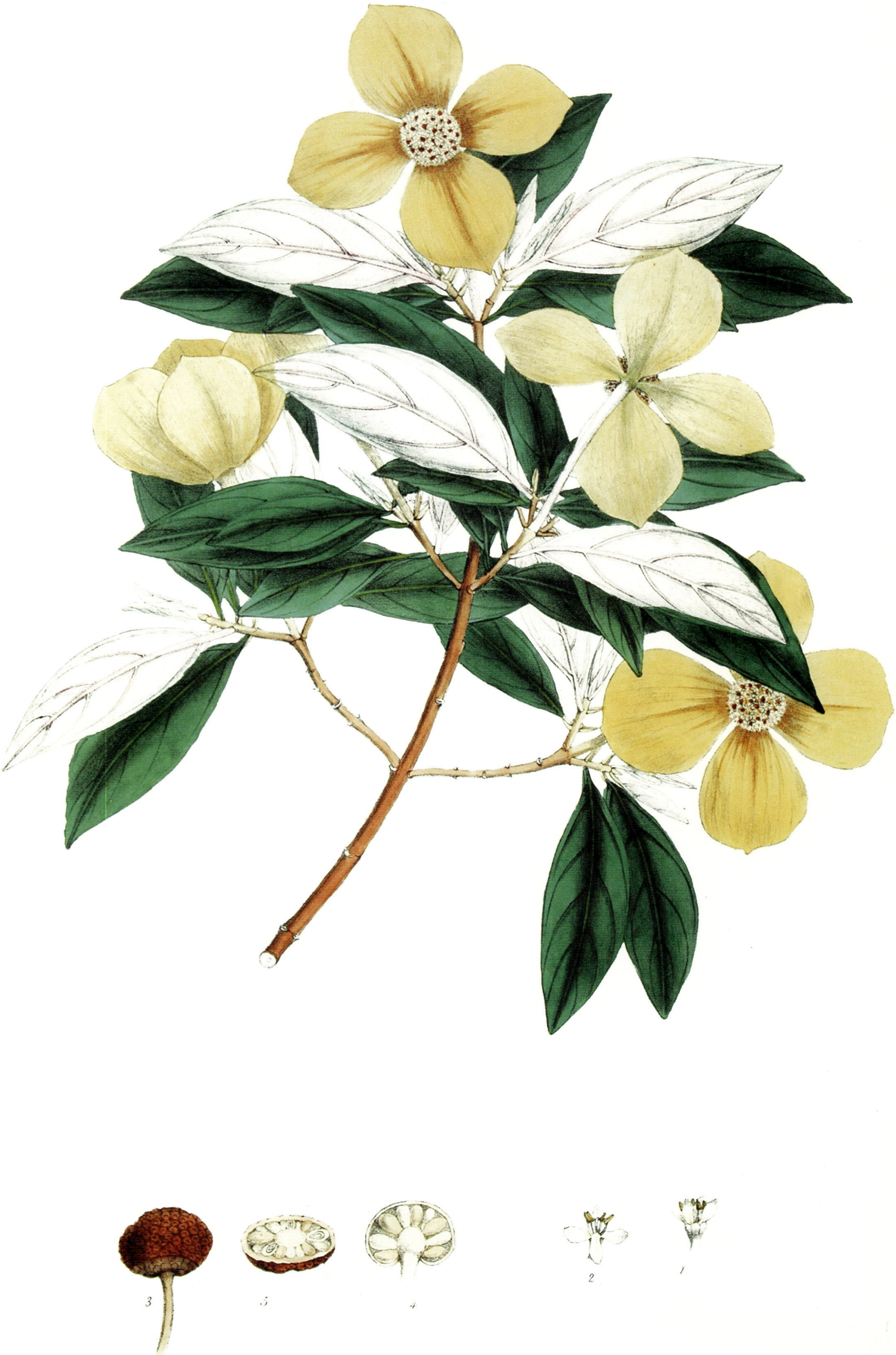

Benthamidia capitata

Himalayan Strawberry Tree, Evergreen Dogwood, Himalayan Flowering Dogwood
Benthamidia capitata
Family: Cornaceae
Synonym: *Cornus capitata*

A deciduous tree, 10–24 m high, found in the Himalaya from Kashmir to Bhutan, at an altitude of 1,100–2,400 m. It has smooth greyish-brown bark and leathery, entire, elliptical-oblong, minutely hairy leaves which are dull greenish-blue above, pale beneath. The tree is especially noticeable when in bloom during May–June; the flowers are greyish-yellow, very small, and closely packed into hemispherical heads, which are surrounded by 4 or 5 white or cream ovate, petal-like bracts. The fruits are composite, strawberry-like, round, fleshy, yellow to pink, with numerous round brown stones. The fruit pulp is yellow and slightly sweet; it is eaten and also made into preserves. The tree is common in Kumaon and Shimla where it is called Bamaur, and often grown as an ornamental tree.

From N. Wallich, *Plantae Asiaticae Rariores*, 1832.

or companion, teaches the cub to kill human beings, and such cubs, when grown up, teach the vice either to their own cubs or to their mates, and so the practice never dies out amongst the tigers of that district. In short I contend that, unlike the case of the poet, the tiger is sometimes born a man-eater and sometimes made one. This theory will account for some districts never being entirely free from man-eaters. In all cases it appears the man-eater shows greater cunning than usual. I have no evidence to show that the tiger turns man-eater for any particular reason. There is abundant evidence to show that the man-eater is physically not different from the ordinary tiger, that age has nothing to do with the question, and that the theory about the manginess of his skin is mere fiction. There is evidence to show that districts in which the tiger has plenty of his natural food are just as much infested by man-eaters as in those districts where the natural food is less abundant. I am afraid my paper is of a mere negative character, and I leave it to some other members to evolve some theory for us on the subject.

Statement showing the number of persons and animals killed by a man-eating tiger during the year 1888, in the territory of Bansda.

Name of village.	Name of person killed.	Age.	Sex.	Caste.	Date of death.	Animals killed. Cows.	Bullocks.	Goats.	REMARKS.
Ambabari	Ardia Bapudia	30	Male	Kukna	15-12-88	1	...	...	
Tadpada	Ganda Kalia	13	Do.	Koli	18-12-88	...	...	...	
Waghai	1 Kesu Punio	60	Do.	Kathis	31-7-88	...	...	...	The tiger was killed by Mr. R. Gilbert, Solicitor, Bombay, whereby the locality has been relieved of the terror.
	2 Tolia Bablia	45	Do.	Kukna	22-10-88	...	...	...	
Sadad Devi	...	...	...	...	...	5	6	...	
Wati	...	...	...	...	...	4	1	...	
Khambla	1 Jivla Kasa	37	Male	Warli	2-11-88	...	...	...	
	2 Radio Bhil	36	Do.	Koli	28-1-88	...	...	...	
Sitapur	...	...	...	...	...	1	2	...	
Mahuvas	Bai Radgi	30	Female	Kukna	4-11-88	...	...	...	
Charanwada	...	...	...	...	...	1	...	...	
Godhbari	...	...	...	...	...	...	2	...	
Kharjai	...	...	...	...	...	3	5	8	
Manpur	Budhia Natha	50	Male	Kathia	12-11-88	...	1	2	
Dhakmal	...	...	...	...	...	1	2	7	
			7 m. 1 f.						
			8			16	19	17	

JHAVERBHAI NATHOOBHAI, Dewan of Bansda. *30th April 1889.*

From *JBNHS* Vol. 4, No. 3, 1889.

FROM MAMTA & ZUBIN MEHTA, AND SHAAN
DALLAS, TX

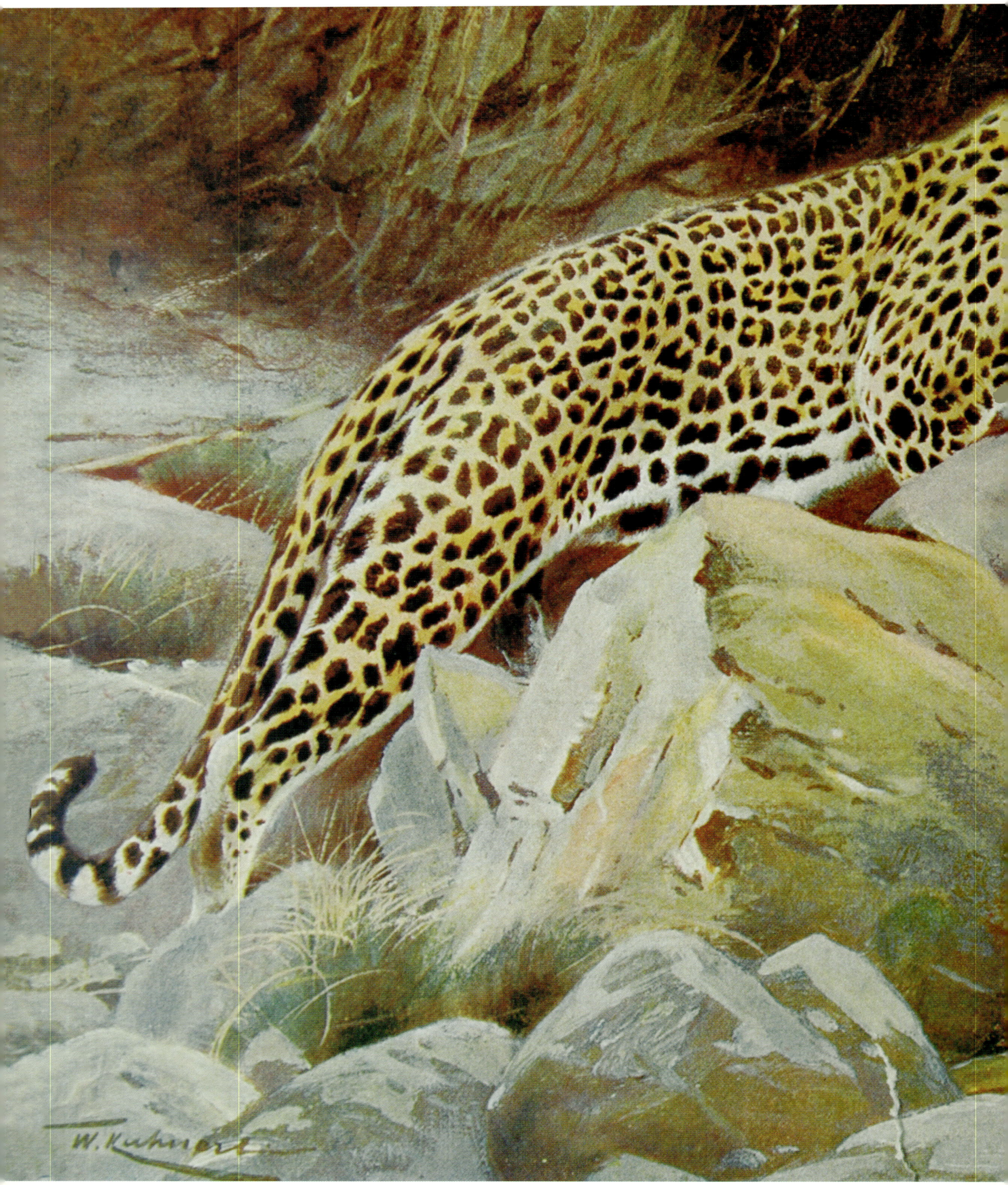

In loving memory of my wife Vijaya
– B.G. Deshmukh, IAS Retd.,
President, Bombay Natural History Society

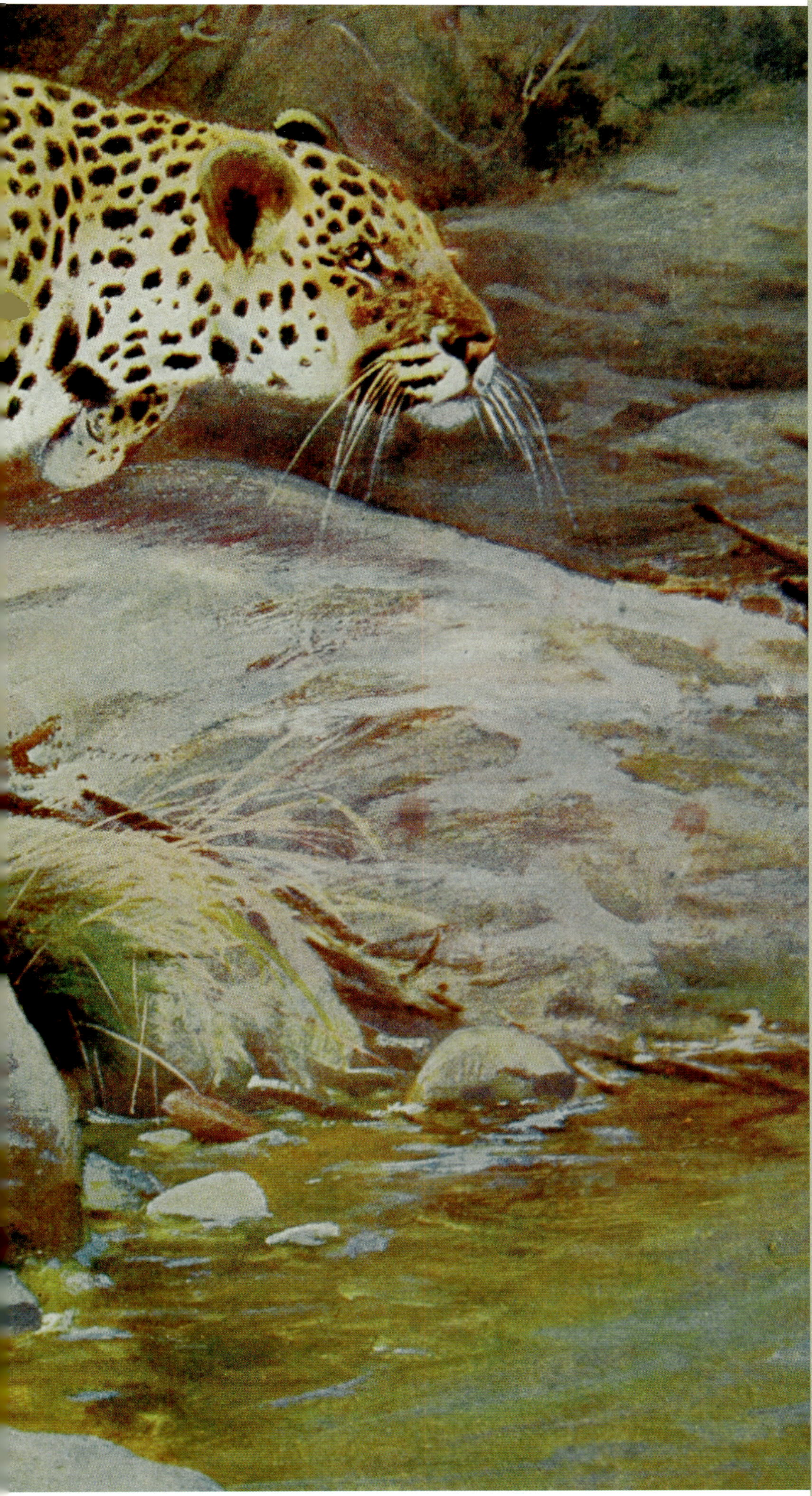

Leopard
Panthera pardus

This sleek, shorthaired big cat, found all over India, has a bright coat marked with small rosettes. Lighter coloured and smaller animals are found in the drier and more open parts of the country. They are often found near human habitation, as is the case in the vicinity of Sanjay Gandhi National Park, Borivali, Mumbai. These animals are usually smaller than those living in thick forests. Leopards are opportunist hunters. Their diet consists mainly of ungulates and monkeys, but rodents, reptiles, amphibians, fish, and village dogs are also eaten. In India the most favoured prey of the leopard is chital. The leopard stalks its prey silently and at the last minute pounces on it and strangles it with a quick bite, also dislocating the spinal column. It can also chase its prey at great speed, running up to 58 km per hour. It often hides its kill in dense vegetation or takes it up a tree. The leopard is known for its climbing ability and is often observed resting on the branches of trees during the day and descending head first. The gestation period is 90–105 days; a litter has 2–4 cubs but usually only one or two survive infancy.

E.P. Gee writes in *The Wildlife of India*: "I have seen leopards in the daytime, but one sees them more often at night when one is travelling along roads in a motor vehicle. Once during my early tea-planting days a friend and I were coming home from a dinner party and saw three different leopards in the same night. The first one sat in the middle of the road facing us, and we had to scare it away. The second one had a freshly killed goat in its mouth." He further states: "Leopards very rarely become man-eaters, but when they do they are a very deadly menace. I once knew of a leopard which became a terrible scourge in a district, regularly killing and eating young boys and girls of about six to ten years of age – till it was shot by a friend. For some leopards occasionally kill and eat human-beings simply because of their omnivorous [?opportunistic] habit, without acquiring a permanent preference for human flesh as man-eating tigers do."

Painting by William Kuhnert, from Lydekker et al., *Harmondsworth Natural History*, 1910.

Pagal Kutta and a Man-eating Panther

George Hogan Knowles

My wife's little dog, a small black-and-tan terrier called Zoe, had just taken up its position before the front door in the verandah, when, suddenly, outside in the garden – to the left front of the extensive mansion – there came a peculiar howl, like a wolf's, yet more of a guttural sound, not only uncanny but extremely puzzling! The howl was immediately followed by sharp barks from some prowling jackals. Instantly, a cry from our sleeping apartments announced that our little son of six months had been startled and awakened from his sleep, for the slats of the closed Venetian doors of his room had been left wide open for fresh air. The sharp, burring yelps of the terrorized jackal pierces stone walls, and is ever – to man as well as to his flocks and herds – a warning of danger that carries far from village to village on a moonlit night.

Into her bedroom my wife rushed, while we two men, calling to the dog to get her inside the room, hastened to our own apartments to seize our guns. But we had hardly crossed the threshold of the doors leading into our rooms when an agonizing yell from the little dog in the verandah, and what sounded like a struggle for life or death, brought us hurrying back to the front door with sticks in our hands. Mough Sing, our veteran head of the guard, had already reached the spot with his iron-bound "lathie", or long stick, in his hands. We had rushed out, prepared to see anything – a panther, a hyaena, or a wolf at least! But, in the outlook of the ill-favoured premonition that had haunted us a few moments back, it appeared to be something even worse than those satanic quadrupeds, for the poor little terrier was struggling in the deadly embrace of a ghastly cadaverous-looking red creature, somewhat larger than a jackal.

"Pagal kutta!" (mad dog) shouted Mough Sing to us. "Take care, your honours," he shouted, as the mad dog, foaming at the mouth, as we could now see in the dim patch of lamplight thrown out from the hall, turned upon us. But in an instant the sporting little terrier was on the rabies-smitten creature again, an act on the brave little dog's part that saved Mr. H. and myself from being bitten.

Then my wife came rushing out, and Mr. H. and I fell back into the hall to warn her and to keep her back, for nothing could be done to save her little dog from its death struggle with the mad one. Poor Zoe had already been bitten, and with her opponent in the clutches of hydrophobia, no one dare grapple! The mad dog's eyes gleamed red, like burning coals of fire, while the verandah was covered with the froth of rabies, and over the distracting scene – my wife in the greatest distress over her little dog, and the child crying out, indoors – that awful, demoniacal howl outside rose up again, drowning for a

George Hogan Knowles was the author of *In the Grip of the Jungles* and *Terrors of the Jungle*, from which this excerpt is taken. Here he describes "night horrors of the jungle", when a rabid dog seizes their pet dog and more terrible things start happening under the shadow of darkness ….

Crimson-backed Flameback, Greater Flameback
Chrysocolaptes lucidus stricklandi

This handsome bird, which Jerdon calls the Red Woodpecker (*Picus ceylonus*), is the Sri Lankan counterpart of the Greater Flameback, with darker red upper parts, longer and paler bill, white throat, broader black stripe through eyes, and narrower white supercilium. The male has a red crest, female has a white-spotted black crown and crest. Jerdon writes, "I have figured this interesting species of woodpecker from the island of Ceylon, where it was procured by Lord A. Hay and where it appears to be not uncommon, replacing the *Picus bengalensis* of the Indian continent, to which it assimilates in structure, belonging to the same division *Brachypternus*." Resident throughout Sri Lanka, it is found in a mixed crowd with other woodpeckers, drongos, and other insectivorous birds. It usually visits large trees in search of insects hidden below the bark and also forages on the ground among fallen leaves. The nesting season may be any time from August to March. Its call is a series of very noisy, rising short trills, shrill and sharp notes. It frequently drums on tree trunks, making a sound like an African tom-tom in a milder form. In *The Common Birds of Bombay* EHA says: "The Red Woodpecker having a rather weak bill saves itself the labour of making a hole in a dead branch of a tree by burrowing into the nests of tree ants" but here he must be referring to the Rufous Woodpecker (*Celeus brachyurus*) which has a shorter bill.

From T.C. Jerdon, *Illustrations of Indian Ornithology*, 1847.

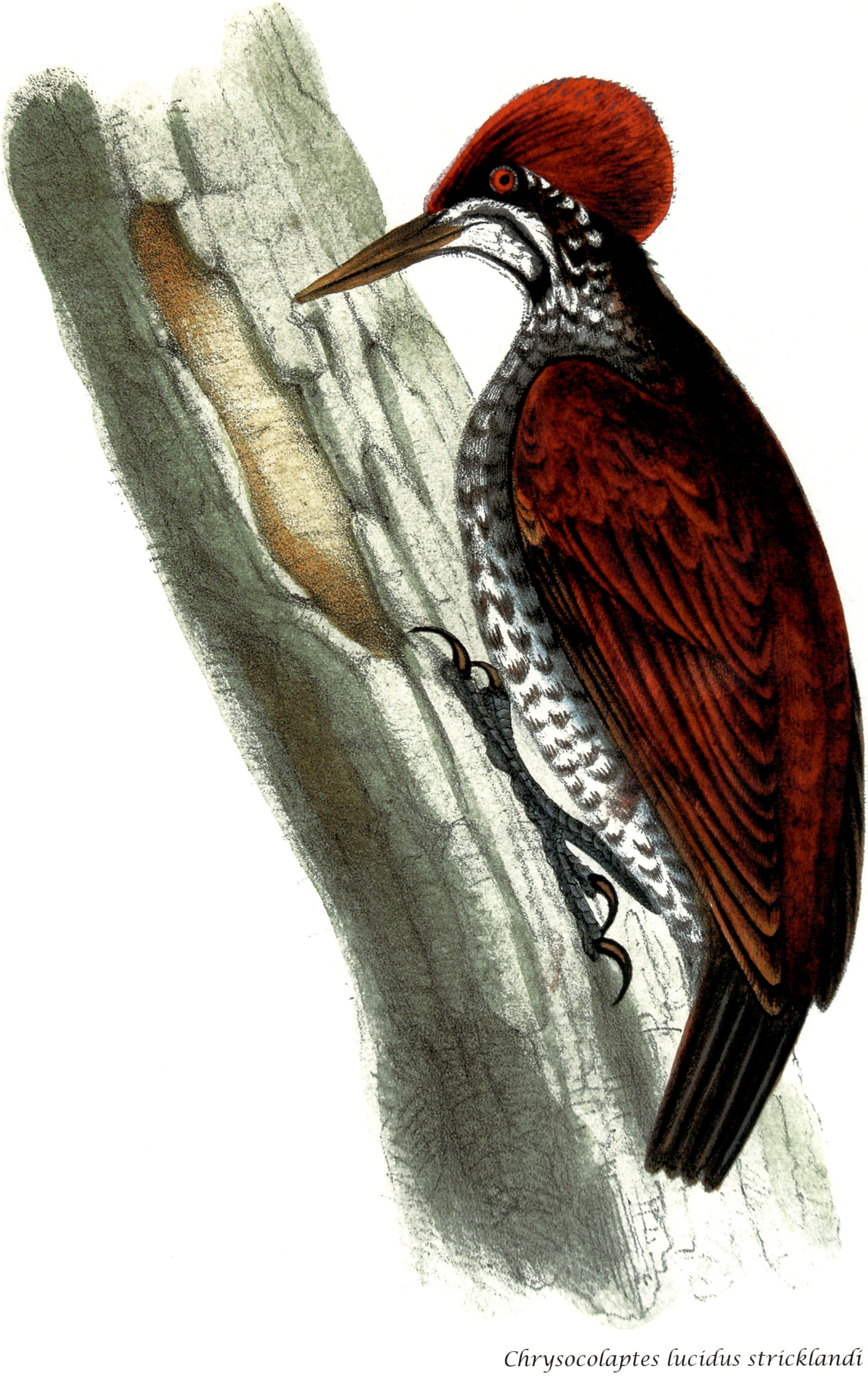

Chrysocolaptes lucidus stricklandi

Courtesy
IndusInd Bank

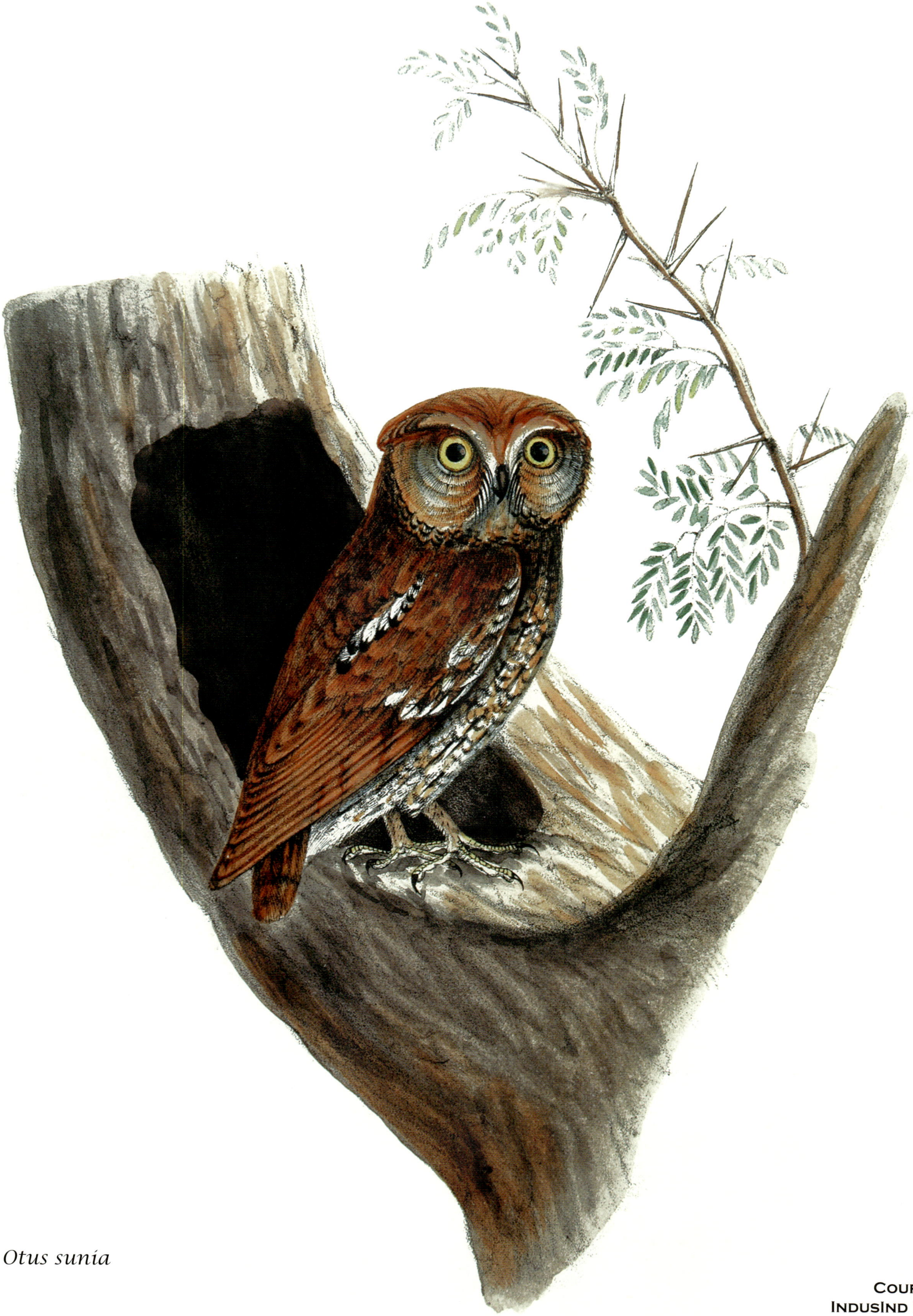

Otus sunia

COURTESY
INDUSIND BANK

Oriental Scops Owl
Otus sunia

The colour of this small owl varies from rufous to grey, with heavy black streaks and white patches below and little streaking above. The eyes are yellow. According to Jerdon: "The Indian Scops Owl is found occasionally throughout all India and Ceylon, in forests and well wooded districts, but is not very common. The first specimen I ever procured was found dead outside my house at Madras, and had probably been killed by the crows, as it was in a good case, and very little injured. I have obtained it both from the Eastern and Western Ghats, but not from Central India. It is also found in the Himalayas, and extends into Burmah, Malaya and China. It has a low mild hoot, which is often heard soon after dark. All I have examined had eaten insects. It breeds, according to Hutton, in hollow trees, laying three pure white eggs on the rotten wood in March." (*Birds of India*, Vol. I.)

From T.C. Jerdon, *Illustrations of Indian Ornithology*, 1847.

moment the desperate yaps and pathetic cries of the little dog, the shouts of the guard, and the ceaseless yelps of the agitated jackals, in the wake of the howling demon.

Shaking off the terrier with a sudden rush, the mad dog flew at Mough Sing, but two men of the guard – for they had all now collected – brought their long-reaching bamboo poles crashing down on that wretched mad dog's head. Lanterns were brought quickly. Poor little Zoe lay panting, and could not be handled, as she was covered with the mad dog's saliva, but as soon as possible the poor little thing was put to sleep, for within ten days she would have been a victim of that dreaded disease rabies.

The mad dog was some village "pariah", which had run away in terror – as animals always do when the first attack of madness comes on – and had unfortunately found its way into the bungalow in the second stage of rabies, which makes the doomed creature fly at every living thing it sees.

Now, but for the sharp yelps of the jackals, the night seemed to mourn silently over the loss of our faithful little dog, who, undoubtedly conscious of some calamity approaching, had taken up its particular watch, and had saved us all, for had not the terrier been in front of the open door, that mad dog – attracted by the lamplight – would have rushed inside the hall and flown at us, biting one or all of us for a certainty.

"Hazoor," said Mough Sing to us, in a hushed voice, "that howl outside, near the garden gate in front, is no wolf. It is a man-eating panther! I know that particular camouflage from experience in my own village, situated in as wild a place as this. In cleverness," said the old veteran, "the man-eating panther can beat the man-eating tiger. The village folk," he continued, "are not afraid to turn out at night, with their iron-bound 'lathies', to drive away a wolf, or to surround him and kill him. Now the panther knows this – how!" exclaimed the old veteran, curling his grey moustaches, "'Bhagwan Janai' (God knows). But the man-eating panther often finds out that a wolf's howl usually brings men out, whereas, knowing that his natural roar has the opposite effect – the people locking themselves up in their houses – he tries to imitate the wolf's howl, in the hope of a chance arising, should the people come out after him, of his being able to grab a straggling man or a boy. The resemblance of the panther's camouflage to the wolf's howl is sometimes perfect," concluded the old man, "but, thank goodness!" he exclaimed, "man-eating panthers are rare."

And in this old Mough Sing was right. Such panthers are rare, but worse, when they are met with, than man-eating tigers.

We could not, however, credit what the old jamadar said, and being anxious to follow up the howl – not only to shoot the animal for everybody's safety, but to investigate, if possible, the matter of the camouflage – we were obliged to leave my wife – though after her sad loss she was feeling sorely distressed – in charge of the trusted guard. Armed with our guns loaded with ball cartridges, Mr. H. and I crept out into the moonlight, after the barking jackals, for we knew that these jungle guardians would be following the animal – whatever it was – that had howled so uncannily.

Cold was the night as, silently in single file, we crept under the shadows of the avenue trees down the drive, to the left front of the bungalow. Then, leaving the garden gate in front of us, we turned due east, to our complete left, and stalked across a stretch of rising ground, covered with crops of grey rock and quartz stone, beyond which the deep forest of sal trees rose up, a few hundred yards in front of us. Over the bare sheen of moonlight oblique shadows lay, as if the glinting rocks had cast off their dusky sheaths of the evening past.

Slowly we crept along, examining the shadows, as we approached the yelping jackals. There were three of these creatures, and they were moving now close to the edge of the forest. Suddenly, inside some undergrowth, there came the distinct sound of an angry rush at the jackals. Gasping and barking in terror of their lives, the jackals darted past our immediate front and disappeared into a dark stretch of forest to our left; and then the jungle guardians became remarkably silent. They had possibly received the fright of their lives and had gone to prowl round some other human habitation.

For a long time we stood in the shadow of a tree, waiting and watching, but not a sound broke into the gleaming silence of the night. Then suddenly – some distance away inside the forest – we heard the hellish howl once again, as though, with the sudden pain of a flesh wound, some giant monster of heathen fable were giving unearthly tongue. Panther or wolf, whatever it was, had also gone for some reason.

We returned to the old factory mansion to find that all was well and that my wife had retired. The morning star seemed to be twinkling already over the dawn of a new day and but a couple of hours' rest lay before us.

"Hazoor," said Mough Sing, as we were about to retire, "that man-eating panther has gone. He must have chanced to cross the tracks of the mad dog, for the animals dread rabies as men do, and, particularly, all the members of the feline species are not only quick to sense hydrophobia, but even give the direction taken by a victim a very wide berth. But," he said, "we shall hear more about this panther."

And true enough! It must have been about seven o'clock in the morning when a poor beggar-woman, dumb and deaf, was brought in front of the house by Mough Sing, the jamadar. A crowd of frightened people from villages close by had collected. Edging the forest, about a quarter of a mile to the north of the bungalow, ran a river, and a rough public road – that passed through the factory premises – crossed this river at right angles. The crossing was the only ford that could be conveniently used by the people and their cattle.

The poor, sobbing woman explained with her hands – while we helped her with interrogatory gesticulations – that her son of about twelve years of age, who had been

Himalayan Honeysuckle, Flowering Nutmeg, Himalaya Nutmeg, Pheasant Berry, Elisha's Tears
Leycesteria formosa
Family: Caprifoliaceae

This is a native of the Himalaya, found from Pakistan to southwest China and Myanmar in shrub and shady forests, from 2,100 to 3,000 m. The name *formosa* means beautiful. It is a deciduous shrub with hollow, upright stem, 1–2 m tall. The leaves are opposite, ovate to broadly lanceolate, long-pointed with entire or toothed margins. The flowers appear in May–June, on 5–10-cm-long pendulous racemes. Each mildly scented flower is small, pale pinkish-white, funnel-shaped, 5-lobed, subtended by a purple-red bract. The fruit is a soft purple-black berry eaten by birds. Some people relish it though it is slightly bitter. The hollow stem is made into whistles and flutes.

From N. Wallich, *Plantae Asiaticae Rariores*, 1832.

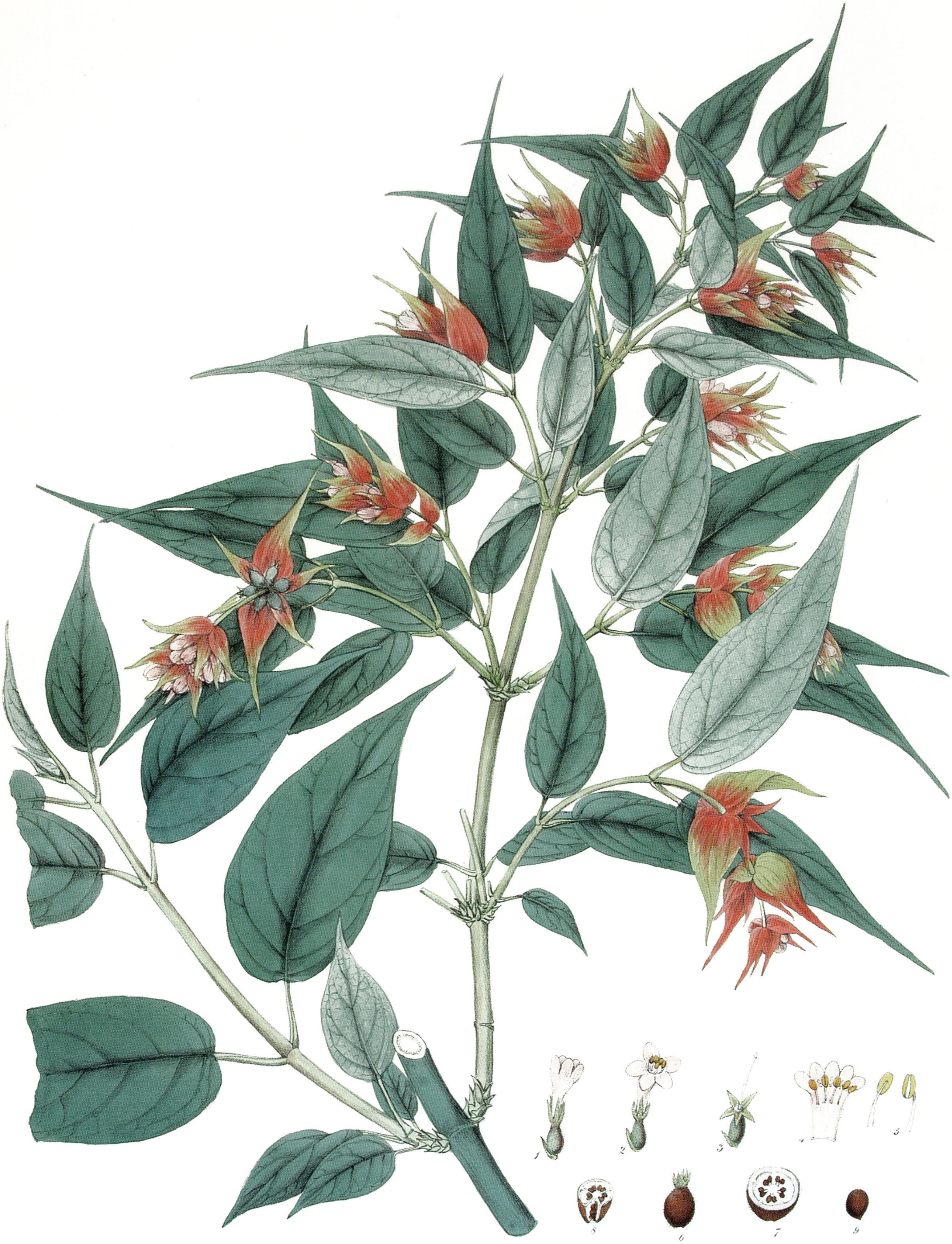

Leycesteria formosa

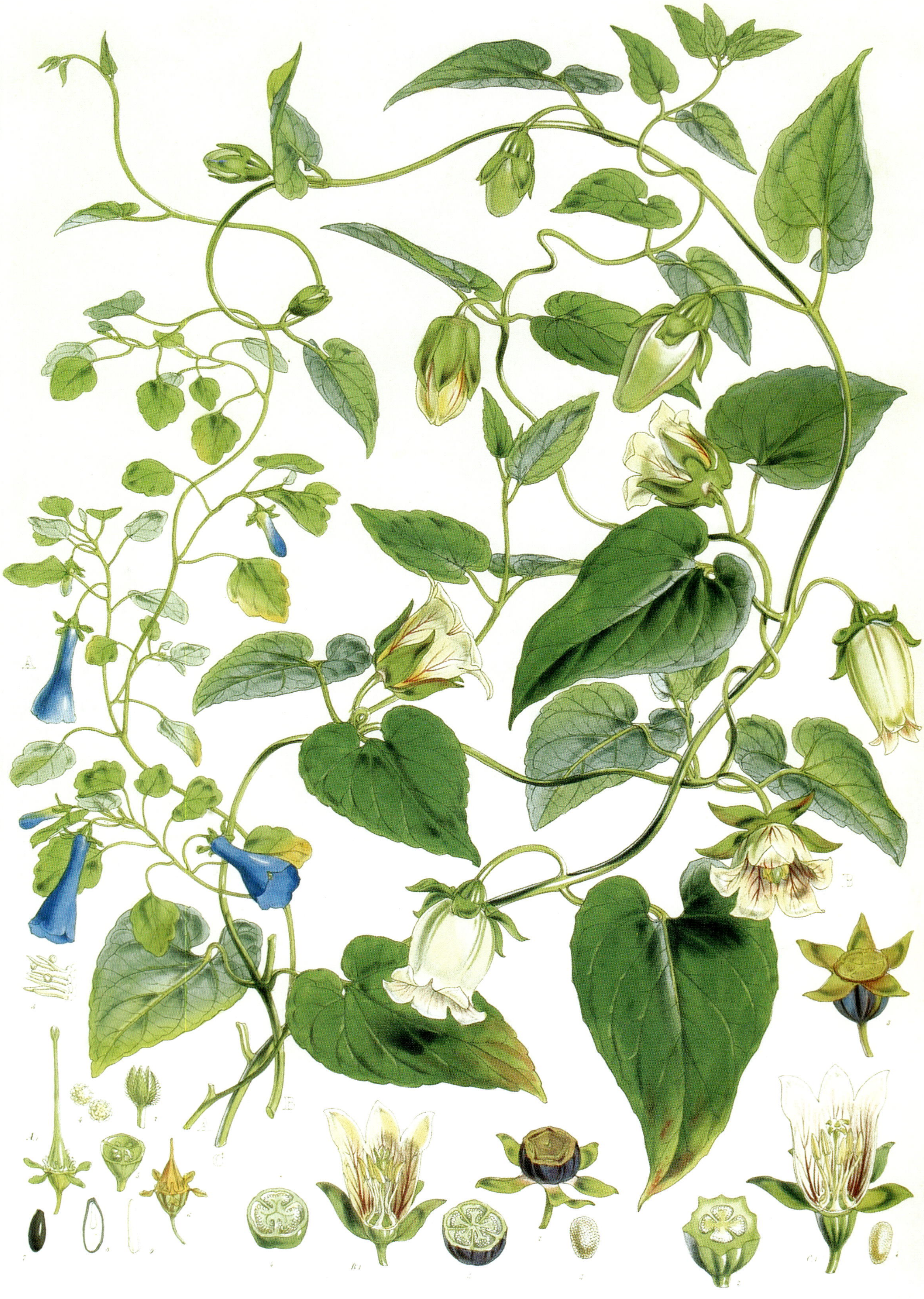

Codonopsis gracilis,
C. javanica, C. inflata

COURTESY
KANTILAL CHHOTALAL

A *Codonopsis gracilis*
Family: Campanulaceae
Synonym: *Leptocodon gracilis*
B *Codonopsis javanica*
Synonym: *Campanumaea javanica*
C *Codonopsis inflata*
Synonym: *Campanumaea inflata*

Codonopsis are strong-smelling perennial plants with large tuberous roots, twining stems, opposite or alternate leaves, pendulous flowers, usually solitary, and beaked capsule. They are called Bonnet Bellflowers. The genus includes 55 species endemic to eastern Asia. They are found variously from 1,000 to 3,000 m altitude in the Darjeeling, Sikkim, Bhutan, Assam, and Arunachal Pradesh hills.

Of *C. gracilis*, Hooker writes: "Nothing can exceed the beauty and delicacy of this little plant which is of rare occurrence in Sikkim, inhabiting water courses in very dense shaded woods, and covering bushes with its pale, translucent, membranous foliage and pale blue flowers. Like all its congeners it is full of milky juice and exhales a peculiar strong and very disagreeable odour when bruised .... This and the two other species figured with it are very singular plants, all so closely allied in many important botanical characters, that I do not doubt their belonging to one genus, but all presenting such important differences in structural character that many botanists doubtlessly separate them. Thus the present species has flowers that seldom arise from the axils of the leaves, their pedicels being adnate with the stem above them."

C. javanica and *C. inflata* have yellow flowers with purple veins.

From J.D. Hooker, *Illustrations of Himalayan Plants*, 1855.

walking in front of her, had just been carried off by a big animal, which she tried to explain was a tiger. That she had run straight up from the river to entreat us to go in search of the boy. She and her son were apparently on their way to a jungle fair, and had come from a distant village on the other side of the river.

Accompanied by Mough Sing and the beggar-woman, Mr. H. and I started off at once, armed with our guns. The ground over which we had crept in the moonlight being close at hand, we passed that way, and we were not surprised to find, on a muddy path, the pug-marks of a very big panther! We began to wonder – in our agitated and hurried inspection of the ground – whether after all there was truth in what the old jamadar had said, he, of course, being completely satisfied that the howl we had heard during the night was the man-eating panther's camouflage. Though we found no tracks of a wolf, we knew that wolves were not as particular as tigers and panthers in travelling along soft paths, and the ground between the pathways, being too hard and dry to show a spoor impression, it was quite possible that a wolf had also happened to pass that way. We had, therefore – looking at the matter regardful of reason – to leave the camouflage in doubt, with a strong suspicion, however, in favour of it.

The road as it led up just across the shallow water of the ford on to rising jungle ground – the river from this point flowing south – was flanked on each side by an embankment ten feet high. The road ran due east, and the poor woman rushed dementedly to the northern side of the road-cutting, pointing out the deep claw marks and the signs of a drag up the muddy embankment, up which her son had been hauled by the panther, after the poor boy had been knocked down by the brute. The panther had leapt upon him suddenly – to judge by the tracks – from the opposite embankment of the cutting. This apparent man-eater must have been lying in wait, hidden in the heavy undergrowth, at the edge of that opposite embankment, for the first opportunity that presented itself – a single man or a woman, or two helpless persons, such as the deaf and dumb woman before us, and her son.

The rumour of the tragedy having spread, a number of men from a small neighbouring bazaar arrived at the "ghat" or ford with sticks in their hands, to help in the search. Our guard had arrived too, consisting of six hefty men, and after sending away the poor, bereaved woman in charge of one of the guard, we began tracking operations, but in this we were grievously disappointed.

After going a few paces from the embankment up which the scramble had taken place, there was no trace of any kind of the panther and his victim. We thought we should at least find bloodstains and pieces of torn cloth from off the boy's loin wrap, but there was nothing! True the ground was hard, and though cut up by ravines, even the bottom of these and the beds of water-courses being crusted with flat rock and stones, we hoped to find traces somewhere to enable us to form up line and move systematically in a given direction. It is a rocky country – Chota Nagpur – even dense forests being strewn with great boulders and huge stones, which makes anything like tracking most difficult in the dry season.

The lifting of the boy, however, so completely, without even the slightest trace being left behind – beyond the drag up the embankment – looked more like the work of a tiger, but the tracks of the panther were unmistakable! We had to agree with Mough Sing, an experienced woodcraftsman, that a big, man-eating panther like this one would shoulder a boy of twelve years and take him away many miles in a few hours.

For a great distance round we searched the forest, peering into and penetrating every ravine we came upon, but it seemed hopeless, and eventually we had to give up the search. On arriving back at the bungalow, late in the afternoon, Mr. H. found that he had to leave for headquarters immediately on urgent business, and my wife and I were once again left alone in that wild region, cut off from all social life – "the world forgetting by the world forgot" – feeling acutely the loss of our one and only little dog.

That evening, accompanied by one of the men of Mough Sing's guard – who was a fair shikarry (hunter) – I went out again after the man-eating panther, with a calling goat as bait, in the hope that I could induce the panther to show himself for a rifle shot. I had accounted for many a panther over a goat as bait, and being sufficiently familiar with the method – to prevent the goat from coming to harm – I decided to resort to this method, in order to try to rid the neighbourhood of the monster that was a nightmare to everybody.

In the heart of the jungles the shikarry and I selected a spot that we considered the most likely cover for a man-eater, and fixing a rough machan up on a tree, we took our places for a quiet watch, a crowd of villagers coming up to tether the goat below our perch, after we had screened ourselves sufficiently for the goat not to see us. In the management of this, great care has to be taken, as, if the goat sees or hears the shooter on the tree, it refuses to bleat – knowing it has human company – and there is no hope of seeing "spots" unless by mere chance the panther happens to pass that way.

The villagers, a crowd of ten men, had strict orders to keep close together, to climb up trees some distance behind our perch (after they had tethered the goat) and to sit quietly until we called out to them, or imitated the belling of a sambur.

Watching the moon hastening – up through the gnarled depths of black forest – and listening breathlessly to every leaf crackle, as the quavering bleats of the terror-stricken goat in full cry penetrated far into the night, the shikarry and I sat motionless, hour after hour, with an occasional imperceptible movement – turn about – to relieve an aching limb! But beyond the strain of being constantly on the alert, no signs of an approach of a dangerous carnivore broke in upon our long wait, to give an edge of excitement to our growing disappointment.

Dagger Tree
Pajanelia longifolia
Family: Bignoniaceae
Synonyms: *Bignonia multijuga, Pajanelia multijuga, Pajanelia rheedi*

A medium or large (up to 30 m) sparingly branched erect, evergreen or briefly deciduous tree, found in the Malabar region, in the Khasi hills, the Andamans, in the Sylhet region of Bangladesh, Myanmar, Thailand, Malaysia, and northern Sumatra. The bark is grey becoming shallowly fissured as the tree ages. The pinnate leaves are opposite, crowded at the ends of the upright branches, large leaf scars being visible on the pale grey twigs; the young leaves are violet. The leaflets are elliptic, acuminate. The large, wide flowers have a soapy odour, and appear in very large terminal panicles. The corolla tube is a dull greenish-red, the lobes yellowish and crisp at the edge. The hanging pod is massive, dagger-shaped, with a longitudinal ridge on either side and contains elongated seeds with translucent, slightly curved wings. The timber obtained from this rapid-growing tree is excellent and many compare it with teak. E.J.H. Corner writes in *Flowering Trees of Malaya* (1940), "It appears to be night flowering and bat pollinated, but we have no information whatever concerning its habits."

From N. Wallich, *Plantae Asiaticae Rariores*, 1832

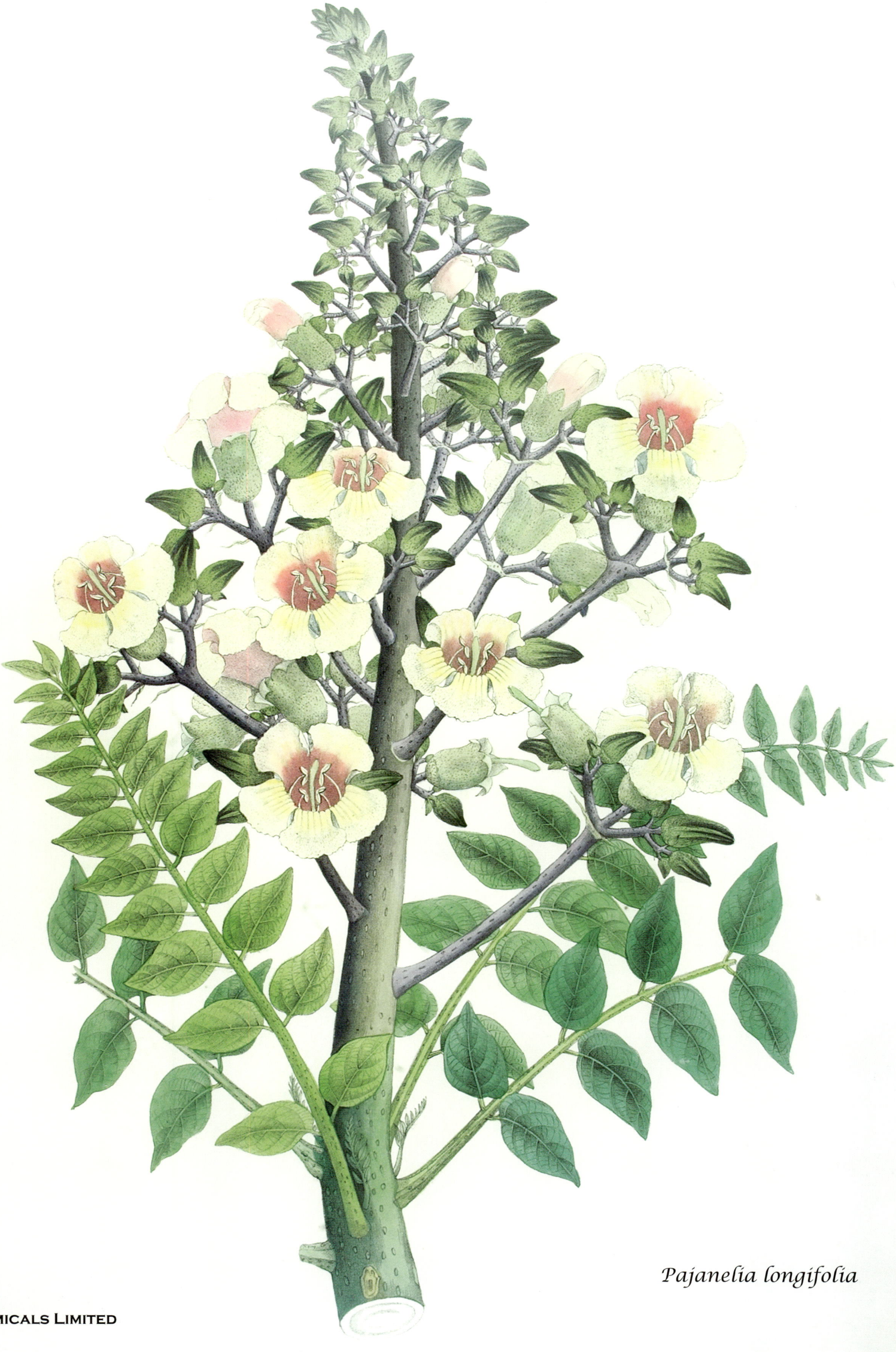

Pajanelia longifolia

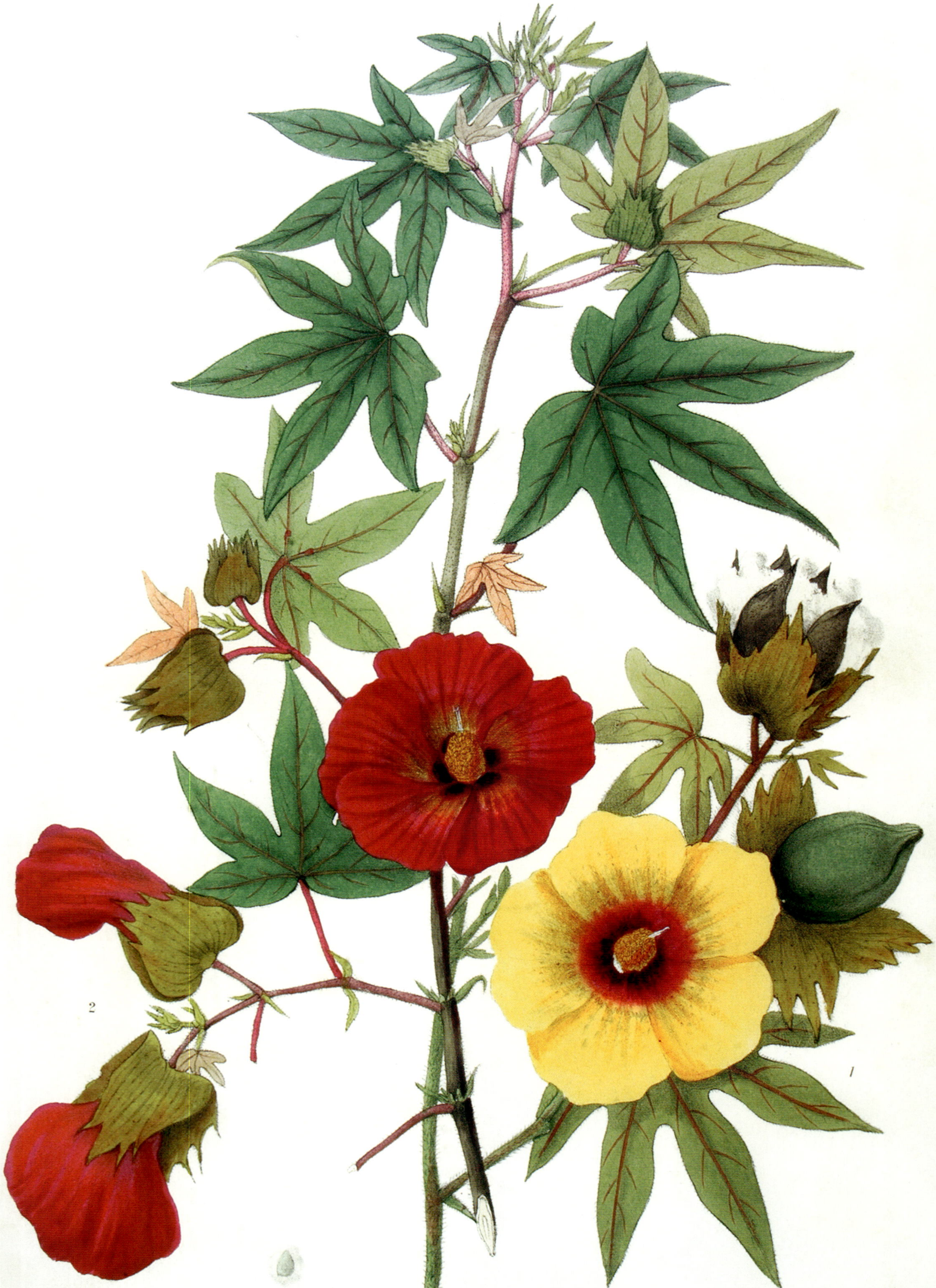

Gossypium herbaceum,
G. arboretum

In loving memory of my husband
Dattatreya Ramchandra Josh
– Suneeta Dattatreya Josh

1 Cotton Plant
Gossypium herbaceum
Family: Malvaceae

This is one of the cultivated cotton species. It is a short-lived, fast-growing perennial shrub reaching 1.5 m height. It has trilobed or palmate evergreen leaves. The flowers are yellow and hibiscus-like, turning yellow-pink when older. The fruit is a green capsule that splits open when ripe revealing the cotton and seeds.

2 Tree Cotton
Gossypium arboretum

This shrub, native to India and Pakistan, attains a height of 1–2 m. The branches have soft hairs and are purple in colour. Linear to lanceolate or falcate (sickle-shaped) stipules are present at the base of the leaf. The leaves have 5–7 lobes and resemble those of the maple tree. The flowers are set on short stalks. The corolla is pale yellow with a deep purple-red centre, and occasionally entirely purple-red, as illustrated here. The fruit is a 3–4-celled capsule, glabrous and ovoid, measuring 1.5 to 2.5 cm across. The seeds are globular and are embedded in the long white cotton.

From J. Forbes Royle, *Illustrations...*, 1839.

We were obliged in the end to give up this attempt as well to shoot the man-eater, and we hallooed to the waiting villagers, who came silently and solemnly up to our machan. The poor goat emitted a low cry of pleasure when it found that, after all, it had not been deserted – left alone to be killed in those awful jungles by some savage animal! It had the pleasurable sensation of being carried home in the arms of the men, who took it in turn to bear its weight.

Then came some blank days, without any information whatsoever, of the man-eating panther.

* * * *

It is about the fifth day after the incidents related that I am riding back to the factory about midday, through the jungle to the east of the bungalow, just beyond the bare, rocky ground crept over in the moonlight by Mr. H. and myself. This stretch of jungle is cut off from the main forest that extends further east, by the river already mentioned, which passes north of the bungalow and the factory, and then, swerving round to the south, cuts down through the forest on this eastern side.

Mounted on a sturdy Australian waler, with a loaded revolver – .450 bore, six-chambered – in the pommel of my saddle, I am riding at a walking pace along a stony path about the width of two men walking abreast. The path is a vista cutting through high tree forest. A wind through the trees is scattering yellow leaves all round me; no other sound disturbs the silent jungle.

Suddenly, pricking its ears with a start, and gazing with head erect into the jungle to its right front, my horse comes to an abrupt halt and begins to snort apprehensively. I grab my revolver and try to cock it, but owing to the alarming restlessness of the horse it is impossible. I can neither see nor hear anything! But, trembling in his limbs, the horse seems to be following something with his eyes, for he now plunges to the left side of the pathway, curving an arched neck round completely to his right, and panting and blowing with long-winded stentorian snorts! The next instant, simultaneously with a rushing sound in the dry undergrowth, the waler – almost unseating me – leaps forward! I see for a second the hesitation – as it seems – of a huge, spotted, doubled-up creature almost below my stirrups, and then I am flying down that pathway trying to guide over the protruding rocks the bolting waler, who is in a mighty panic.

Out into the open we come at a mad gallop! I hear behind us a roar of disappointment, as the horse plunges over the rocks of that bare stretch of ground between the forest and the eastern side of the bungalow. In another moment – in the garden in front of the mansion – I am turning and twisting the waler about in circles to try to stop him!

My wife, hearing the clattering of the horse, hastens out into the front balcony, and I find the jamadar, Mough Sing, there with some of his "guard". They had come rushing up to the bungalow on hearing the panther roar. Apparently I had had a narrow escape!

Quickly Mough Sing and I hold a consultation. So near the bungalow, in that strip of jungle! We must collect all the men we can, and have a drive at once (with tom-toms) before this brute of a panther moves elsewhere! Two or three men of the guard, along with some factory peons, are sent off immediately to broadcast in the neighbouring bazaar the news of the panther and the urgent call for beaters.

It is a mid-week afternoon and the bazaar is thronged with tribes from various distances, crowding round the small shops to barter their jungle produce, and to buy cloth-pieces and foodstuffs.

The Doctor Baboo – our only hope and stay in the jungles as medical adviser – in charge of the local Government Dispensary, has heard the news. Bravely flourishing a revolver in his hand, he is walking up the drive, talking loudly to an admiring crowd behind him. He is accompanied by his assistant, the humble Compounder Baboo – who thinks his master a prodigy of all the learned professions under the sun – and by several big "lathials", wearing yellow turbans, whom I recognize as the Raja's guard of honour, for a small local Raja also resides in the weird jungle settlement.

"Sir," says the Dr. Baboo, sensationally costumed, with khaki Jodpur breeches bulging below a gaudy blazer of some Bengali cricket club in Calcutta, and with a small black bowler cap on his head, cocked jauntily to one side, "I shall help your honour to shoot this leopard of unheard-of pretensions, if your honour will seat me on a high tree. The Raja Sahib," continues the Dr. Baboo, "sends his 'salaams' (best wishes) to your honour, and begs you to command his services. He is, however, unable, owing to unwieldy size, to present himself in person, but he begs to place at your honour's service his fearless guard," concludes the Dr. Baboo, as he sweeps a hand across the front of the Raja's yellow-turbaned men, standing at attention and saluting, with broad grins on their faces!

"Shabash!" (splendid) exclaims Mough Sing, curling up his moustaches sarcastically, "we have now no fear of any anxiety, and the 'bagh' (leopard or panther) must have already ceased to smile."

"Heingh!" (heigh) exclaims the Dr. Baboo, sounding a formal inquiry, as with open mouth he screws up his eyes at Mough Sing in an all-knowing manner.

"Hazoor," sniggers the jamadar, putting his hands up to the little medical grandee in perfunctory respect, "what I mean to say is, your honour knows where the heart lies."

"Hei-ngh!" exclaims the Dr. Baboo again – this time in a long-drawn tone of puzzlement, as if he had an idea he had suddenly slipped somewhere into deep water. The Bengali distrusts the up-countryman and vice versa, and the well-known loyalty of the latter to British employers is generally resented by the Bengali Baboo.

Heartspotted Woodpecker
Hemicircus canente

A small black-and-buff woodpecker with a short, thin neck, large crested head, and a stumpy rounded tail. In the male the forehead, crown, and crest are black; the back is black with a broad buff band on either side marked with heart-shaped black spots. The throat, foreneck, and sides of the neck are creamish-white; the rest of the underparts are dusky olive and black. The female is similar but with forehead and crown buffy-white; rest of the underparts dusky; and crest black. Found in pairs or small groups on high trees a considerable distance from the ground, where they make their nest-holes, in the forests of the Malabar coast of southwestern India and the Western Ghats and east to Madhya Pradesh, Chhattisgarh, and Orissa, the northeastern hill states, and Bangladesh.

From T.C. Jerdon, *Illustrations of Indian Ornithology*, 1847.

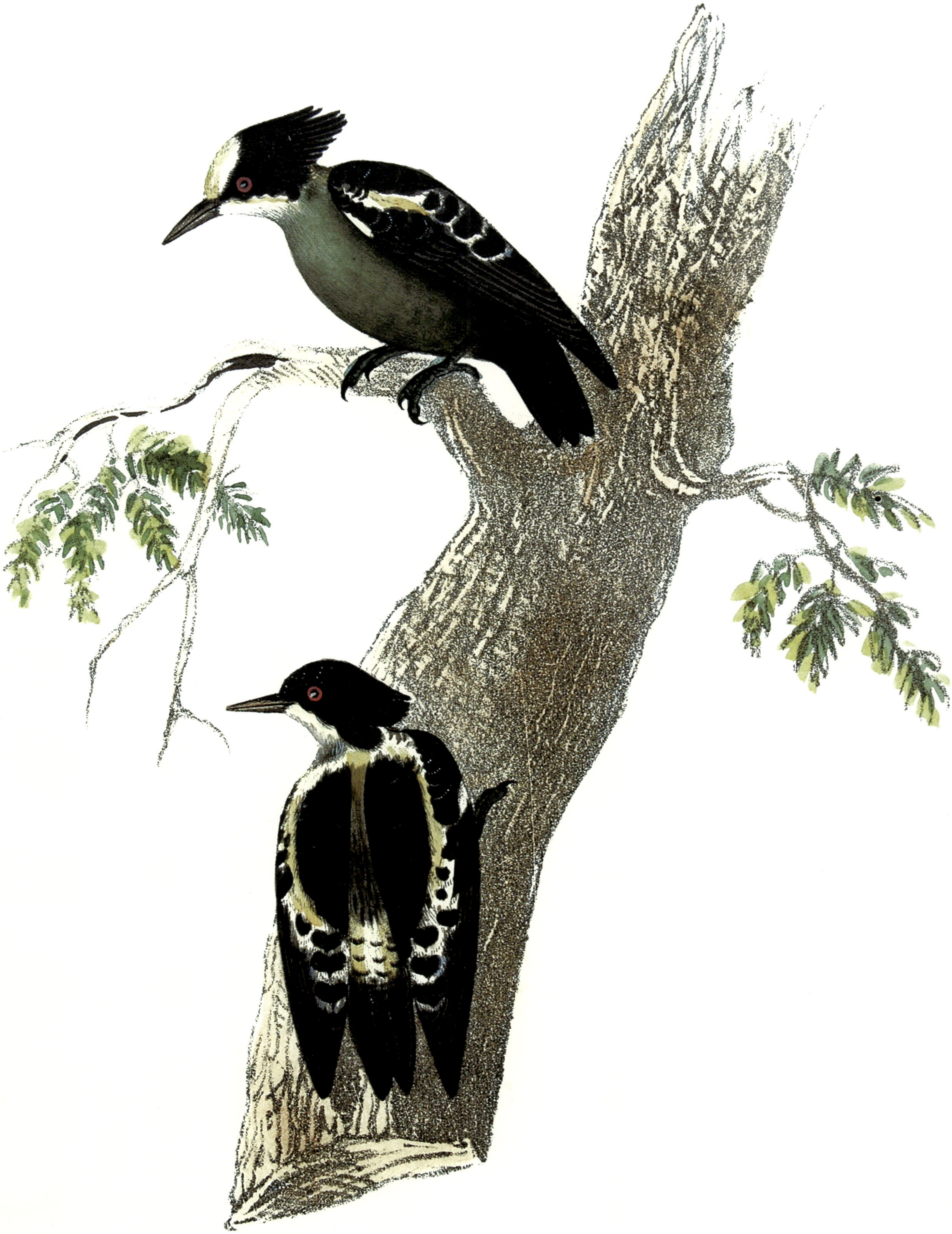

Hemicircus canente

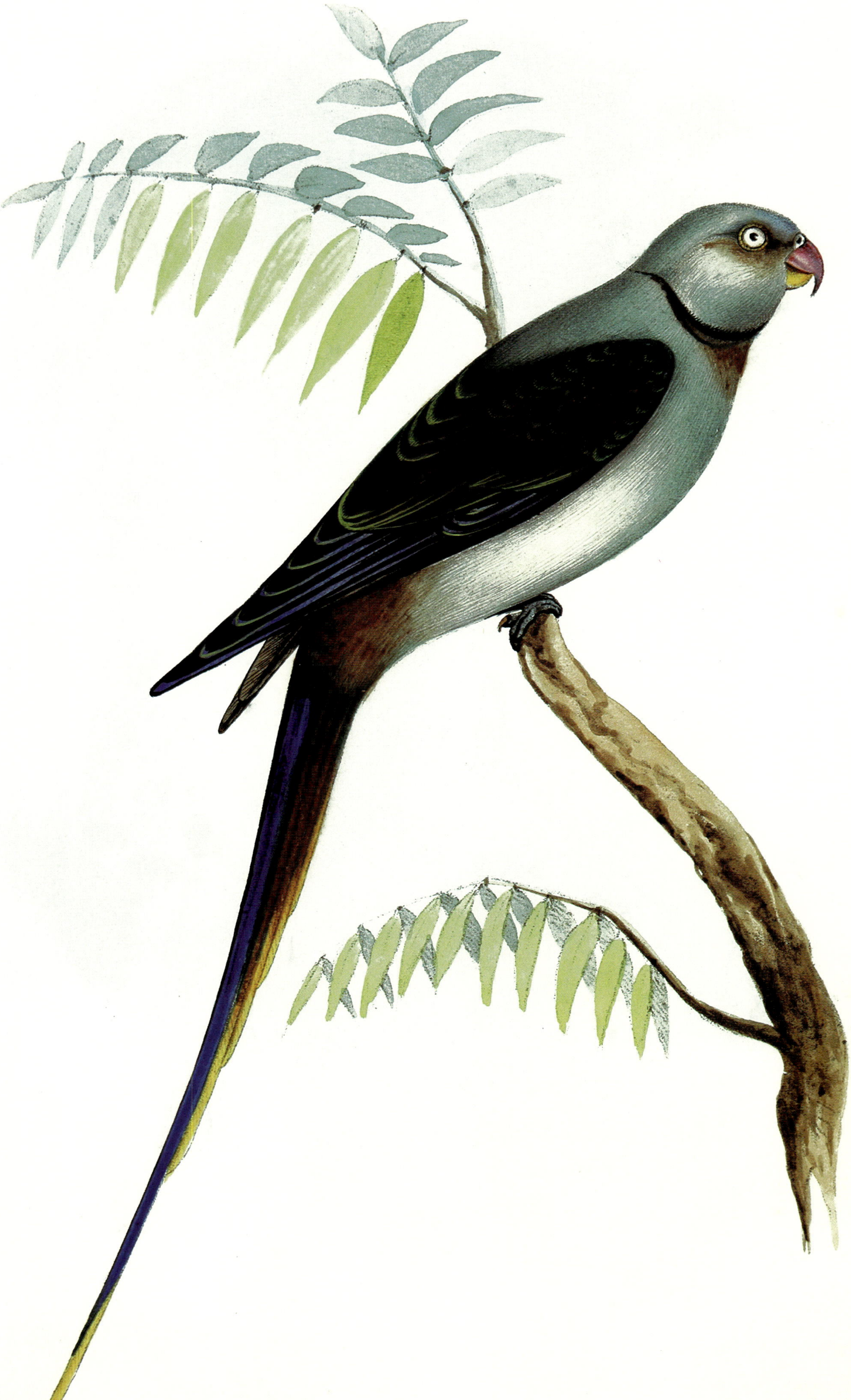

Psittacula columboides

COURTESY
FARIDA & SHYAM MARIWALA

Blue-winged Parakeet,
Malabar Parakeet
Psittacula columboides
Synonym: *Paleornis columboides*

This parakeet is greyish-green in colour, with brilliant blue-green and black collar, conspicuous blue wings and tail, the latter with yellow tip. The female lacks the green collar behind the black neck-ring. Jerdon states: "Previously to obtaining this bird, I had heard from several sportsmen of a blue parakeet said to be common in the depths of the forests of Malabar. I first procured it at Trichoor, and afterwards in various other localities on the west coast, and on the sides of Neilgherries up to a height of [1,700 m]. As Colonel Sykes has it in his catalogue, it probably extends along the range of Western ghauts for some distance, but I have not heard of it having been procured in any other locality. ... It keeps entirely to the depths of the forests, and frequents only the loftiest trees. Its flight is very rapid and elegant, and it associates in small flocks. Its cry though similar in character to the harsh call of the common parakeet, is much more mellow, subdued, and agreeable."

From T.C. Jerdon, *Illustrations of Indian Ornithology*, 1847.

And now we move on to join the large crowd of beaters that have collected. Armed with big sticks, bows and arrows, with many drummers dotted about there are over two hundred men present with almost as many "pariah" dogs jumping about with excitement at the promise of a hunt.

The crowd of beaters having been carefully instructed, Mough Sing and I, accompanied with the Dr. Baboo and his faithful compounder, and by the "lathials" of our own and the Raja's guard, proceed to skirt round the jungle marked down for the exciting drive. We know that in close formation the tribesmen will, ordinarily, be quite safe from the panther's attack.

Two machans near the ford of the river are speedily erected, and Mough Sing and I now place the Dr. Baboo and the compounder on a safe tree, at the further western corner of the drive, opposite my machan, the beat being so arranged as to move in a line from south to north. I leave two good "lathials" with the Dr. Baboo and his companion, to sit with them on their tree, as the Baboos are now shivering and chattering with fear, and, placing myself on my own tree, I instruct Mough Sing to order the drive to commence. The beaters have to come up about half a mile, hugging on their eastern flank the river, along whose banks men have been posted on top of trees as "stops", to make a noise and keep the panther from cutting across into the forest beyond the river.

Mough Sing, with a strong guard of men, has taken up a position covering some deep ravines some distance beyond my tree – on my righthand side – and the drummer with him, now sounds the muffled signal for the hullabalooing to start; and the drive commences!

And – ye little gods and apples – what a bedlam! If ever, with terror and excitement, a fanatical crowd of aborigines had taken leave of their senses, it was this one!

A mighty roar, as if a great river had burst its dam, opens with its many scenes the final performance of this panther hunt, drowning for a few moments the salvo of booming drums of all sizes, as these tom-toms are belaboured. Then the blowing of horns and the firing of crude muzzle-loading guns – loaded with powder only – lent by the Raja, and such weapons as we pulled out from the factory store of by-gone arms, drown the first ebbing roar of the crowd, and awaken the jungle ghosts; for the first to appear, followed by his many wives, is a large mongoose, who stands up on his hind-legs, scratches his ear to make sure he is awake, then listens to the tumult! Then, glancing round at his wives, innocently sniffing the air – as if he suspected them of some foul plot to disturb his serenity – he moves on, and the wives follow, gliding by like streaks of silvery grey.

Through the continuous din, now intensifying with mightier uproar, whenever anything – from a hyaena to a tree lizard – is seen, and now subsiding, there falls suddenly upon my ear panic-stricken exclamations of: "My God, my God!"

At the extremity of a narrow vista down which I can see from my machan, I am able to distinguish wild arm movements, and to my horror I see the flashing barrels of a revolver sweeping round in every direction! Then a remonstrant voice rings out: "Han! Be careful of that weapon – don't do that! – han – han!"

It relieves me considerably to hear the rebuke. It is the voice of one of the Raja's guardsmen, who is trying to stop the panic-stricken Dr. Baboo from swinging his revolver recklessly about, regardless, whatsoever, of the appearance of any animal! The sight of this, with the exclamations of "My God, my God!" rising above the voice of the guard's remonstrance, comes as a sudden comedy, to relieve for a few moments the tense excitement; notwithstanding the danger of the loaded revolver being pointed at everything, from a passing crow above one's head, to a twittering bird in the bushes below. The most solemn would have been induced to commit himself to a smile.

Suddenly, to my left front, there comes a crackling sound of dry leaves and three or four jackals emerge. A hare at that moment giving them a start, these jungle guardians – with their tails between their legs – jump about in the most comic fashion, as if they were completely nerve-racked! Then, standing to pant, and looking at one another in ridiculous wonderment at the oncoming of the tom-toms, and the mad shouting, they sneak away, passing suspiciously under my tree. The next moment a huge boar, followed by several sows, dashes across my direct front.

Then suddenly, somewhere to my immediate left – along the river bank – a couple of jackals commence to bark. It is a serious alarm cry; and, "click, click", mechanically the hammers of my gun are pressed down by my thumb to full cock, as for a better view I stand up slowly in my machan! Then comes suddenly an uproar from the "stops" along the river bank! The men are clapping their hands to make a noise, and shouting out: "Bagh hai, bagh hai" (the panther – the panther!). The brute is apparently trying to break cover that way. With my gun loaded with "lethal" ball, I look anxiously around! And now down the main drive – an entangled mass of creepers and dense undergrowth – shots are fired, and shouts arise as if hell were let loose! Suddenly I see something – moving fast – a long belly trailing the ground – something dark – huge. It might be a black panther gliding between the shadows! At that instant there rings out again into the air: "My God, my God!" But I am barely conscious of the Dr. Baboo's exclamations of terror. The fearsome-looking dark creature – as suddenly as I glimpsed him – has disappeared behind a ridge to my left front, a ridge covered with shrubs, densely packed beneath some high trees! My eyes are desperately scanning the screening undergrowth, and the beaters, shouting like maniacs, are coming nearer every moment. The men will soon be in the way, making the shooting most difficult – zounds!

The very air trembles with excitement, and the jungle is alive with moving game. Birds are fluttering wildly over my head, and I am expecting at any second to get a

Dombeya palmata
Family: Sterculiaceae

This shrub is named after J. Dombey who was an English doctor and naturalist in the West Indies. The leaves are palmate, resembling the castor oil plant, with 7 lobes, smooth and glossy. The greenish-white scentless flowers appear in large terminal corymbs, during September–November.

From N. Wallich, *Plantae Asiaticae Rariores*, 1832.

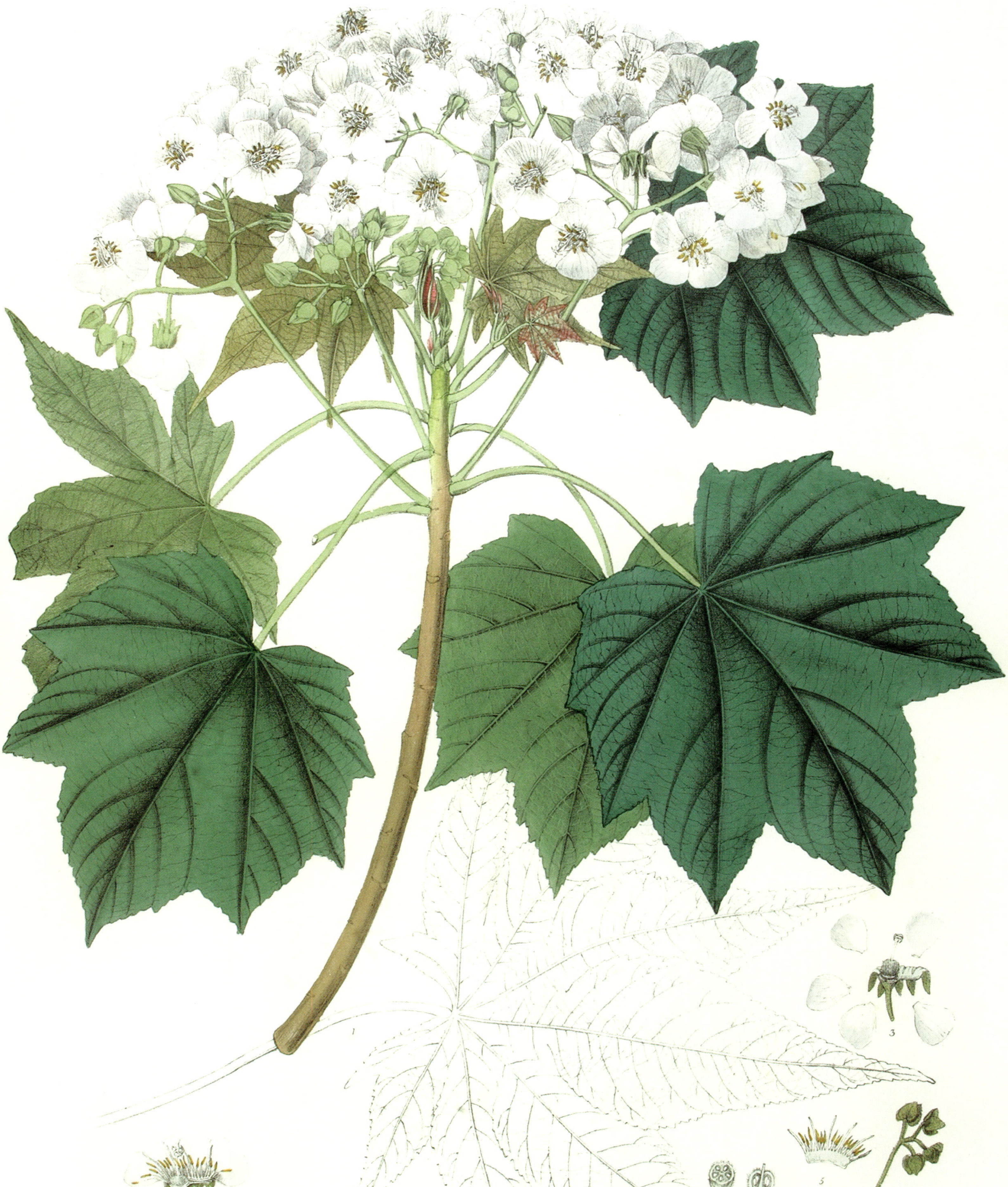

Dombeya palmata

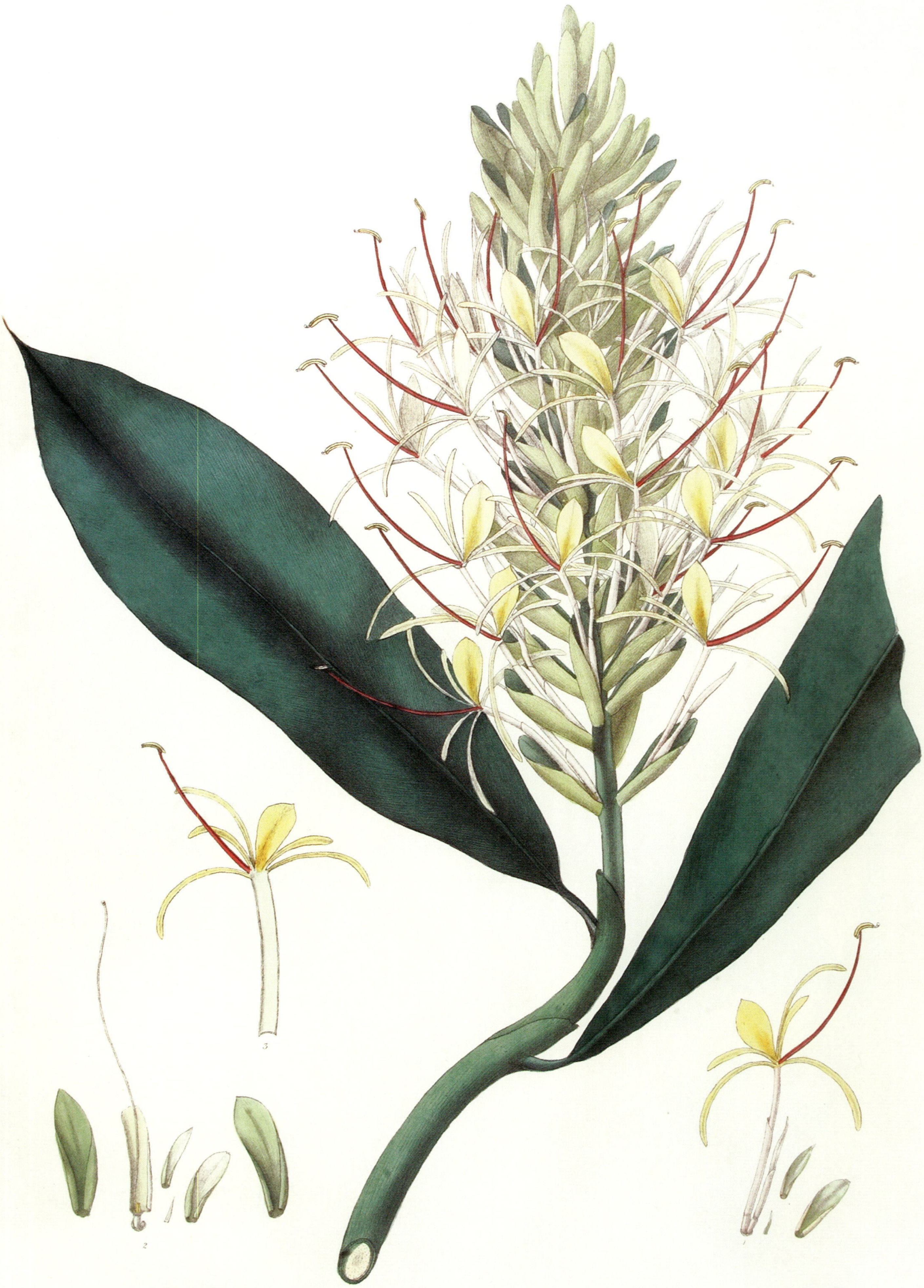

Hedychium gardnerianum

Himalayan Ginger Lily, Kahli Ginger, Yellow Ginger, Wood Ginger
Hedychium gardnerianum
Family: Zingiberaceae
Synonym: *H. speciosum*

Nathaniel Wallich called this plant "the queen of the genus". It is a native of the eastern Himalaya found from central Nepal to Sikkim up to 1,900 m. In Nepal it is an important element of the Indrajatra, the festival that takes place in the early spring, when the living goddess Kumari and other deities are decorated with crowns and garlands made from its flowers. The species was named after Edward Gardner who represented the East India Company at the court of Kathmandu. It is a perennial herb with leafy shoots growing from a large branching rhizome. It grows to 1.5 m tall with long bright-green leaves clasping the tall stems. The fragrant red and lemon yellow flowers appear in dense spikes towards the end of summer.

From N. Wallich, *Plantae Asiaticae Rariores*, 1832.

flying shot at the panther across my direct front. But all on a sudden there comes from nowhere, as it seems a furious rush, and a convulsive, ear-splitting roar – where is it? "Houngh, houngh" – great Scott! at the foot of my tree – beneath my machan the brute is passing – under my very feet!

To get in even an instantaneous snap shot, is utterly impossible. I swing round like lightning, but the savage man-eater has crossed – slipped into cover again – now on my righthand side, where the ravines slice the ground – where the stalwart guard of "lathials" are acting as "stops"!

Numbers of the beaters are now up the trees, and, in order to draw their attention, I shout to the braver and more reckless ones, as they emerge into close view. "Back, back," I wave them and make desperate signs to them to climb up trees to give me a chance to fire! I can depend upon the brave guard. They are sure to drive the man-eater out of the ravines, back towards my machan for a certainty – and back the brute may come roaring at any moment!

Scarcely have I anticipated that likelihood, when on my right roar follows roar – terrific, savage roars again: "Houngh – oungh, oungh!" Then come sounds of a turmoil. Bang, bang on the ground crash down the iron-bound bamboo poles, wielded by the big, tall men of the guard, as they accompany every loud crash of their swinging weapons with stentorian shouts of solid determination in order to drive the panther back – out of those deep ravines!

But the brute has charged the plucky "lathials"! He is at bay now – hedged in – surrounded by a bedlam of tom-toming and yelling human beings. An exclamation of human pain – "aouw!" – suddenly arises. Someone of the guard has been mauled, and the stalwart men give vent immediately to mighty shouting. "Maro–maro, sala!" (beat him off – beat the brute off!) resounds Mough Sing's voice as he shouts to his men.

Consequent upon the action of the guard, in obstructing the panther's escape, suddenly – to the accompaniment of tumultuous shouting – back comes the roaring man-eater, rushing savagely through the jungle on my right. He is nearly out of the cover; I am expecting him at any moment to cross the strip of open ground I am covering when, suddenly – my God! – to my horror, down from a small tree on my right, scrambles a frail, turbaned youth – the sweeper boy of our bungalow! How did he get there? – who allowed him to leave the house, against my strict orders that none of the house servants were to join the beaters?

The poor boy, terror-stricken, has rushed towards my tree – yelling to me wildly to save him – and out of the cover at that instant lollops the roaring demon man-eater! He sees the boy in front of him instantly. Midst the terrible, distracting bedlam of shouting that now rages, upon that terrified youth – without a second's hesitation – turns that huge panther in murderous fury, claws, fangs and deafening roars numbing to sudden

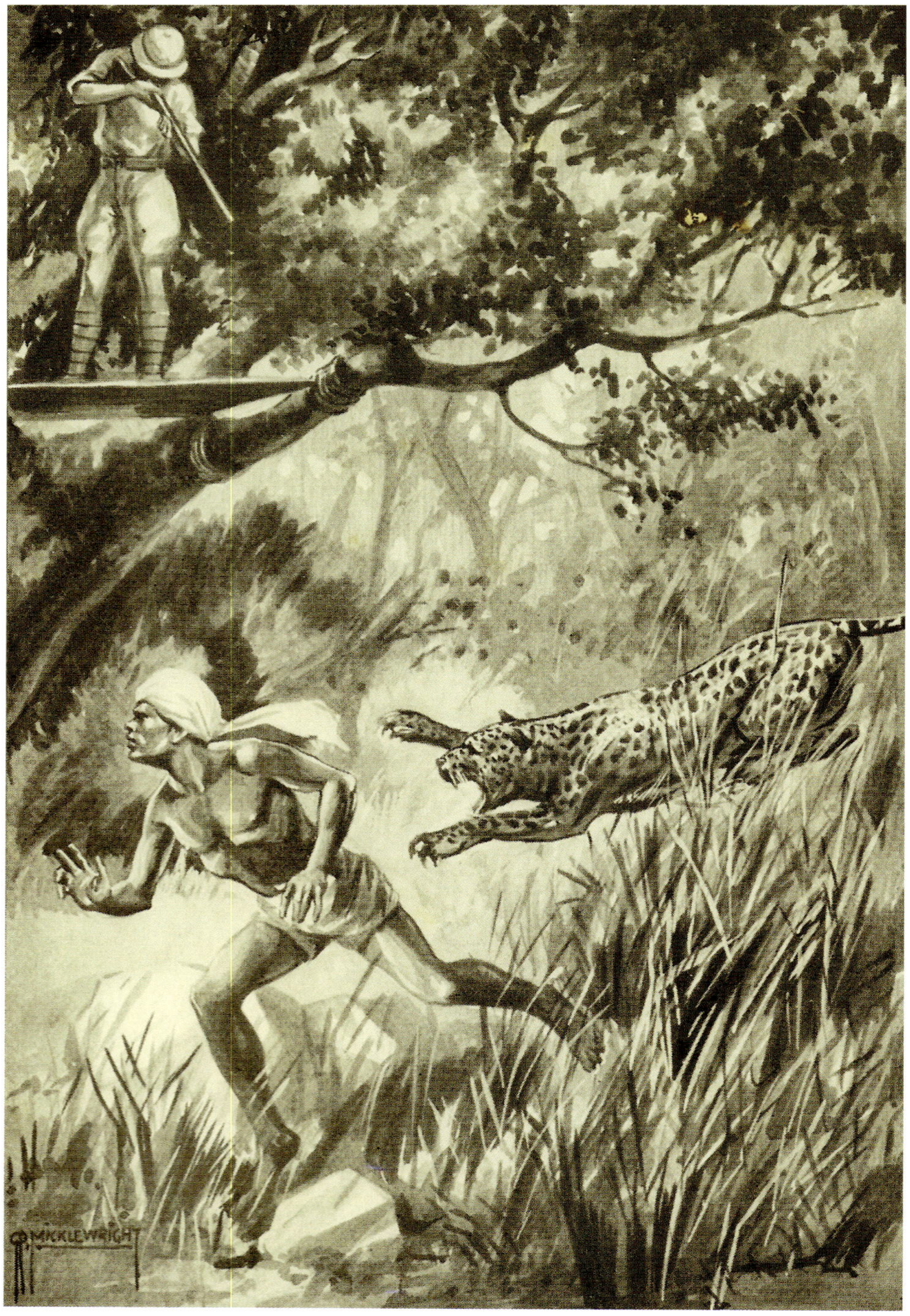

The spotted "man-eater".

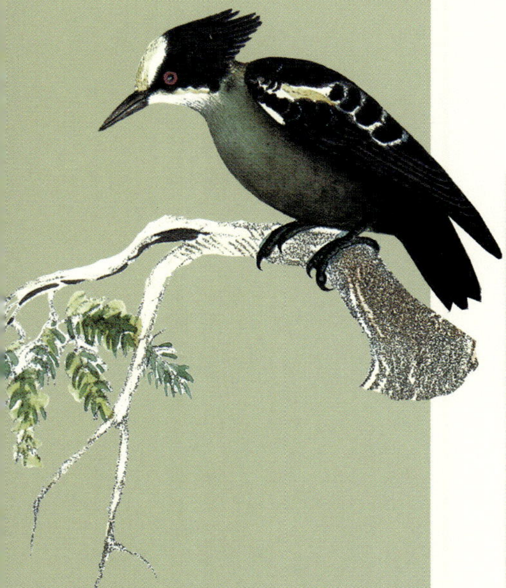

silence the terrorized onlookers, who hold their breath for a few seconds, expecting to see the boy torn to pieces!

Now across my front they come; the poor victim running for his life, with the man-eater close behind him. The boy is leading by about three yards only, and yelling desperately for help. Annoyance and anxiety have upset my nerves, and my gun is shaking – shaking even at the shoulder!

No words can paint my feelings. With bated breath and a prayer on my lips, I am taking aim. "If I can only get steady for half a second!" My bead is on that roaring mass of spots, claws and fangs. "Yes – thank God!" I breathe out, soliloquizing, "I am steadier. A touch and go," I hear myself saying, "I must not miss." The savage brute has reached the heels of that poor boy – another moment – but I pull!

"Bang!" There is a leap to one side, a gasping roar – a blind rush – a collapse, with a thud against the ridge down which the man-eater had come; and – dying groans follow!

* * * *

"Thank God!" I exclaimed.

It was a lucky shot in the nick of time.

Mough Sing brought the youth of seventeen years before me, catching him by the ear! The youth had sneaked away, he said, from the bungalow, and, stalking into the forest, had climbed up the small tree near me. The panther on its return, seeming to be coming straight for his tree, the boy got into a panic and behaved in the foolish manner as described.

The Dr. Baboo – still exclaiming: "My God!" – was already giving first aid to the "lathial" of Mough Sing's guard, who had had his left arm and shoulder mauled when the panther rushed the jamadar's unflinching line. The wounds might have been much worse, as the attacked man's comrades were a little late in coming to his rescue.

Then, hearing the great shouts of triumph from the teeming crowd of beaters and others that had now collected to see the dead man-eating panther – which measured in the flesh seven feet six and a quarter inches – the servants of the bungalow – sent by my wife to inquire after me – came hurrying down. A little later I greeted my wife, who came to meet the cheering multitude. The vast procession was headed by dancing tribesmen, who performed amusing evolutions in front of the carriers, who bore on their shoulders the ends of the stout pole from which – with its four legs lashed up – hung that dreaded man-eater.

Late into the night the jungle tribesmen were busy dividing up – for the extraction of oil for rheumatism – the carcass of that panther; each man – and many a woman – carrying away a tiny piece of bone for a charm, to be worn around the neck!

From George Hogan Knowles, *Terrors of the Jungle*, 1932.

Newly Born Elephant Calf Saved from a Hungry Tigress

S. Eardley-Wilmot

THE herd of elephants, finding food, shade and water in the vicinity, spread themselves at ease over the country with no intention of travelling during the ensuing night. One amongst them there was who had roamed uneasily during the day, avoiding her fellows, seeking vaguely for retirement in some spot sheltered and secure. Annoyed by the busy idleness of the herd, she wandered to a distance, and as evening fell found herself alone; at her back a perpendicular rock, still festooned with moss and ferns, before her a tangle of bamboos from which issued stems of straight trees whose leafy crowns gave promise of shelter from the midday sun.

As morning dawned with tropical haste a bear came sauntering past the spot, self-absorbed, as is the custom of his tribe, thinking solely of the good fortune of the past night when he had found abundance of fruit, longing for his lair where he might doze and drowse through the daylight hours. To him came suddenly a loud hissing sound, followed by the thud of a heavy blow on the earth. He leapt to one side with an agility remarkable in one so clumsily built, and saw within a few yards an elephant standing with its back to a rock, at its feet a huddled mass which he suspected to be a new-born calf. He did not stay to investigate. With a bound he hurled himself into the underwood, and, as the noise of his headlong flight died away in the forest, the elephant commenced to rumble deeply with a sound like the purrings of some gigantic cat, undecided whether to be pleased or angry.

The calf lay extended at his mother's feet, under the shelter of her head and trunk, as yet too weak to rise; from time to time she waved the flies away with her trunk or gently kicked the soft earth so that it fell in light showers on his body. On such occasions her foot struck the ground so near to the reclining calf that one could not but wonder at the extreme accuracy of her movements, an accuracy which in the human race is only acquired by the constant application to some one handicraft. She had stood thus for some hours and would so stand for many more, without food save such as she might reach from her post of sentinel; not until the calf was able to take nourishment would she move from the spot, lest harm should befall the helpless young. It was not, in fact, till the commencement of the second day that the calf uttered

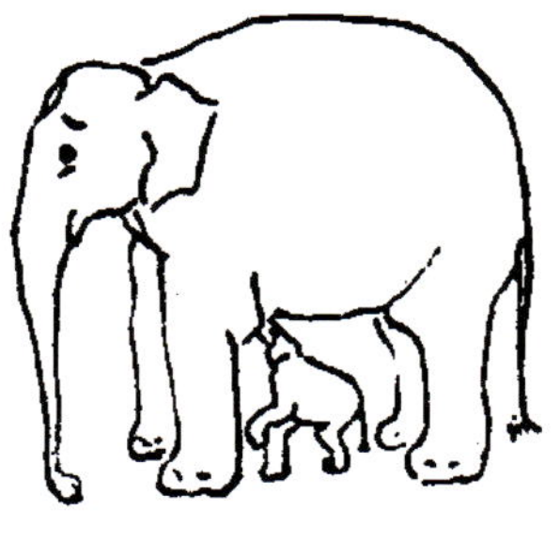

The author describes the birth of an elephant calf in the jungle and the tender care and protection provided by the mother. In a tragic turn of events, another mother, a tigress, tries to prey on the calf to feed her hungry cubs but gets crushed by the furious mother elephant.

Nun's Orchid, Nun's Hood Orchid, Red Crane Orchid
Phaius wallichi
Family: Orchidaceae
Synonyms: *P. tankervilleae, P. grandiflora, P. grandifolius*

Phaius, frequently spelt *Phajus*, are terrestrial orchids. This spectacular plant is found from 1,700 to 2,000 m in Himachal Pradesh, Sikkim, Meghalaya, and Chhota Nagpur, also in Sri Lanka, Malaysia, and Myanmar. It has thin pleated leaves and produces erect flower-spikes with numerous showy, long-lasting flowers during March–April. The pseudobulbs are conical-roundish, sheathed by the leaf bases. The 3 or 4 leaves are up to 90 cm long, 20 cm broad, elliptic-lanceolate, with prominent longitudinal nerves. The stout, erect inflorescence arises from the base of the pseudobulbs, and is up to 150 cm tall with 10–20 fragrant flowers ranging in colour from pink to burgundy, rose red, orange, and yellow tones. The flowers last for 6 to 8 weeks.

From N. Wallich, *Plantae Asiaticae Rariores*, 1832.

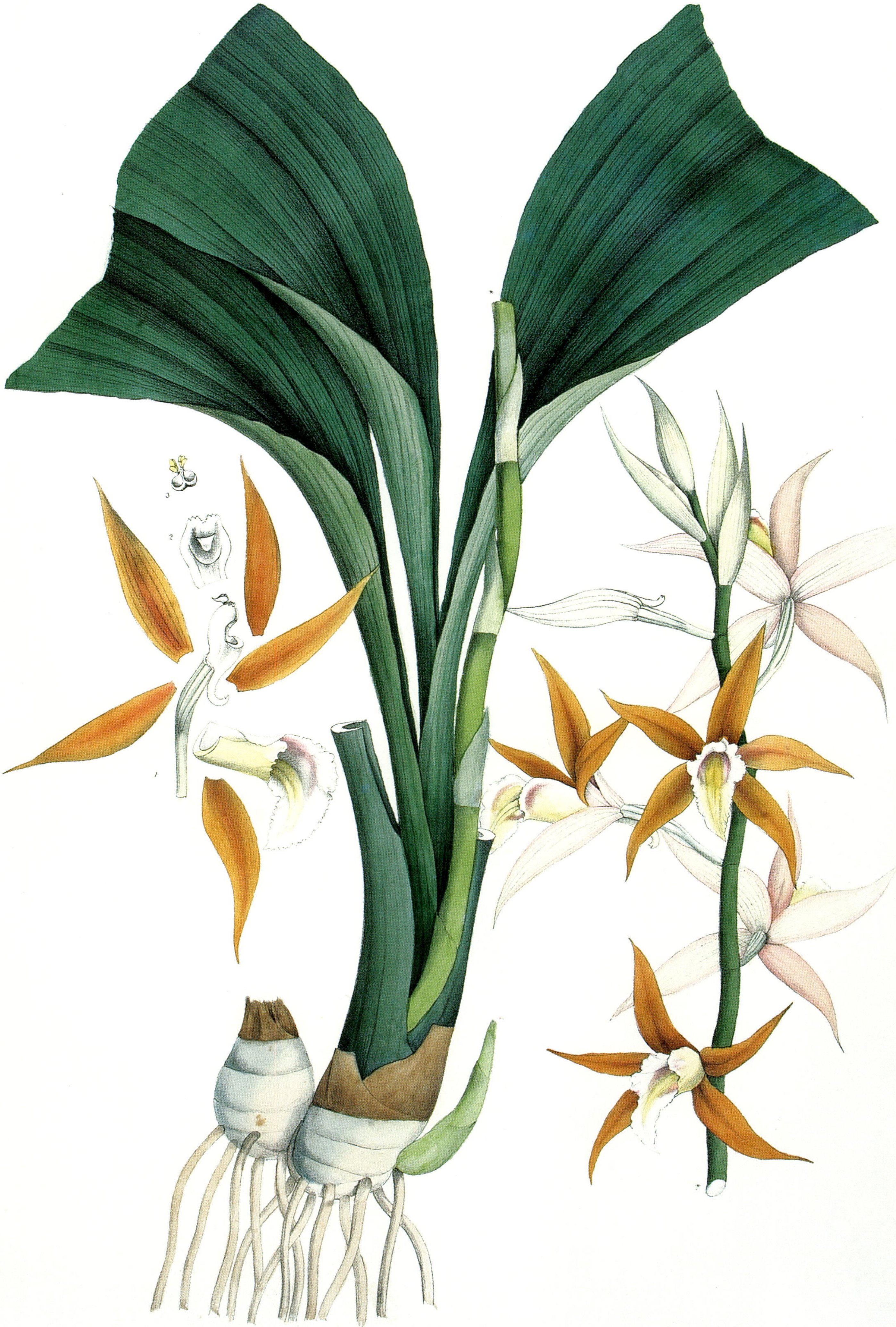

IN LOVING MEMORY OF OUR PARENTS
SMT. MITRABAI & SHRI ANAND LAXMAN CHANDAVARKAR
– MANGALA & ASHOK ANAND CHANDAVARKAR

Phaius wallichi

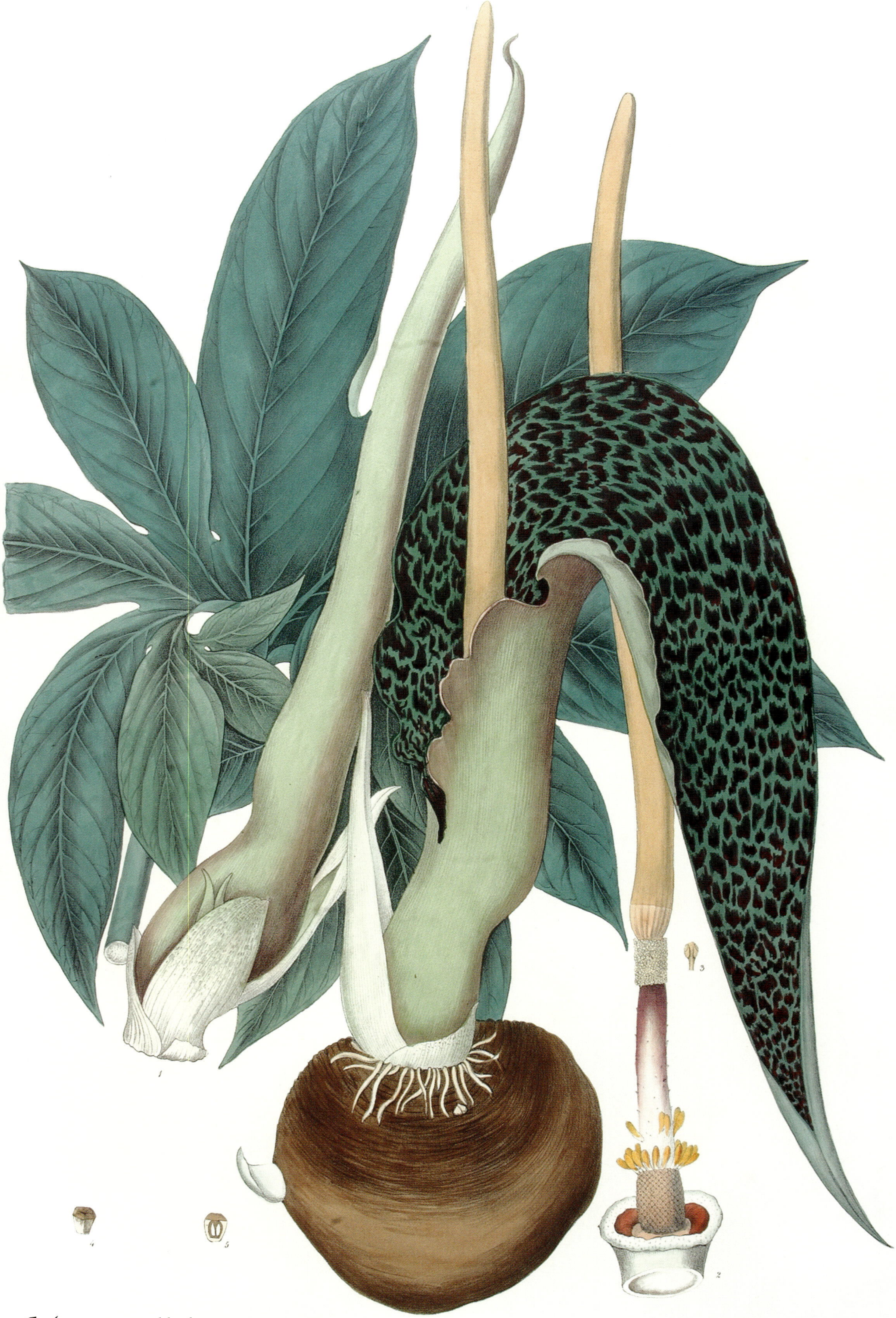

Arisaema guttatum

In loving memory of my sister,
Vijaya Hemchandra Purandare
– Yashwant Dattatreya Joshi

Voodoo Lily
Arisaema guttatum
Family: Araceae
Synonyms: *Sauromatum guttatum, S. venosum, Arum guttatum*

This is a very stately Arum that flowers in March–May, with the inflorescence almost sessile on the tuber. It is found from Himachal Pradesh to central Nepal and southeast Tibet, on rocky mountain slopes and river valleys, from 1,000 to 2,000 m. It has a disagreeable smell. The spathe is very loose and leathery, sometimes 60 cm long, hanging from below the middle of the stem and trailing on the ground. The outer surface is of livid purple while the inner is bright green, covered with innumerable purple angular spots. The root is supposed to be a strong poison.

From N. Wallich, *Plantae Asiaticae Rariores*, 1832.

grunts and murmurs of discontent, and then his mother raised him to his feet by aid of trunk and forefoot, and so guided his trembling limbs and supported him till he found the udders pendant from the breast and gained strength as he satisfied his hunger.

And so the days passed in feeding and sleeping, the mother always within sight or hearing of her young, always ready to face any odds in its protection, returning to its side whenever any suspicious sound reached her ears, or when the wind brought notice of the approach of any man or any beast that might be harmful, till, at the end of ten days, the calf was able to walk, to follow his mother for short distances at a time, then lie down to sleep while she satisfied her hunger. She, still full of maternal care, moved persistently, yet slowly, in the direction followed by the herd, hoping soon to overtake them and so profit by the protection their numbers afforded, and she had indeed nearly attained her object without mishap when an adventure occurred which might have altered the whole course of this narrative.

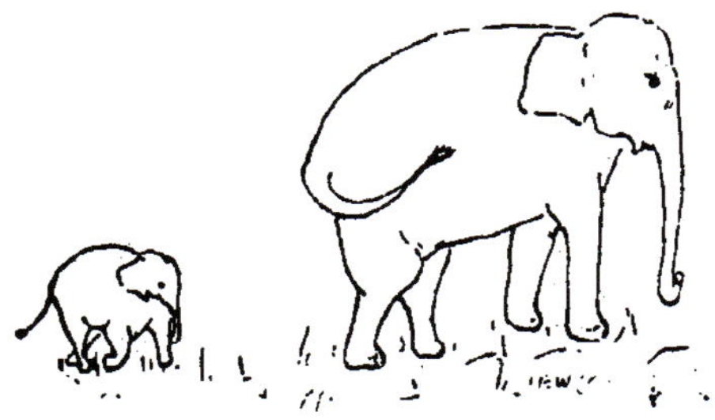

One morning, aware of the proximity of the herd, and intending to join them during the day, the elephant was grazing on the bamboo shoots and grass, not far from the spot where her calf was lying, when her sensitive nostrils gave notice of danger. She waved her trunk in the direction of the wind, inhaling deeply, then strode hurriedly to where her calf was hidden, standing over him as he lay. The next moment a tigress appeared on the scene, lean and anxious looking, seemingly pressed by hunger, evidently, too, the mother of young cubs whose increasing demands for food had to be satisfied. In point of fact the animal was starving: for days she had been unsuccessful in the hunt, and she must eat or die, and with her her helpless family. She had scented the presence of the young calf, and thus it was that nature in its cruelty had ordained that the love of these mothers for their offspring should lead up to a struggle which must end in the slaying of one family, yet resulting in no apparent good.

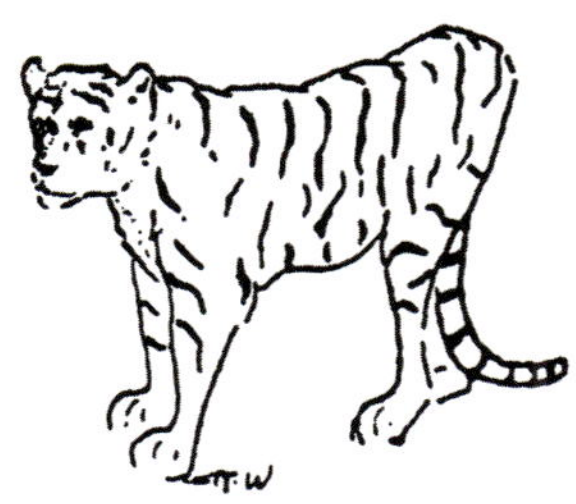

The tigress walked slowly round the elephant, continually decreasing the distance between them, her object being to drive off the mother and then kill the defenceless calf. The elephant also turned slowly, keeping her head to the foe. Soon the tigress made a rush towards the calf, and was met by a kick from the elephant's hind leg, which staggered her for the moment and made her resume her prowl in the hope of finding

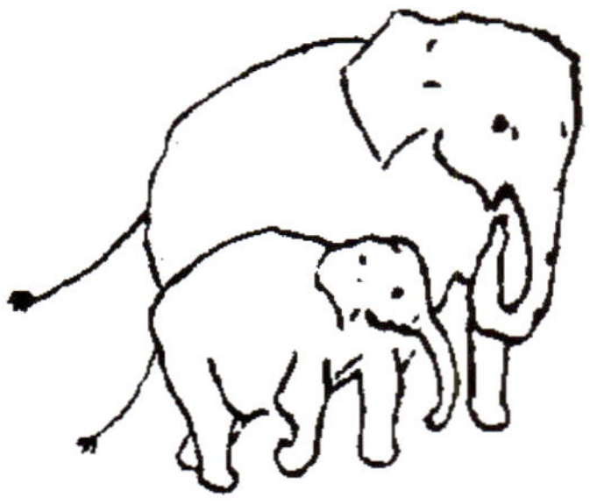

some better opening for attack. The duel proceeded in silence, and gradually an arena was formed in the forest, a circle of trampled grass where no concealment was possible. Several times the tigress made onslaughts, only to be repulsed, and, at the last, savage at her ill-success, she bounded at the elephant's head, clasping her with armed fore-paws, burying her fangs in the soft flesh at the base of the trunk; scratching wildly at fore-legs with her hind claws, hoping that she could drag the elephant to earth and inflict such painful injuries that she would desert her young. The elephant flinched under the attack, and then stood firm. Maddened with pain and fear she knelt and tried to crush the adversary with her weight, but the tigress held firmly, and, with the purchase gained, pulled violently to one side in the attempt to throw the elephant. She, despairing, rose to her feet, rushed blindly forward in the desire to get rid of this rending, biting torment, and so by good fortune hurled herself full against the stem of a stout tree. She felt the tigress crushed against her bony forehead, and the grip relaxed; again she butted with her full strength, and the enemy fell limply at her feet. Then, possessed solely by the fury of battle and of pain, she stamped on the still writhing body, crushing it to a pulp, kicking it till but a muddy mass remained to show her victory. At last, streaming with blood, trembling with excitement, she returned to her calf, feeling it over with her wounded trunk, satisfying herself that it had suffered no harm.

That evening, the elephant, scarred with many wounds, now plastered with clay to keep off the attacks of flies, rejoined her tribe after many days of danger and tribulation, and felt once more in the safety of home. But the tigress's cubs were calling for their mother, till their feeble cries died away in a whimpering protest as the weakness and pain of starvation gradually overcame them.

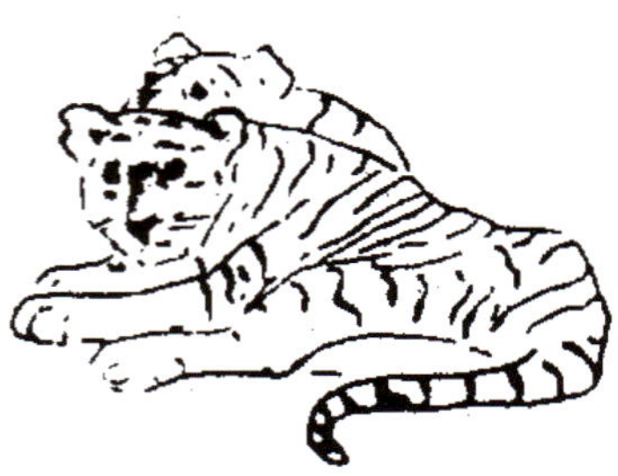

From Sir S. Eardley-Wilmot, *The Life of a Tiger, and the Life of an Elephant*, illustrated by Iris Eardley-Wilmot, 1933.

Chestnut-tailed Starling, Malabar White-headed Starling, Greyheaded Myna
Sturnus malabaricus blythii
Synonym: *Pastor blythi*

Considered by some a separate species, this white-headed race is resident in the southwestern peninsular region of India. It is chiefly arboreal, feeding on berries and insects and descending to flower bushes in search of nectar, constantly chattering, and moving in flocks. The long yellowish bill is dark at the base, the legs yellowish orange. The upperparts are grey, with the snowy-white hood extending over neck and breast reaching mid-belly; the lower belly and flanks are rufous, with chestnut tail.

From T.C. Jerdon, *Illustrations of Indian Ornithology*, 1847.

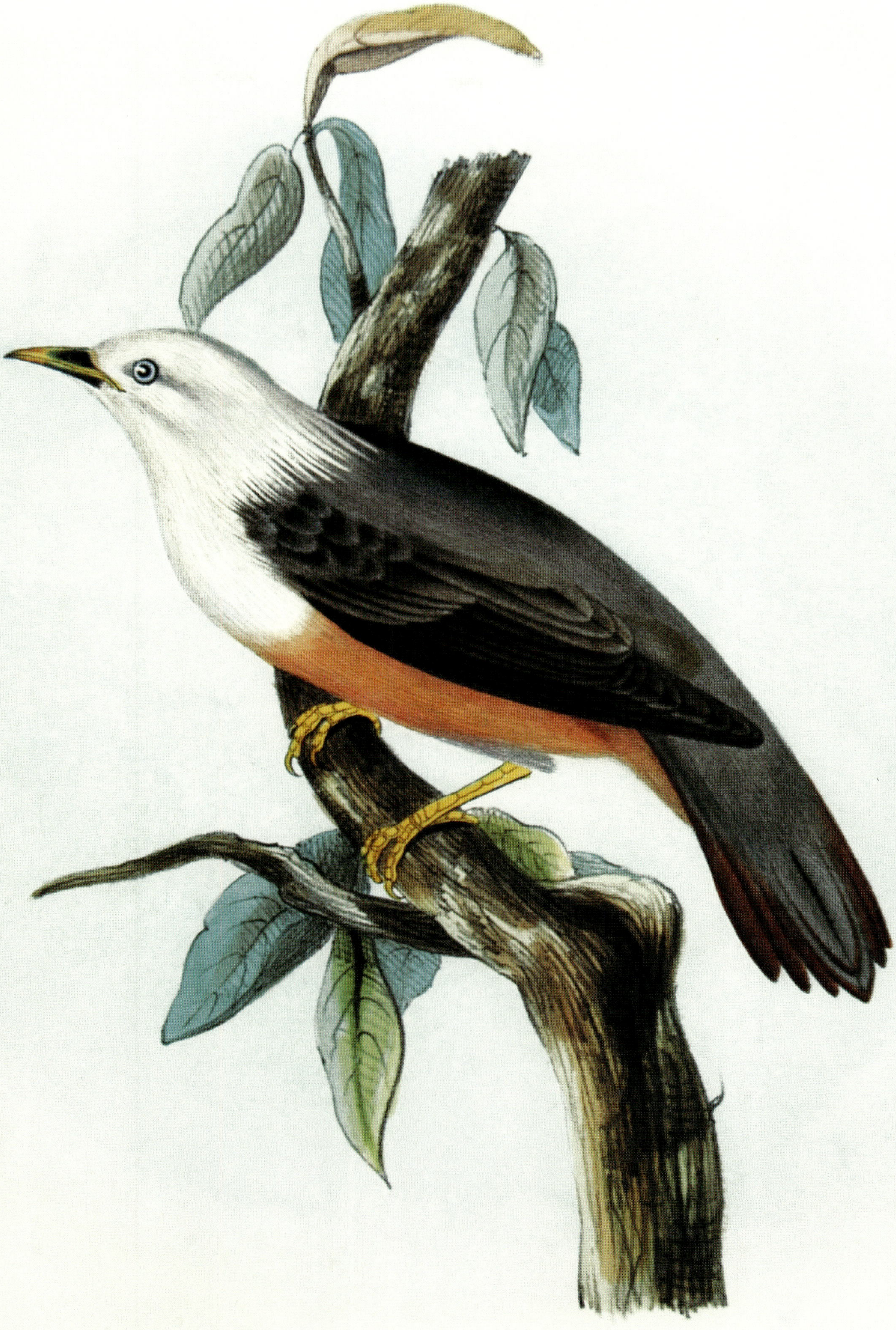

Sturnus malabaricus blythii

Courtesy
Mahindra & Mahindra Ltd.

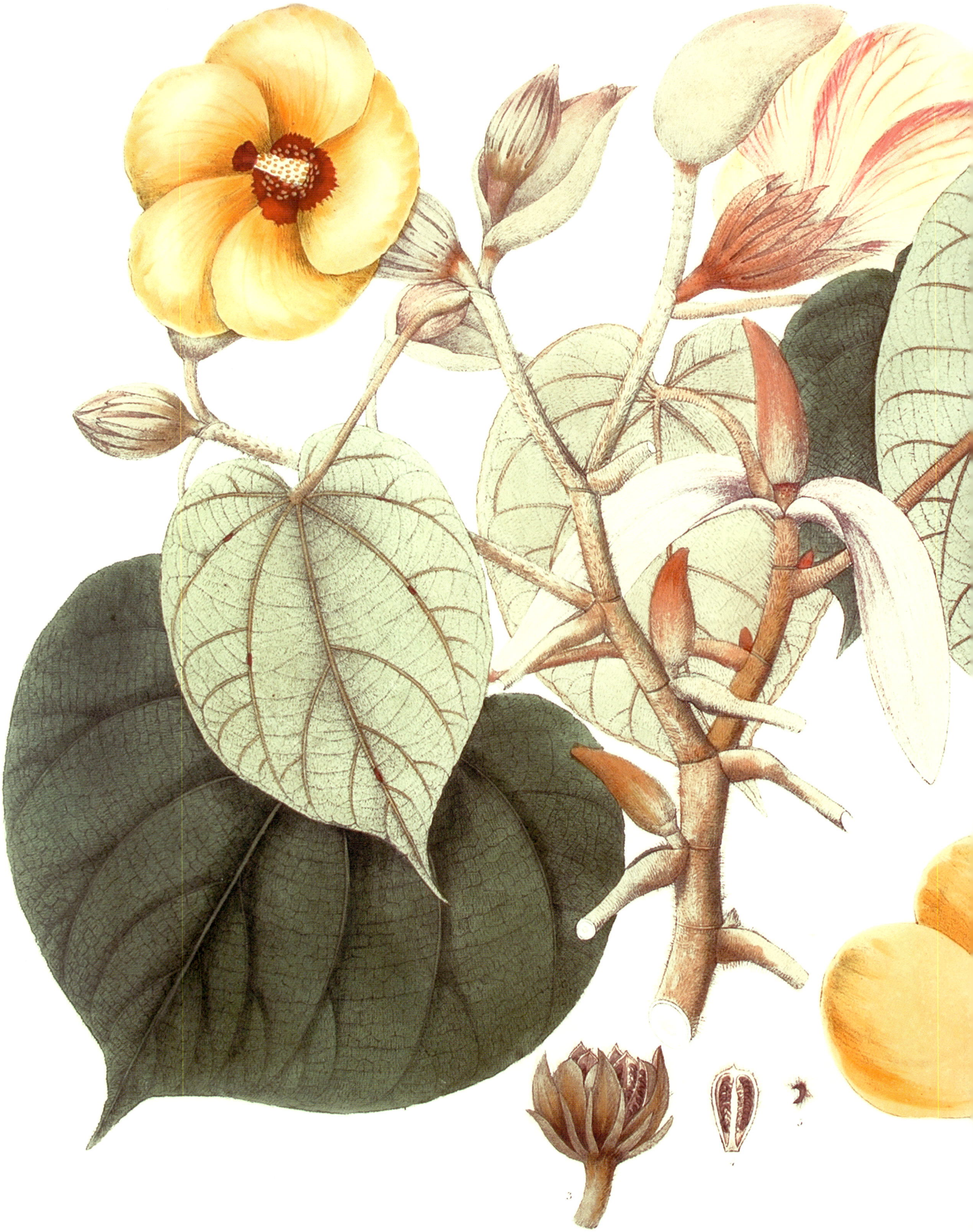

Inserted in fond memory of
and as a mark of gratitude to
the late Father H. Santapau, S.J.
– Dhun Panthaki

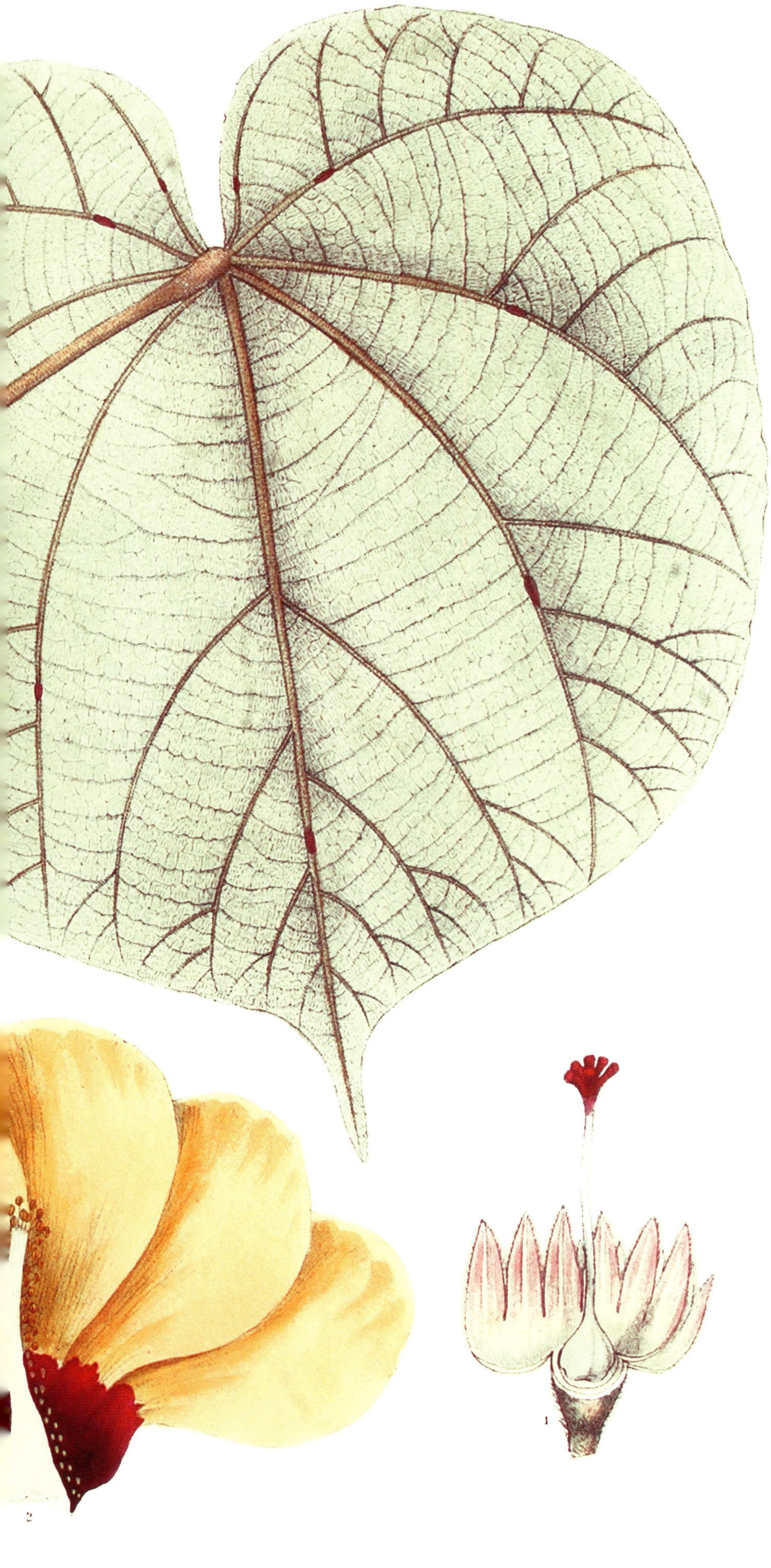

Largeleaf Rosemallow

Hibiscus macrophyllus

Family: Malvaceae

This tree, found in India, Indonesia, and Malaysia, grows up to 15 m tall. The young stems, petioles, and pedicels are hispid. The leaves are circular, cordate, 20–60 cm in diameter, smaller and narrower distally, with veins on the lower surface. Both surfaces are loosely pubescent, the margins subentire, the apex acute or acuminate. The petioles are often as long as the blades. The flowers appear during March–May, are usually solitary, the petals yellow with purplish-red basal spot, fading to yellowish-orange. The capsule is ellipsoid to obovoid, 2–3 cm long, the seeds reniform.

From N. Wallich, *Plantae Asiaticae Rariores*, 1832.

How the Rhinoceros Got His Skin

Rudyard Kipling

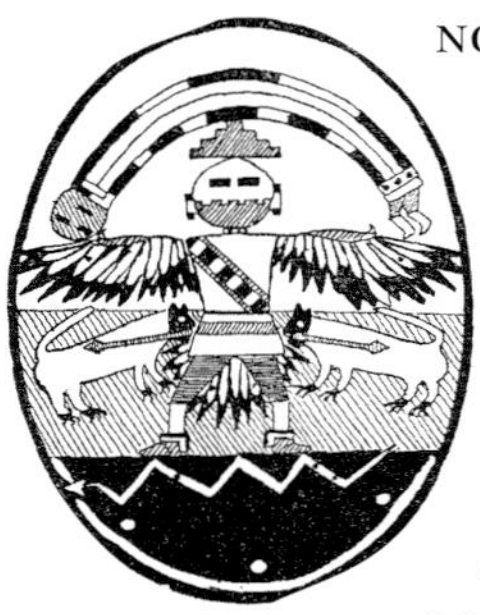

NCE upon a time, on an uninhabited island on the shores of the Red Sea, there lived a Parsee from whose hat the rays of the sun were reflected in more-than-oriental splendour. And the Parsee lived by the Red Sea with nothing but his hat and his knife and a cooking-stove of the kind that you must particularly never touch. And one day he took flour and water and currants and plums and sugar and things, and made himself one cake which was two feet across and three feet thick. It was indeed a Superior Comestible *(that's* Magic), and he put it on the stove because *he* was allowed to cook on that stove, and he baked it and he baked it till it was all done brown and smelt most sentimental. But just as he was going to eat it there came down to the beach from the Altogether Uninhabited Interior one Rhinoceros with a horn on his nose, two piggy eyes, and few manners. In those days the Rhinoceros's skin fitted him quite tight. There were no wrinkles in it anywhere. He looked exactly like a Noah's Ark Rhinoceros, but of course much bigger. All the same, he had no manners then, and he has no manners now, and he never will have any manners. He said, "How!" and the Parsee left that cake and climbed to the top of a palm-tree with nothing on but his hat, from which the rays of the sun were always reflected in more-than-oriental splendour. And the Rhinoceros upset the oil-stove with his nose, and the cake rolled on the sand, and he spiked that cake on the horn of his nose, and he ate it, and he went away, waving his tail, to the desolate and Exclusively Uninhabited Interior which abuts on the islands of Mazanderan, Socotra, and the Promontories of the Larger Equinox. Then the Parsee came down from his palm-tree and put the stove on its legs and recited the following *Sloka,* which, as you have not heard, I will now proceed to relate –

Them that takes cakes
Which the Parsee-man bakes
Makes dreadful mistakes.

And there was a great deal more in that than you would think.

Because, five weeks later, there was a heatwave in the Red Sea, and everybody took off all the clothes they had. The Parsee took off his hat; but the Rhinoceros took off his skin and carried it over his shoulder as he came down to the beach to bathe. In those days it buttoned underneath with three buttons and looked like a waterproof. He said nothing whatever about the Parsee's cake, because he had eaten it all; and he never had

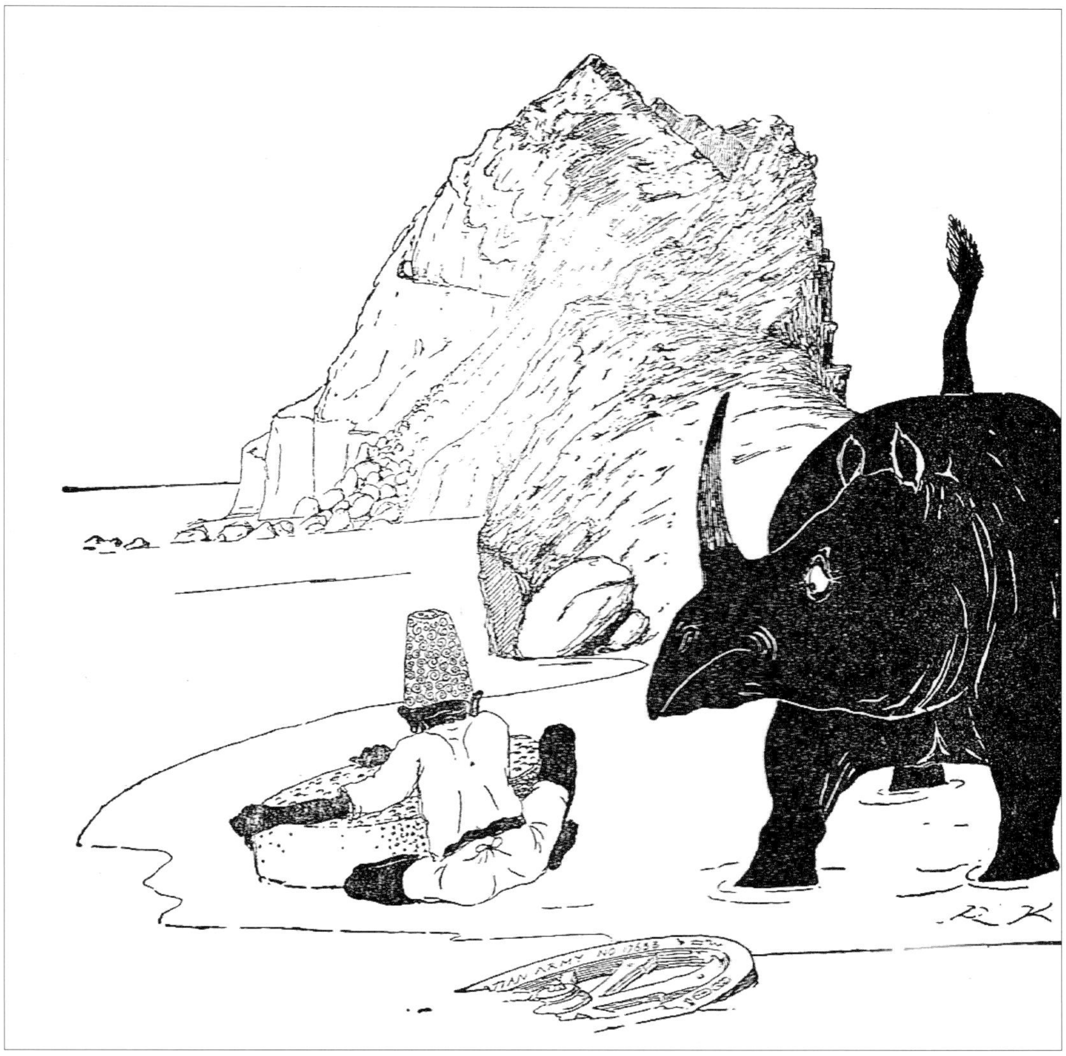

THIS is the picture of the Parsee beginning to eat his cake on the Uninhabited Island in the Red Sea on a very hot day; and of the Rhinoceros coming down from the Altogether Uninhabited Interior, which, as you can truthfully see, is all rocky. The Rhinoceros's skin is quite smooth, and the three buttons that button it up are underneath, so you can't see them. The squiggly things on the Parsee's hat are the rays of the sun reflected in more-than-oriental splendour, because if I had drawn real rays they would have filled up all the picture. The cake has currants in it; and the wheel-thing lying on the sand in front belonged to one of Pharaoh's chariots when he tried to cross the Red Sea. The Parsee found it, and kept it to play with. The Parsee's name was Pestonjee Bomonjee, and the Rhinoceros was called Strorks, because he breathed through his mouth instead of his nose. I wouldn't ask anything about the cooking-stove if I were you.

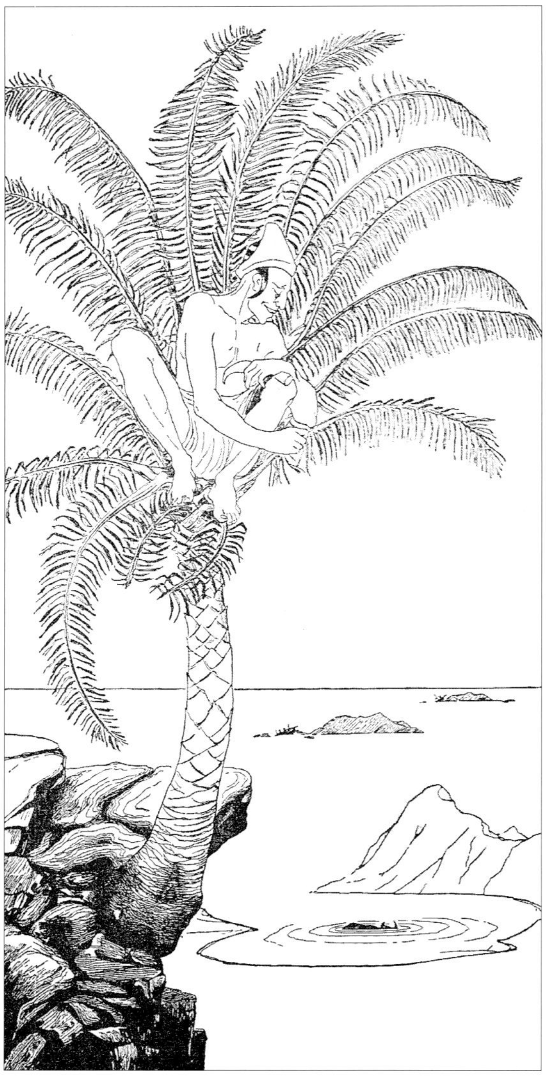

THIS is the Parsee Pestonjee Bomonjee sitting in his palm-tree and watching the Rhinoceros Strorks bathing near the beach of the Altogether Uninhabited Island after Strorks had taken off his skin. The Parsee has rubbed the cake-crumbs into the skin, and he is smiling to think how they will tickle Strorks when Strorks puts it on again. The skin is just under the rocks below the palm-tree in a cool place; that is why you can't see it. The Parsee is wearing a new more-than-oriental-splendour hat of the sort that Parsees wear; and he has a knife in his hand to cut his name on palm-trees. The black things on the islands out at sea are bits of ships that got wrecked going down the Red Sea; but all the passengers were saved and went home.

The black thing in the water close to the shore is not a wreck at all. It is Strorks the Rhinoceros bathing without his skin. He was just as black underneath his skin as he was outside. I wouldn't ask anything about the cooking-stove if I were you.

any manners, then, since, or henceforward. He waddled straight into the water and blew bubbles through his nose, leaving his skin on the beach.

Presently the Parsee came by and found the skin, and he smiled one smile that ran all round his face two times. Then he danced three times round the skin and rubbed his hands. Then he went to his camp and filled his hat with cake-crumbs, for the Parsee never ate anything but cake, and never swept out his camp. He took that skin, and he shook that skin, and he scrubbed that skin, and he rubbed that skin just as full of old, dry, stale, tickly cake-crumbs and some burned currants as ever it could *possibly* hold. Then he climbed to the top of his palm-tree and waited for the Rhinoceros to come out of the water and put it on.

And the Rhinoceros did. He buttoned it up with the three buttons, and it tickled like cake-crumbs in bed. Then he wanted to scratch, but that made it worse; and then he lay down on the sands and rolled and rolled and rolled, and every time he rolled the cake-crumbs tickled him worse and worse and worse. Then he ran to the palm-tree and rubbed and rubbed and rubbed himself against it. He rubbed so much and so hard that he rubbed his skin into a great fold over his shoulders, and another fold underneath, where the buttons used to be (but he rubbed the buttons off), and he rubbed some more folds over his legs. And it spoiled his temper, but it didn't make the least difference to the cake-crumbs. They were inside his skin and they tickled. So he went home, very angry indeed and horribly scratchy; and from that day to this every rhinoceros has great folds in his skin and a very bad temper, all on account of the cake-crumbs inside.

But the Parsee came down from his palm-tree, wearing his hat, from which the rays of the sun were reflected in more-than-oriental splendour, packed up his cooking-stove, and went away in the direction of Orotavo, Amygdala, the Upland Meadows of Anantarivo, and the Marshes of Sonaput.

This Uninhabited Island
Is off Cape Gardafui,
By the Beaches of Socotra
And the Pink Arabian Sea:
But it's hot – too hot from Suez
For the likes of you and me
Ever to go
In a P. and O.
And call on the Cake-Parsee!

From Rudyard Kipling, *Just So Stories*, 1902.

Clouded Leopard
Neofelis nebulosa

The Clouded Leopard is a medium-sized cat, 55 to 110 cm long (like a small Leopard) and weighing between 15 and 23 kg. It is found in the eastern Himalaya, northeast India, Southeast Asia, and southern China. Its coat is pale yellow-brown with large, irregular-shaped markings, each dark brown/black around the edge and lighter within. It is this "cloud" pattern that gives the cat its name. The undersides and short, stout legs are spotted and the head and neck streaked with black/dark brown. The Clouded Leopard is a tree dweller and has almost squirrel-like agility. It can hang by its hindlegs using its long tail for balance, and runs head first down tree trunks. It is generally a solitary creature. Females give birth to a litter of up to 5 cubs, after a gestation period of 85 to 93 days. It is estimated that less than 10,000 Clouded Leopards exist worldwide and their number is declining. Habitat loss due to widespread deforestation, and poaching for use in Chinese medicine are the main causes of the decline.

From *JBNHS* Vol. 37, No. 4 (Supplement: *The Wild Animals of the Indian Empire and the Problem of their Preservation*, Part II).

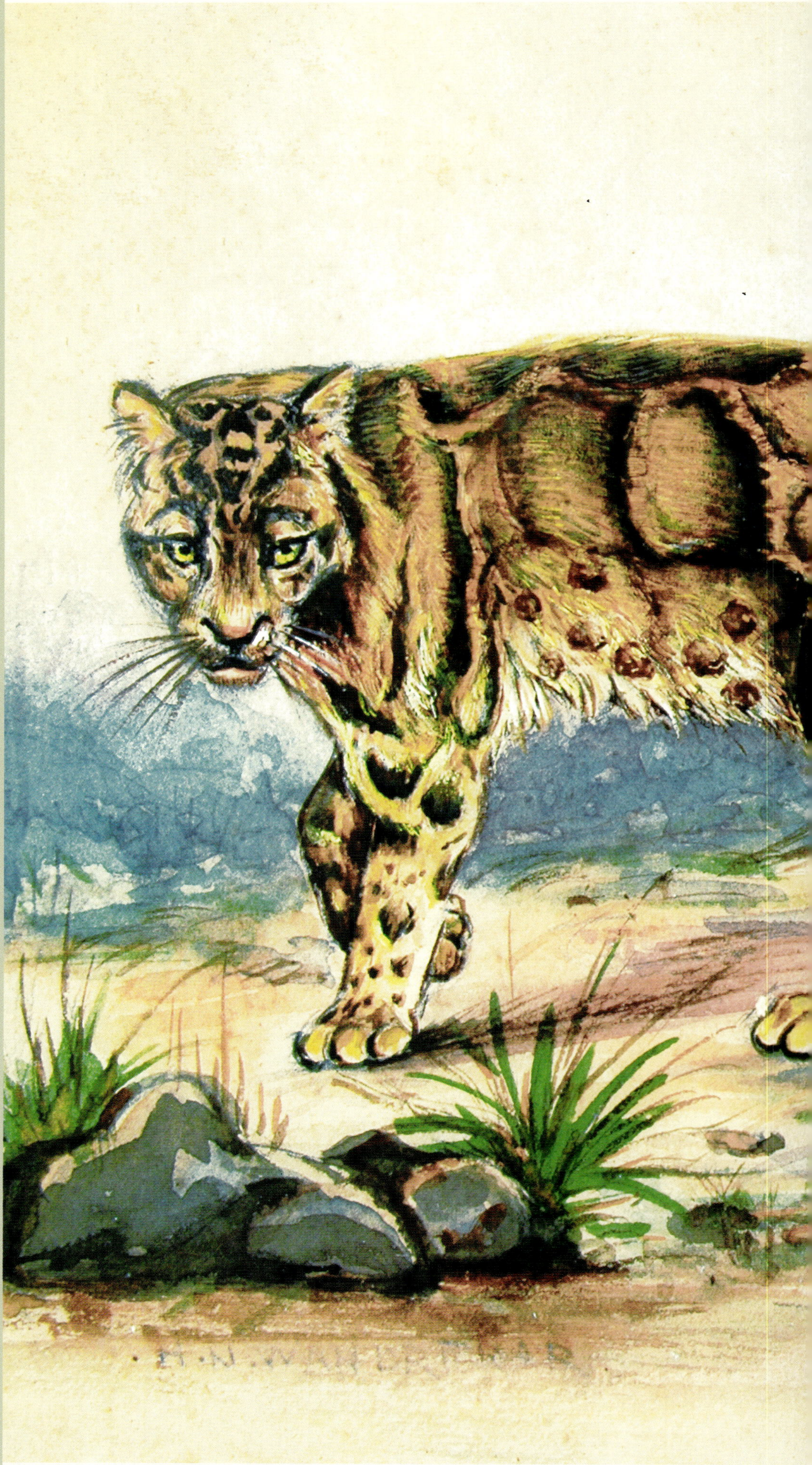

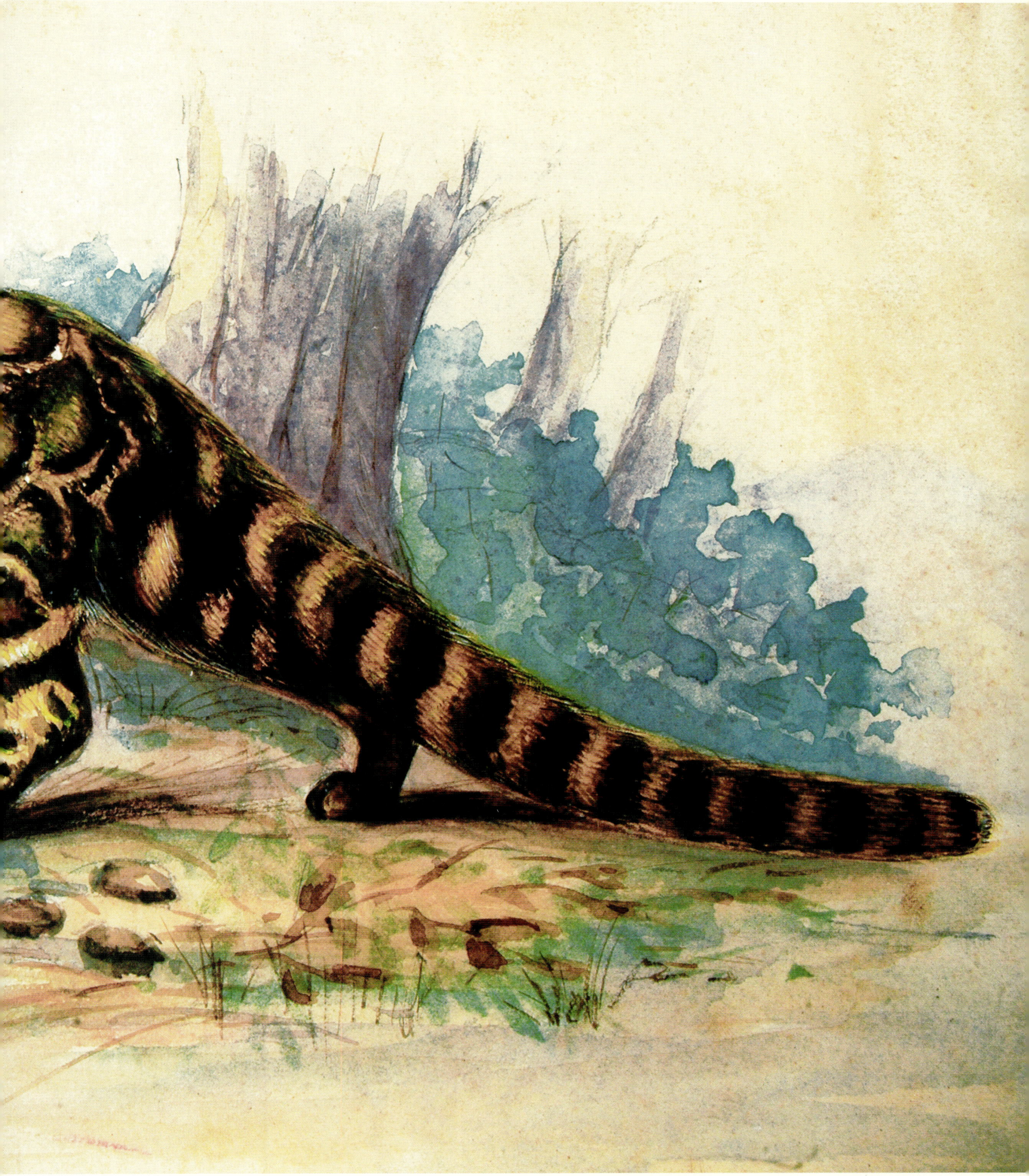

COURTESY HSBC
THE HONGKONG AND SHANGHAI
BANKING CORPORATION LIMITED

Wounded Bear Charging up a Tree

It is well known that bears are good climbers, and that they are in the habit of ascending trees after bees' nests and jungle fruits. That they charge readily enough if encountered on foot is familiar to all sportsmen. The incident I am about to relate of a bear charging up a tree at its adversary is, I think, uncommon, although I remember on a previous occasion seeing a wounded bear charge to the foot of the tree from which it was fired at, and rear itself up against a ladder on the top rung of which was seated Colonel Bowie, now Inspector-General of Police, Central Provinces.

On May 17th, 1890, I was beating a rocky hill in the Central Provinces, and was posted in a Saj tree (the Black Eque), when a she-bear galloped by at a few yards' distance. I shot her, and she rolled about on the ground after the manner of bears, complaining bitterly. I fired the second barrel at her as she lay on her back, which appeared to act like a tonic, as she jumped up, charged straight at the tree, and at once began to climb it. So quickly did she come, that, although I never reloaded quicker in my life, she reached my legs just as I got one barrel reloaded. She was on the point of seizing my leg with her teeth when I fired and dropped her. My hat fell off in the scrimmage, and on reaching the ground she tore the inside out. She then made off at a slow walk; I reloaded and hit her with both barrels, knocking her down once, and, though all the shots but one she had received were well forward, she was still able to walk, and was finished by a shot from my companion, Captain Burton, Royal Fusiliers, who was posted about 100 yards from me. This was a case which one occasionally meets with in sporting prints, appropriately headed "Extraordinary Tenacity of Life in a Bear". The shot in the tree I found had hit her inside the head of the humerus, breaking the scapula, and so close was she that the hair on her face and chest was singed with the flash, and the wads were driven into the wound: I found them resting against the head of the humerus. The express bullet had split up on the scapula, and only a small fragment had entered the cavity of the chest. My mala was tied on the lowest branch of the tree, and I was sitting 13 feet from the ground. The stem of the tree was 3 feet 6 inches in circumference, two feet from the ground, or about 14 inches in diameter. The bole of the tree was perpendicular with only one small shoot at a distance of 5 feet from the ground. My legs hung down alongside the tree, so that my feet were at a height of 11 feet 6 inches. The highest claw marks of the bear were 11 feet 6 inches high, alongside my right foot. The above details are given from actual measurement.

In order to carry away with me an accurate record of the incident, I tied the defunct bear to the tree, and with the able assistance of Captain Burton took a photograph, with the result depicted opposite. Although, of course, the bear was dead, the picture gives a correct representation of the tree. The bear, however, is too low, as her highest claw marks were at the point my right foot touches.

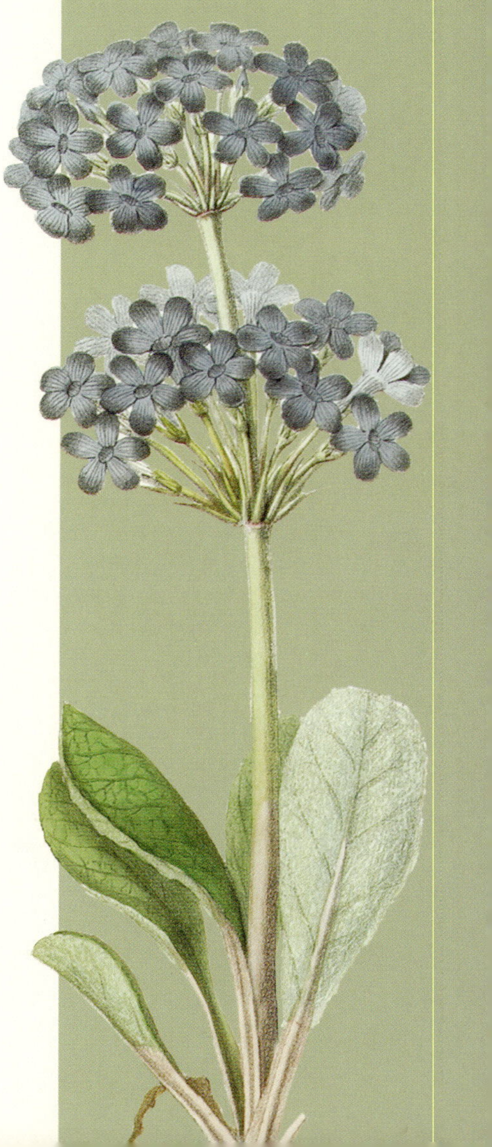

John Duncan Inverarity was born in Bombay in 1847; he was the son of Dr. Inverarity, friend and brother-in-law of Jonathan Duncan, the Governor of Bombay. Educated at Cheltenham, he returned to Bombay in 1869, where by the sheer force of his character and the brilliancy of his intellect he soon established himself as one of the leading lights of the Bombay Bar. It is not as a great lawyer, however, that we would speak of John Duncan Inverarity, but as a big game hunter and naturalist. It was in the early days of his career that Mr. Inverarity associated himself with the Bombay Natural History Society. He joined the Society almost >

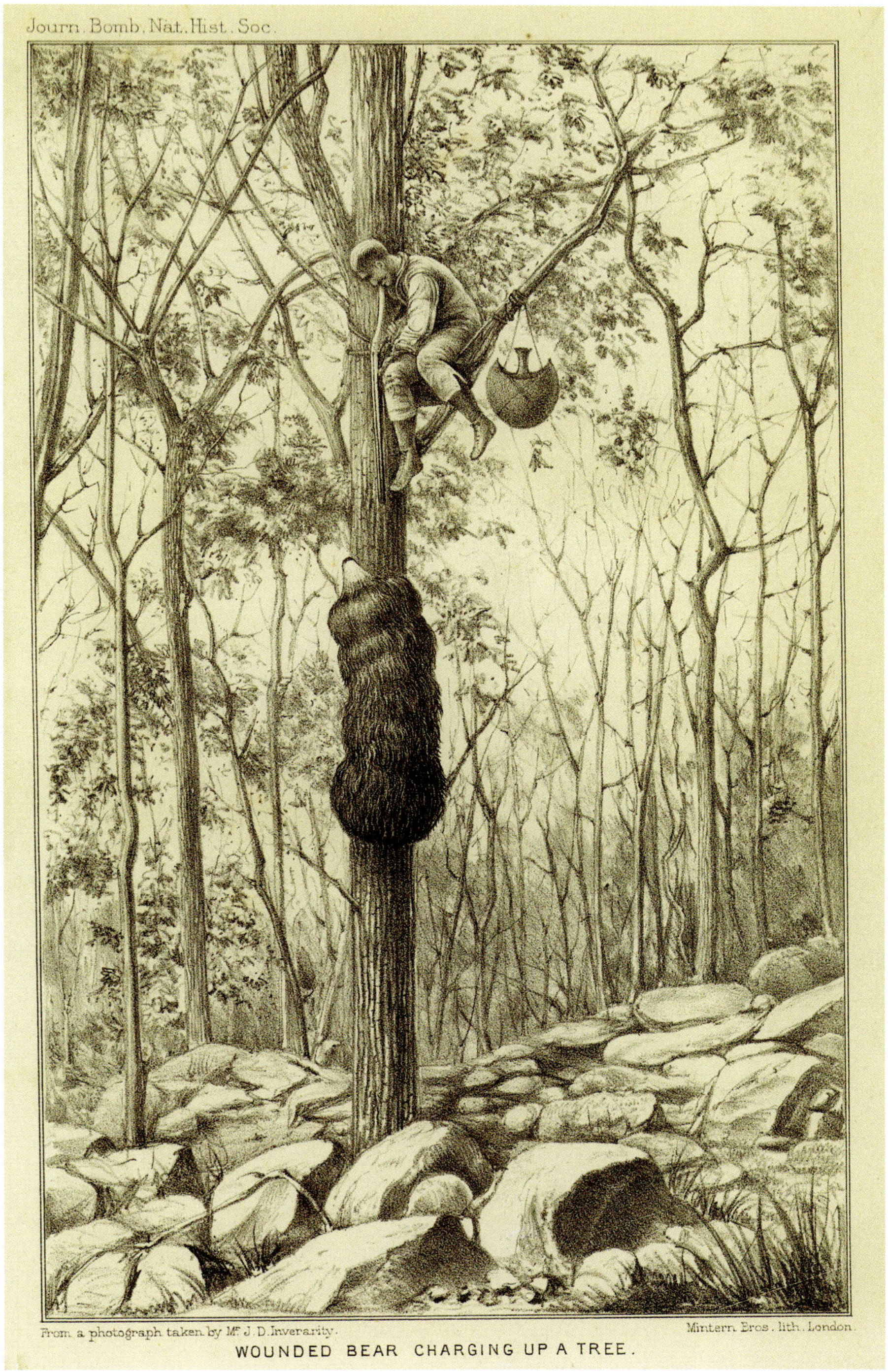

Wounded Bear Charging up a Tree, from a photograph taken by Mr. J.D. Inverarity.

>at its inception in 1886, became member of its Managing Committee in 1891 and a Life Member in 1893, and was elected its Vice-President on April 6th 1897 and continued as such till the time of his death in 1923. He was a valued and constant contributor to the pages of the *Journal*. He was a big game hunter of no mean repute and excelled, not only as a shikari, but as a close and accurate observer of the habits and characteristics of the animals he hunted. Being gifted with a facile pen he was able to include his observations in his writings on sport and jungle life. – From *JBNHS* Vol. 29, No. 3.

As I had other things to think of, I am not sure if I observed accurately the manner of climbing, but she appeared to me to swarm up with her belly close to the tree and her legs well round the bole.

J.D. INVERARITY

From *JBNHS* Vol. 5, No. 4 (1890).

A Day's Sport in Berar

The following account of a day's sport in Berar, some few years ago, when game was more plentiful than it is now, may be of interest.

The writer's duty, as a District Officer, led him one day to the neighbourhood of a deserted village – Dharur – situated on the banks of a mountain stream which, coming down from the Satpurahs, sweeps the base of the hill on which stands the fort of Naruallha. Dharur, in the days I am about to write of, was just lovely in its solitude. Mixed jungle clothed the small valley down which ran the stream on the banks of which the ruins of the village stood, and for about 3 miles of its course both banks of this stream were densely clothed with "Sendi", or the date palm, with here and there small open patches of grass.

On the day of my arrival a "kill" was reported by a herdsman who grazed cattle in the vicinity. The marks on the "kill" – a large female buffalo – indicated that one, if not two tigers had their home in the date palm. As it was impossible for a single gun to command the jungle, a message sent in to Ellichpur soon fetched out two friends – General (afterwards Sir Harry) Lumsden of Punjab Frontier fame, at the time commanding the Hyderabad Contingent, and his Brigade-Major, the late Colonel Hugh Watson. With them came as escort officer Resaldar Beg Mahomed, a typical specimen of our splendid irregular cavalry. After a careful reconnaissance, we selected a position on a mound of earth at the foot of a big mohwa tree, which stood about the centre of one of the larger aforesaid patches of grass in the centre of the Sendi-bund, about two miles up stream. Here the General, Colonel Watson, Beg Mahomed, and the writer stood shoulder to shoulder, with a servant behind each carrying a spare rifle. Up the mohwa tree, in the shade of which we gratefully stood, was posted a Shikari of Colonel Watson's as look-out. The troopers of the escort were placed at intervals along either bank, outside the denser jungle, as stops, falling in with the beat as it came along. The beat itself, mainly composed of hill-men villagers, with a forest elephant in the centre to give direction, formed line at the village, and, provided with tomtoms, horns, and rattles, started about noon. It was the 8th of March, and the sun was tolerably hot. It seemed an age before anything showed up. Presently, putting us on the *qui vive,* well in advance of

Buddleja colvilei

Family: Loganiaceae (also Scrophulariaceae)

Often spelled *Buddleia*, named after Adam Buddle, a botanist from Essex, England, this genus has over 100 species, most of them shrubs but also a few trees. *B. colvilei* is a small deciduous tree with arching branches, found from Nepal to Arunachal Pradesh. The leaves are elliptic-lanceolate, finally serrate, and dark green with rusty hairs below. In June–July the tree bears attractive long drooping terminal spike-like clusters of deep crimson to pink flowers with white centres. The corolla is funnel-shaped with four rounded lobes; the calyx bell-shaped. The capsule is woolly-haired, splitting into two valves.

Hooker found this species in Darjeeling and Sikkim, saying, "It has no rival anywhere for beauty and graceful habits. It is abundant towards the summit of Tunglo, from [3,000 m] to the top [3,300 m], and is also frequent in the Lachen and Lachung valleys, at similar elevations .... This [tree] will probably prove very hardy, as I have found it in very exposed places as well as in woods; and from the abundance of its flowers and its lasting for some weeks in bloom it would be a most desirable addition for our gardens. The plate was made from a sketch of my own."

From J.D. Hooker, *Illustrations of Himalayan Plants,* 1855.

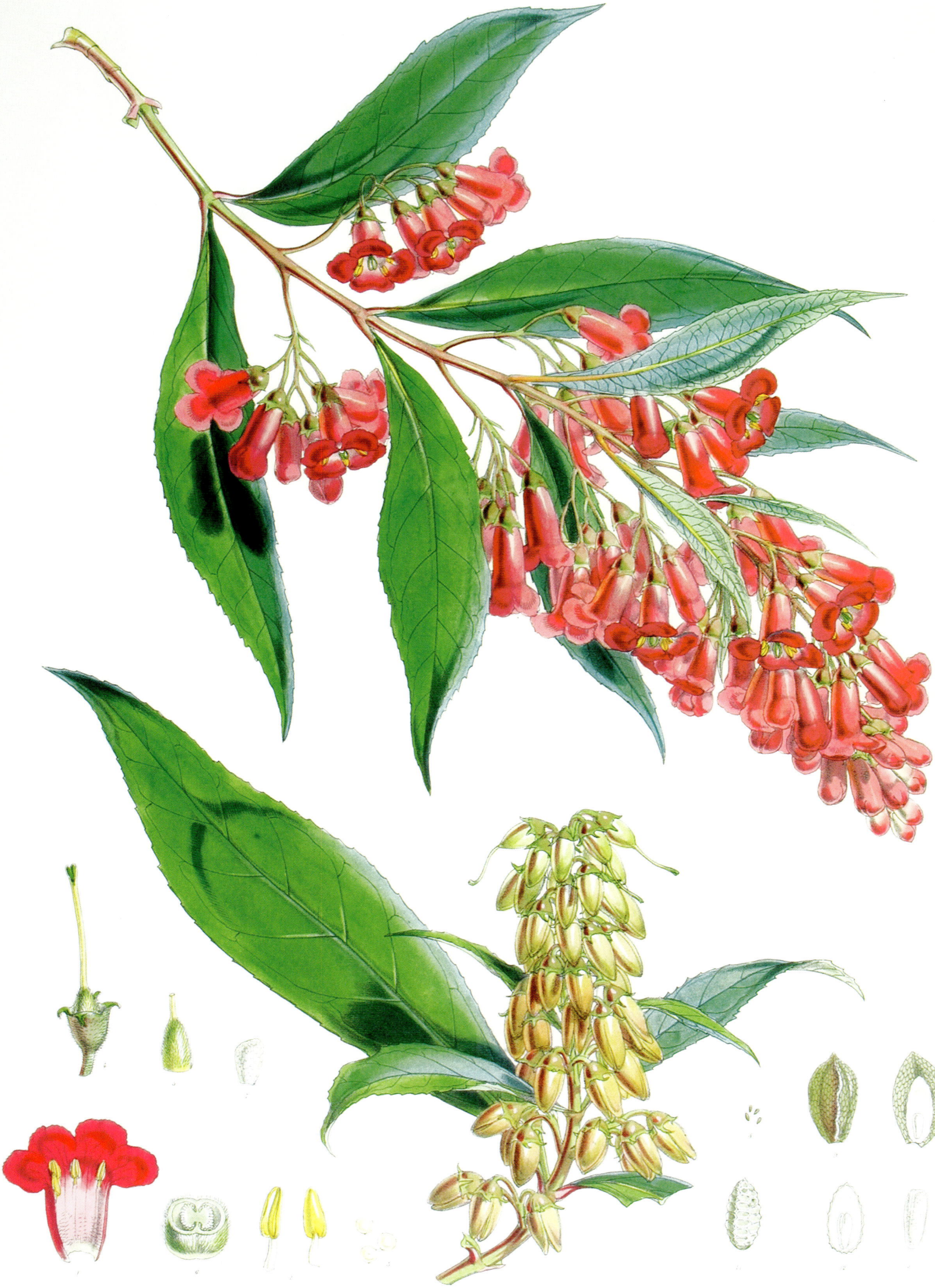

Inserted in fond memory of and
as a mark of gratitude to
the late Father H. Santapau, S.J.
– Dhun Panthaki

Buddleja colvilei

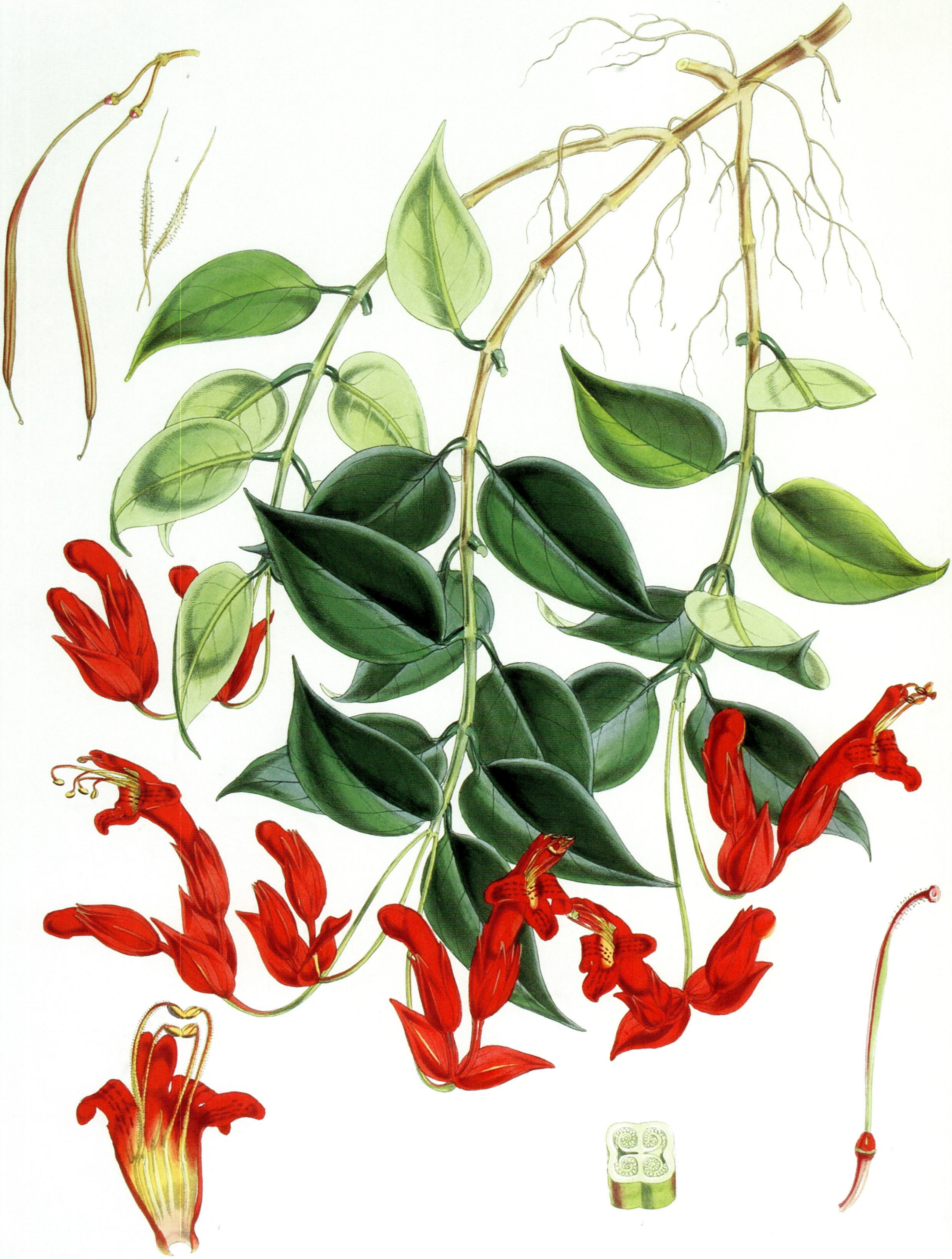

Aeschynanthus peelii

In loving memory of my daughter Jayat
– B.G. Deshmukh, IAS Retd.
President, Bombay Natural History Society

Aeschynanthus peelii
Family: Gesneriaceae

Aeschynanthus is a large genus of ornamental tropical herbs native to South and Southeast Asia. They are usually trailing epiphytes with brightly coloured flowers, most often pollinated by sunbirds. They are called "lipstick plants" as the blossoms of many species have the appearance of a lipstick emerging from a tube. *Aeschynanthus peelii* is found on trees in forested valleys and cliffs by the side of streams in northeast India (Assam, Sikkim), Bhutan, Myanmar, and China at 900–3,200 m. Hooker writes, "This splendid species inhabits a greater elevation and cooler climate than any other known to me. It used to grow on the lofty trees of the Jillapahar, behind Darjiling, before the forests were so thinned that the situation became too exposed for it. It is at the same time one of the most brilliantly coloured species known, the peduncles, pedicels, bracts, and flowers being of the same vivid red colour, and the leaves a deep glossy green above and pale beneath. A very similar plant inhabits lower levels on the Khasia mountains, but bears many flowers in each pair of bracts, and has longer, narrower leaves; it may however be a variety of this."

From J.D. Hooker, *Illustrations of Himalayan Plants*, 1855.

the approaching beaters, came a hare or two; then some jungle cat; and finally peafowl. After all these had passed, there was another seemingly endless interval of suspense – that interval during which, as all shikaris know, strung to the highest pitch of expectancy and excitement, one hears and feels the beatings of one's own heart, just as a steamer quivers to the pulsations of its screw. But the feeling passes away in a second, as the expected quarry comes into view. In our case the view was *five* full-grown tigers "mooning" along close together almost in line, unconscious that their way was barred by man. "Mooning" is the only suitable expression, for they came along quite slowly, heads down, with a bored expression of countenance, as if they felt the heat, and thought the hustling of them by the beat behind most inconsiderate and unkind.

When they first broke into view, about 30 to 35 yards separated us. It was a beautiful and never-to-be-forgotten sight, but the position was in a way critical. There was no time for counsel, but the General, without a moment's hesitation, whispered "Take the big fellow." As a matter of fact, the tigers seemed much of a size, and it required some fineness of discrimination to say on the spur of the moment which was the biggest. But the matter was settled for us, for one of the tigers stopped dead and looked up at us in a startled and surprised fashion. Instinctively we all banged at her (her sex, of course, we only subsequently discovered), and then there was a sight for the gods. It is really impossible accurately to describe, and it was all over in a second or two. But as the sound of our rifles and the roar of the wounded tigress woke the other four tigers up – so to speak – they just raised their heads, looked at us, jumped on one side, and roared in concert. The untouched tigers were evidently momentarily quite undecided what to do – whether to try and get past us, or to face the din and row behind them, which had very appreciably increased as the sound of our shots reached the beaters. While these tigers were thus seemingly debating what to do, the female we had fired at made a bound towards us as if to charge home; but her heart failed her, and she turned sharp to her left and to our right, making for the shelter of the jungle. As she went across us, we gave her the contents of our second barrels, and she fell out of sight into a small nallah. As she did so, the rest turned back in the direction of the beaters. We reloaded rapidly, knowing they would soon be headed and return, but this time prepared for us. Presently I saw a tiger trying to sneak past to our left, and called out "there goes one." I raised my rifle at the same time to indicate the direction to my companions. "Fire if you see him" shouted the General. Our voices had the effect of making this tiger stop and take a good look at us; as she was doing so, I fired. She instantly responded with a roar and charged straight down on to us. I call her "she", as subsequently we discovered that she too was a female, and a very lively and determined female. I failed to stop her with my second barrel, and it seemed to me that my three companions had emptied both theirs equally ineffectively. At this critical moment, when she was within a few yards of us – near enough for us to see her ears back, teeth showing, a savage gleam in her eyes, and blood streaming down one fore shoulder, – when it looked as if one of us in her next bound must be knocked over, our gun-bearers from behind jumped up the mound on which we stood to hand us our second rifles. In their excitement they shoved Resaldar Beg Mahomed and the writer over the crest of the mound, and we, losing our footing, slid down the mound in a sitting posture with our rifles in our hands. For a moment it seemed as if one of us two must be seized, for the tigress was now quite close to us; but in that moment Watson, who was a very steady and cool shot, fired, hit her between the eyes, and in fact drained her. Instead of continuing to come on, she spun round and round twice or thrice and fell dead close in front of us. Had she come on, instead of spinning round, some of us might, possibly would, have been clawed. While this little scene was being enacted, the other three tigers had, it seems, come into view and again turned back. But they could not face the row the beaters were making, and again they turned towards us to force a passage through. One was passing to our left, two to the right; we fired at and wounded the former, and he doubled back on the beaters. As he did so, the other two tigers faced the stop and broke past us to the hills behind. We at once stopped and withdrew the beaters and sent for the elephant to look the wounded tiger up. It would have risked an accident to let the beaters come on with a wounded tiger in their front. The elephant was a very unsteady one; if he saw a tiger, he invariably charged it, and it was difficult to shoot off him. In such jungle, however, we had no choice but to mount him, and we went first to look up tiger number one. As we approached the nallah, where the look-out indicated she had fallen, we heard a growl, followed by a faint attempt at a charge; she was put out of pain immediately. Number three gave us a lot of trouble. Nothing would draw him; he kept retreating from one thick and almost impenetrable patch of "Sendi" to another, and dodged about, till eventually he dodged us altogether; and we had to return to camp with two tigers only, which the General photographed. This ended for us a very eventful day.

The place I speak of is no longer a home for tigers; the jungle has given place to fields with cover barely sufficient for a hare.

KENNETH MACKENZIE, Colonel
Chikaldah, Berar, 28th April, 1893

From *JBNHS* Vol. 8, No. 2.

Man-eating Panther of Seoni

The following are a few notes with reference to a man-eating panther – known to have killed over 20 people in rapid succession and all within a radius of 15 miles of Gunsore village in the Seoni District.

Negro Bean
Mucuna anguina
Family: Fabaceae
Synonym: *M. monosperma*

A large woody climber of the eastern Himalaya found up to 900 m along the southern slopes of Khasi hills in Meghalaya (where it is called Mei-sia-ryntim), Tripura, Orissa, the Deccan plateau, Andamans, Nepa (where its name is Baldhrengra), and Thailand (names: Ma Ba Lai, Ma Mui Yai, Saba Long Lal). The species is found growing in thickets, on hedges, and on the banks of rivers and streams. The leaves have three, nearly egg-shaped leaflets, 7–12 cm long on a common stalk; the main veins and leaf-stalk are generally red. The flowers are purple 3.5 cm long in 6–12-flowered racemes. The flat, oblong pod is 5–7 cm long, plaited, and contains a single, round, compressed dark brown seed. The seeds are eaten as vegetables and are supposed to have restorative power. They are also sedative and expectorant, and given to patients suffering from cough, asthma etc. In Ayurveda the plant is referred to as Kakandola. In some parts of India the seeds are eaten by Hindus during fasts. Both flowers and fruits have reddish-brown, irritant bristly hairs. The hairs on the pods are exceedingly slender, brittle, and easily detached, the fragments readily penetrating the skin producing an intolerable itching; hence at times they are employed for mischievous purposes.

From N. Wallich, *Plantae Asiaticae Rariores*, 1832

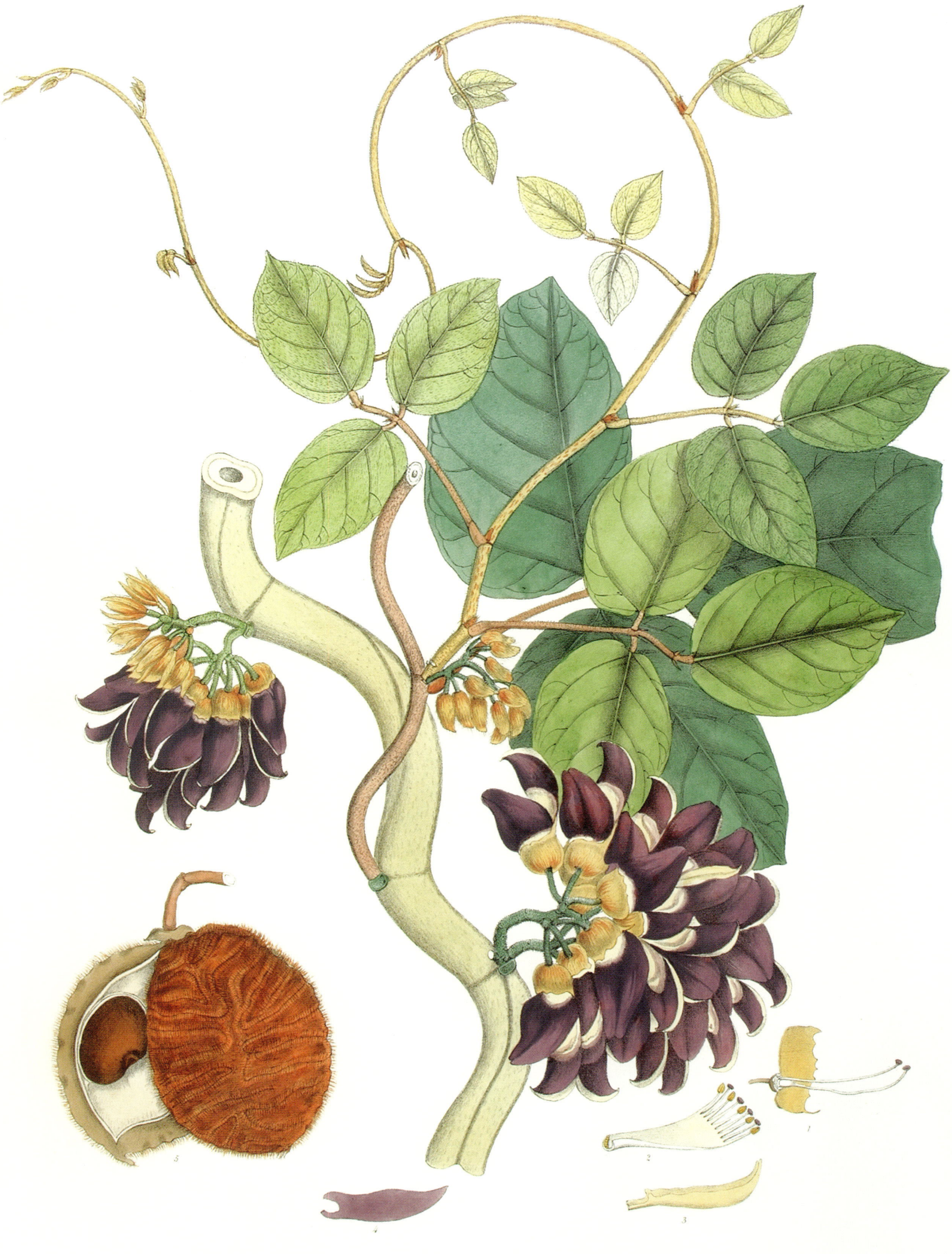

Mucuna anguina

INSERTED IN FOND MEMORY OF
AND AS A MARK OF GRATITUDE TO
THE LATE PROFESSOR P.V. BOLE
– DHUN PANTHAKI

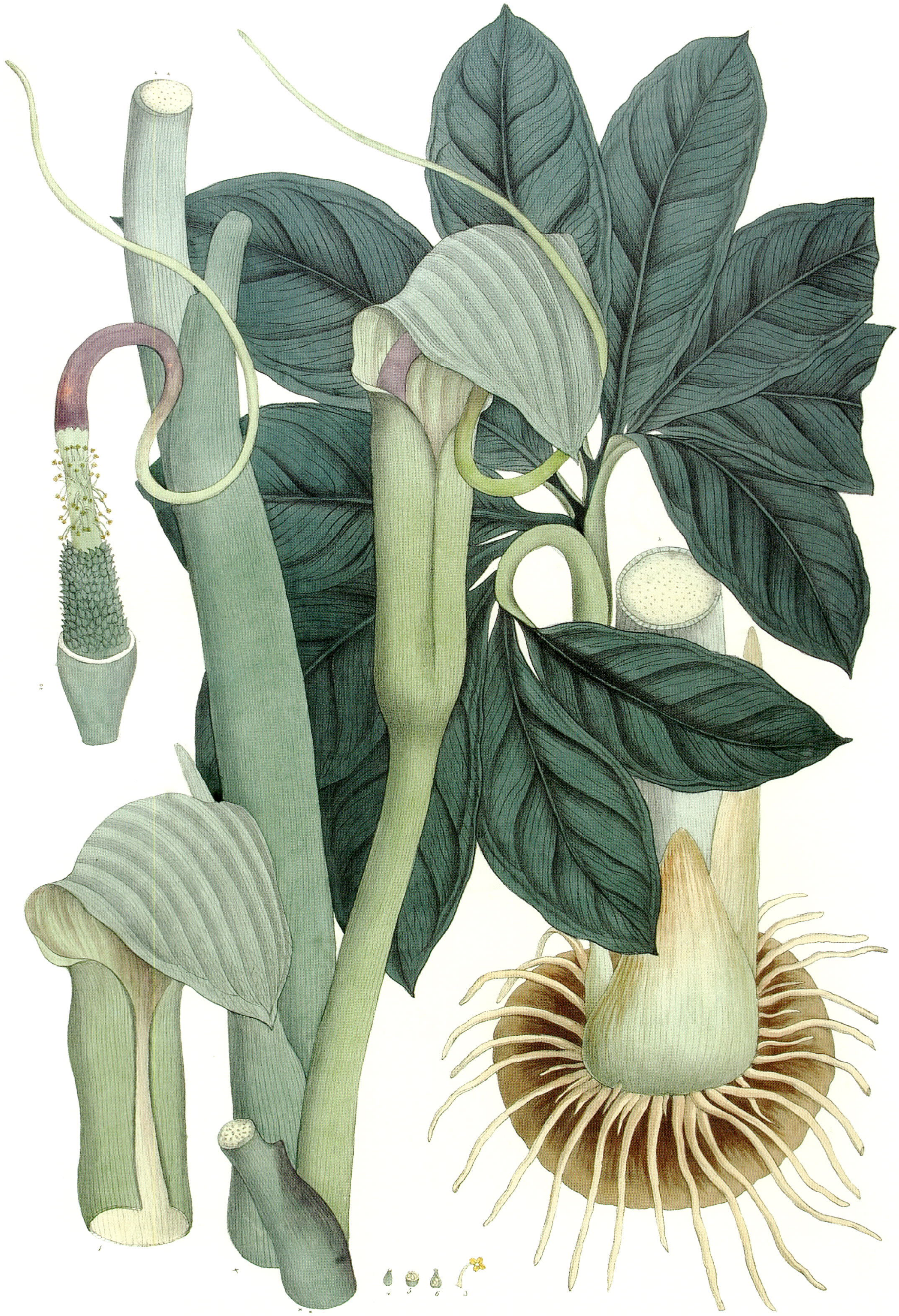

Arisaema tortuosum

In loving memory of Dinesh Balwant Bapat,
from his father Balwant G. Bapat

Whipcord Cobra-lily
Arisaema tortuosum
Family: Araceae
Synonyms: *Arum tortuosum, Arisaema curuatum*

This plant is found in the Himalaya from Shimla to Sikkim and Bhutan. It has multidivided foliage over a tall stem that bears an elegant spathe in May–June. The long spadix is green and points upwards.
The plant contains calcium oxalate crystals that cause an extremely unpleasant sensation, similar to needles being stuck into the mouth and tongue, if eaten raw.

From N. Wallich, *Plantae Asiaticae Rariores*, 1832.

It was not long after my arrival at Gunsore that I came to hear of the panther, so I made it my first opportunity to try and bag him, knowing he was such a source of danger to the surrounding inhabitants.

A kill was reported to me as having taken place on the 23rd of March, *i.e.*, four days before my arrival. On April the 10th another kill was reported. On hearing of this I went out to ascertain as to whether it was true and found the body in a mangled condition with one arm torn right-off. The body was lying in a nullah within 200 yards of a village (Balwara). The villagers stated that the animal must have visited them about twelve at night and dragged away the body from just outside a house and it seems there were four people sleeping outside this house, but that none of them heard the slightest sound. I made arrangements for sitting up that night (*i.e.*, the night after the kill took place). The animal put in his first appearance about 9 p.m., again at 11 p.m., and for the last time at one o'clock. During each of these visits I thought it advisable not to shoot as I could see nothing of the animal owing to darkness, although I could distinctly hear him at the body.

On April the 21st I again heard of a kill (Somnapur village). This time the beast had actually gone into a house and dragged away the body to one side of a village only 100 feet from one of the houses and on an open plain. There were two children asleep in the house where the body was taken from but they were not disturbed.

In order not to be disappointed a second time owing to darkness, I thought it best to try a hurricane lamp which I fastened to a pole about 5 feet off the ground and about 20 feet off the body and I took the precaution to cover up part of the globe of the lamp with a dark cloth so as to allow a dark side for the animal to approach from. At 11 p.m. I noticed the animal having a quiet search round and as soon as he got down to the body I fired. I left a man all night to see that nobody should remove them until I came the following morning. I might remark that the body was only mauled at the chest, neck, and head, and that from these parts a good deal of flesh had been devoured.

The Government reward for the animal was Rs. 50.

W.A. CONDUITT
Seoni District, 21st November, 1901

From *JBNHS* Vol. 14, No. 3.

Man-eating Panther of Dangs

The panther first began man-eating in this district in the rains of 1901; apparently he came from the Dangs, as his first kills were in the villages bordering on those native states. I think he probably took to man-eating on account of the 1900–1901 famine which was very bad indeed in the Dangs, many people dying out in the jungle (I have myself seen skulls of those who tried to get into British territory but died by the way); during the famine the deer and pig in the forests would naturally diminish on account of the Bhils and Konkanis being hard up, and this together with the circumstance of people moving about in the wildest and least frequented parts in search of edible fruits and roots probably led the panther on some extra hungry day to take to man for food. The first man he took was probably dying or asleep in the jungle. I did not hear of the man-eater till last February when my camp was at Mulher in the centre of the valley where he was killing: his beat in the Mulher Valley was about 7 miles long and 3 broad. I remained 10 days at Mulher trying my best by means of "holding the hills", beats, and sitting up over kills to get the beast, but was unsuccessful and could not stay any longer. In November my work took me to Ajmir Saundana within 22 miles of Mulher, and while at Ajmir I heard on the 26th of November that the panther was killing again. I at once went to Mulher, but it was not till the 4th of December that the beast was killed; I should have had to leave on the evening of the 4th as I could spare no more time after him, so luck was with me at last.

On the 3rd, at noon, I heard a boy of 15 had been killed at Wadai, 4 miles from my camp; this boy was most unfortunate. Last year the panther had tried to get him, but only mauled one leg; my wife and I were able to dose the wound with carbolic oil and the boy got well; this time he and one or two others were sitting close to a bright fire on a threshing floor near the village in the early part of the night and the panther came and carried him off: the panther took him about a quarter of a mile to a patch of high grass and brushwood and ate all he could of the head, the flesh of one leg and all his inside; so there was plenty left for the beast to come back for. The body was dragged into deep shade under the brushwood and there was no tree anywhere near to sit up in, so four holes were made in the ground and a wooden post put in each; a cart was then taken off its wheels and lifted on to the top of the posts so as to be about 8 feet from the ground: the cart and posts were then hidden in branches and grass, only a small opening being left to shoot through. An avenue was cut through the brushwood so that I could clearly see the body from the erection, which was about 35 feet from the kill. At 3-30 p.m. I was sitting as quiet as a mouse in the cart and had sent all the men back to the village. At about 4-30 I distinctly heard the panther coughing about a quarter of a mile away. In another hour's time and without any warning the beast growled just *close* behind me; I had been absolutely still and he could not have seen me and so probably he scented me because of my erection not being far enough off the ground. I did not turn or move, hoping he would in a few minutes come round to the kill in front; but he was too cute, and it got darker and darker till at 6-30 I could not see the body at all: there was a crescent of a moon, but it was useless, as the body was in deep shade. At about a quarter to seven I settled to turn up my hurricane lamp and hang it outside on one of the

Sapphire Flycatcher, Sapphire-headed Flycatcher
Ficedula sapphira

Synonym: *Muscicapa saphira*

This small bird with tiny bill, has an orange-rufous throat and breast, and white belly. The breeding male is deep blue above with bright blue forecrown, the non-breeding male has an olive-brown head; both having bright ultramarine wings. The female is olive-brown above with rufous rump and tail. It is resident in the evergreen broadleaved forests of the northeast Himalaya, breeding during March–July at 1,600–2,100 m and descending to the plains in winter. The flashy, jewel-like blue of the wings is conspicuous when the bird swoops on insects in flight.

From T.C. Jerdon, *Illustrations of Indian Ornithology*, 1847.

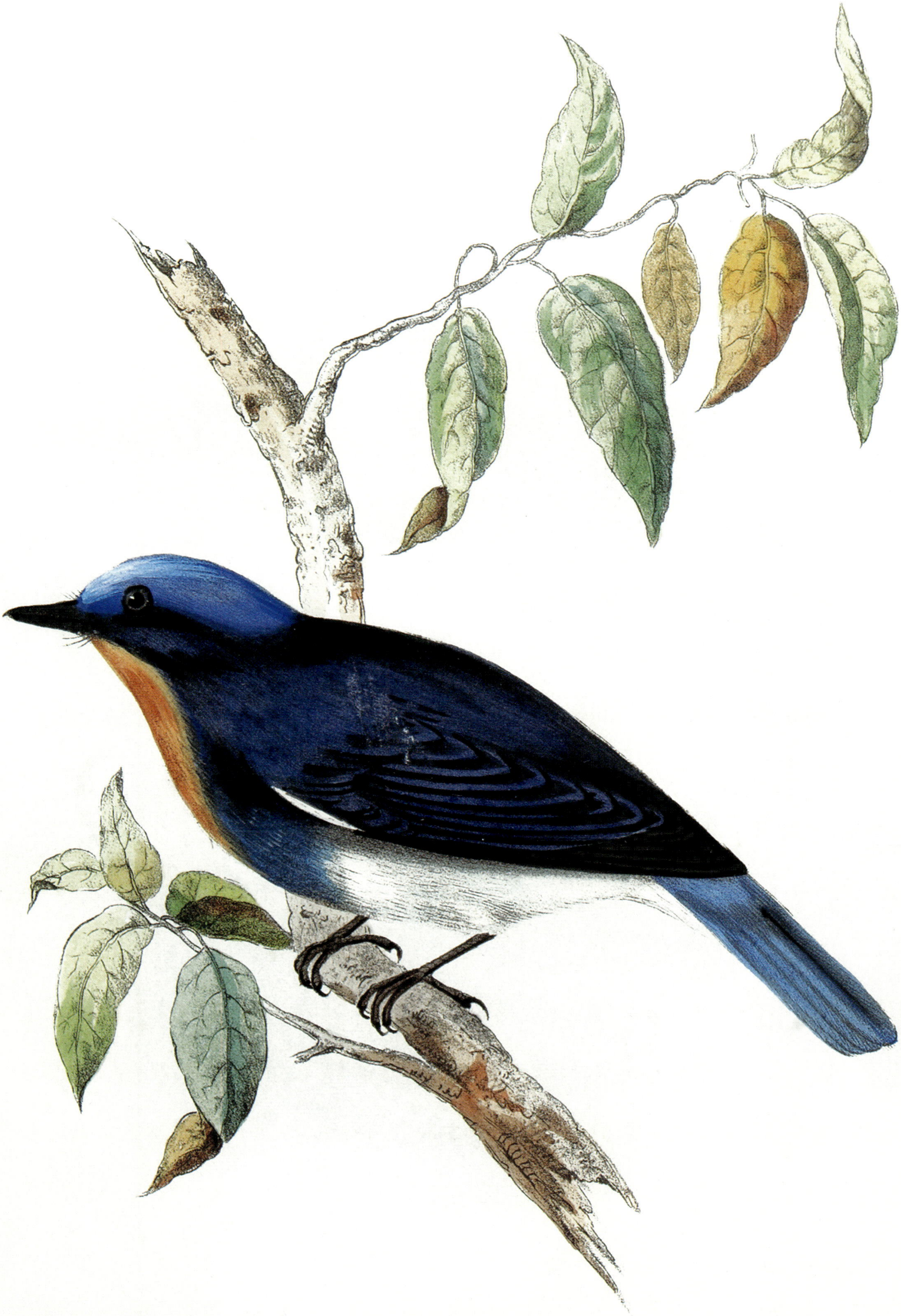

In fond memory of
the late Mrs Lily Sripad Rao,
from her granddaughter
Anusha Babbar

Ficedula sapphira

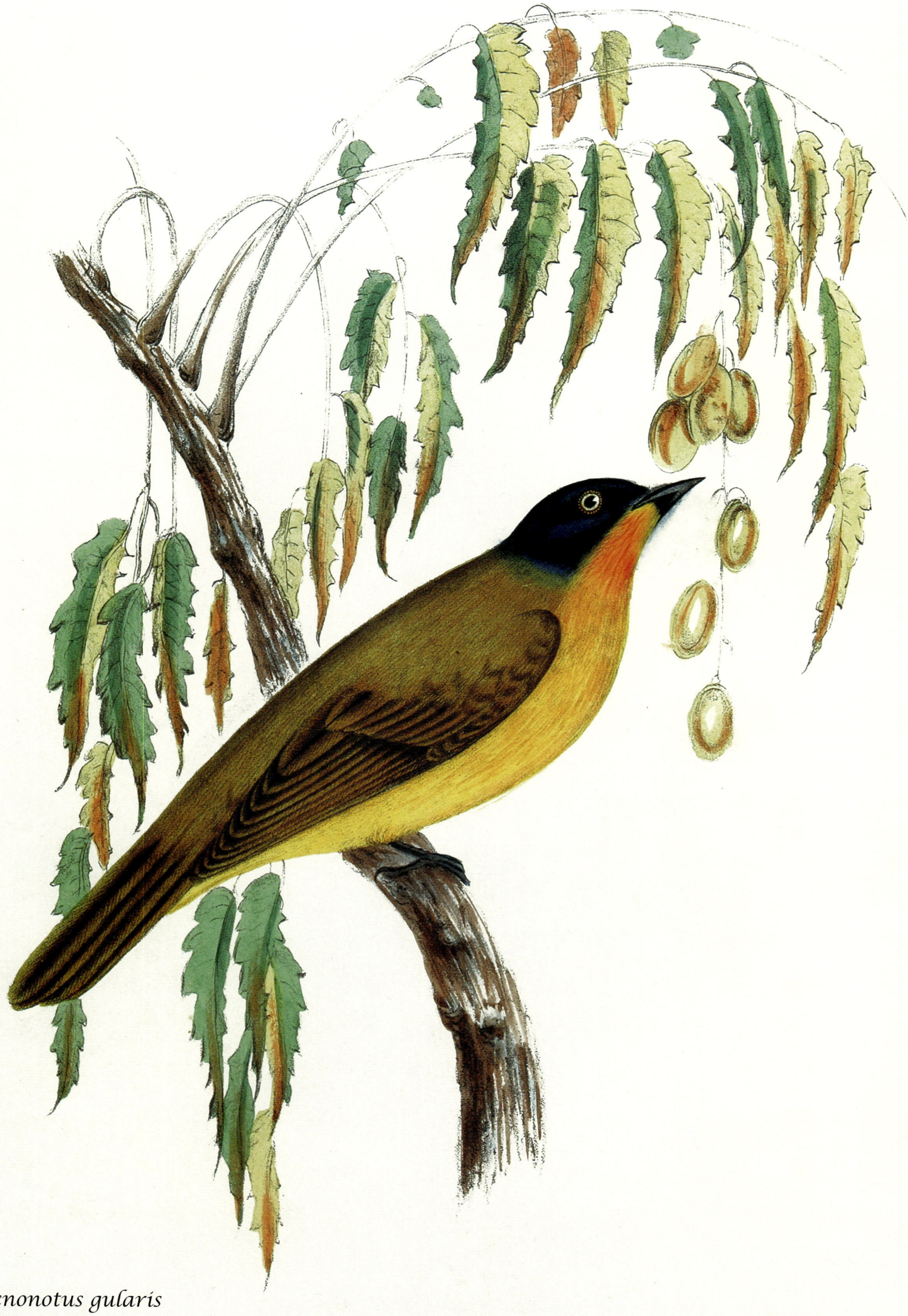

Pycnonotus gularis

COURTESY
CENTRAL BANK OF INDIA

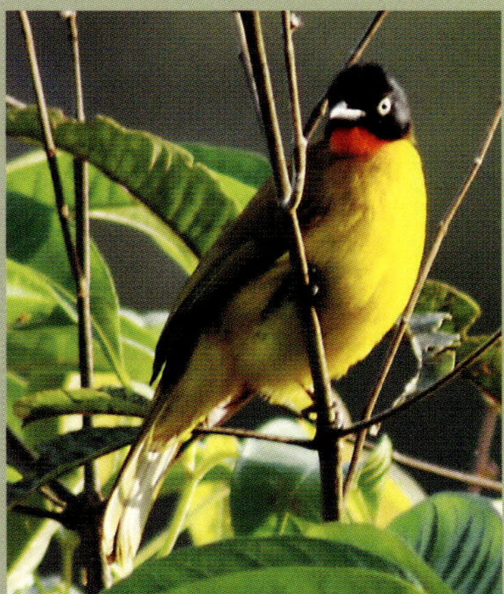

Rubythroated Yellow Bulbul, Flame-throated Bulbul
Pycnonotus gularis
Synonyms: *P. melanicterus gularis, Brachypus rubineus, Rubigula gularis*

Jerdon describes this bird as follows: "Head and cheeks pure glossy-black; plumage above yellowish-olive-green; a small chinspot black; throat beautiful shining ruby-red; the rest of the plumage beneath bright yellow. Bill black; legs greenish dusky; irides light yellow. This pretty species is found in the forests of Malabar, extending from Travancore to North Canara. It is found from the level of the sea to nearly about [700 m] of elevation. It is not common, frequents the more open spots and glades in thick jungle, and is usually found in the neighbourhood of water. It lives in small families, is sprightly and active, hopping about the smaller branches of trees, and uttering now and then its pleasant twitter, much in the manner of the crested Bulbuls. Its food chiefly consists of fruits and berries." (*Birds of India*, Vol. II, Part I.)
The photograph above was taken near Jog Falls, Karnataka.

From T.C. Jerdon, *Illustrations of Indian Ornithology*, 1847.

front posts and about 7 feet off the ground: I did so as quietly as I could and the light shone well on to the body.

I now felt as if my best chance had gone, but settled to sit right through the night till morning. At 8-30, after the moon had set, when looking towards the body I saw a grey mass beyond it which had not been there before, so I knew the panther had come at last and was more thankful than I ever remember being in my life. He looked most weird as of course he had come absolutely noiselessly, and he gradually got nearer and nearer and looked whiter and larger and more like a ghost than anything. When he got up to the body he half turned his side although still looking straight towards me; so I settled to shoot. I had purposely put an expanding bullet in the left barrel of my 450 express and a solid one in the right and meant to shoot both barrels at once (a finger on each trigger) as the lamp invariably goes out with the concussion of the air so that one cannot get in a second shot. I aimed at his body behind the shoulder as far as I could and pulled both triggers; a flash, a loud report, and total darkness, and then I listened and heard the panther crashing down the hill for about 5 seconds, then quiet, and then a groaning growl; so I knew he was hard hit. Nothing more could be done till morning so I went back to the village, and after a fine drink of milk tried to sleep on a bed of chaff thrown down on the ground under the eaves of a hut, but a rat which would try to walk over my face, numerous insects, and cattle and goats a few feet away in the hut hardly allowed me a wink the whole night. At sunrise we started off to the kill: we found blood very soon and tracked very cautiously: the tracking was easy, as there was a lot of blood but there was plenty of cover so one had to be careful. The panther had gone down to the stream bed and after about 15 minutes we saw the body stretched out on some rocks ahead with back towards us. I threw a small stone towards it but there was no movement so I went up to within 10 yards' distance. I then was pretty sure he was dead, but threw a rock at him to make sure, when to my surprise he lifted his head to look at me. I fired two solid bullets into his head and all was over. He was a splendid male 7' 3¼" long and the length of the middle toes in his fore feet showed he was the man-eater of last hot weather. One of the bullets of the night before had struck him in the stomach and the second (the solid one) had broken one of his hind legs.

When first he took to man-eating he would never touch any one unless he (or she) was lying down and unless it was quite dark with no moon or fire or light burning; under these circumstances if no one were sleeping outside he would go right into a hut if the door were open and lay hold of one of those asleep: if he happened to grip by the arm or leg and the victim yelled out then probably the panther would let go and try his luck elsewhere, but if he laid hold at once by the throat so that there could be no cry, then the man, woman, or child would be carried out of the hut before the others sleeping near knew anything at all.

He had got bolder however lately and the kill before the last was in broad-day-light (in Gad forest on the Dangs boundary).

The following list shows the damage done by the panther in the Mulher Valley: he was probably the beast which did the man-eating in the Dangs.

Number of people killed at once	24
Number of people who died subsequently from being mauled	6
Total deaths in Mulher Valley	30

Number of people who were mauled but recovered, 11.

L.S. OSMASTON, Deputy Conservator of Forests
Camp Bhowli, Nasik District, January 1903

From *JBNHS* Vol. 15, No. 1.

A Gaur Attacking and Killing a Man without Provocation in the Central Provinces

As I believe instances of unwounded bison making a determined and unprovoked attack on human beings are rare, possibly the following account may be of interest.

At the end of May last I went out after bison in the Central Provinces. In sending me the permit, about 6 weeks previously, the Forest Officer told me there was a savage old bull at the place I was going to that had recently killed a native. One of my trackers turned out to be the brother of the man killed, and he told me – and other local natives corroborated his statement – that his brother had been quietly cutting grass in a glade in the forest when a bull bison deliberately charged him from the other end of the glade, they said, 150 yards off. The man climbed a bamboo clump to escape, when the bull went for the clump, more or less broke it down, knocked the man out of it and killed him. According to their account it was a most deliberate, savage, and unprovoked attack. I was fortunate enough to get this bull after four days' hard tracking with a shade temperature of 98° to 102°. He did not then appear to be particularly savage. Though wounded and being continually "jumped", in dense jungle where it was impossible to go noiselessly on account of the dead leaves, he only twice showed fight.

As near as I could tell, he was about 12 or 13 years old and had a very perfect and massive head with a spread of 40". He was in prime condition and showed no signs of having been previously wounded but there were not enough men to roll him over so as to examine both sides.

H.V. BIGGS, Lt.-Col., R.E.
Mhow, C.I., September 1907

From *JBNHS* Vol. 18, No. 2.

Red Lotus Tree, Evergreen Magnolia
Magnolia insignis
Family: Magnoliaceae
Synonym: *Mangliletia insignis*

This evergreen tree was found by Wallich "at Sankoo in the great valley of Nepal and on the mountain Sheopur towards its summit". It grows in Assam, Meghalaya, Manipur, Sikkim, Nepal, Myanmar, and southern China between 900 and 2,600 m. According to Wallich, each tree is of vast size, 7.5 to 12 m high, the trunk 1–2 m in diameter. The narrowly pointed, lanceolate leaves are green all year around; they droop gracefully over the stem. They are dark green above and pale and bluish beneath. When in full bloom the tree is one of the most magnificent sights to behold. The flowers are produced in vast numbers among the dark green glistening foliage, The petals are cream or pink-tinged within and pale pink to red on the outside. They are succeeded by oblong capsules, from which the scarlet seeds are suspended by delicate white threads. The buds have a beauty of their own before they open. They are reddish-green and stand erect on the branches like candles.

From N. Wallich, *Plantae Asiaticae Rariores*, 1832.

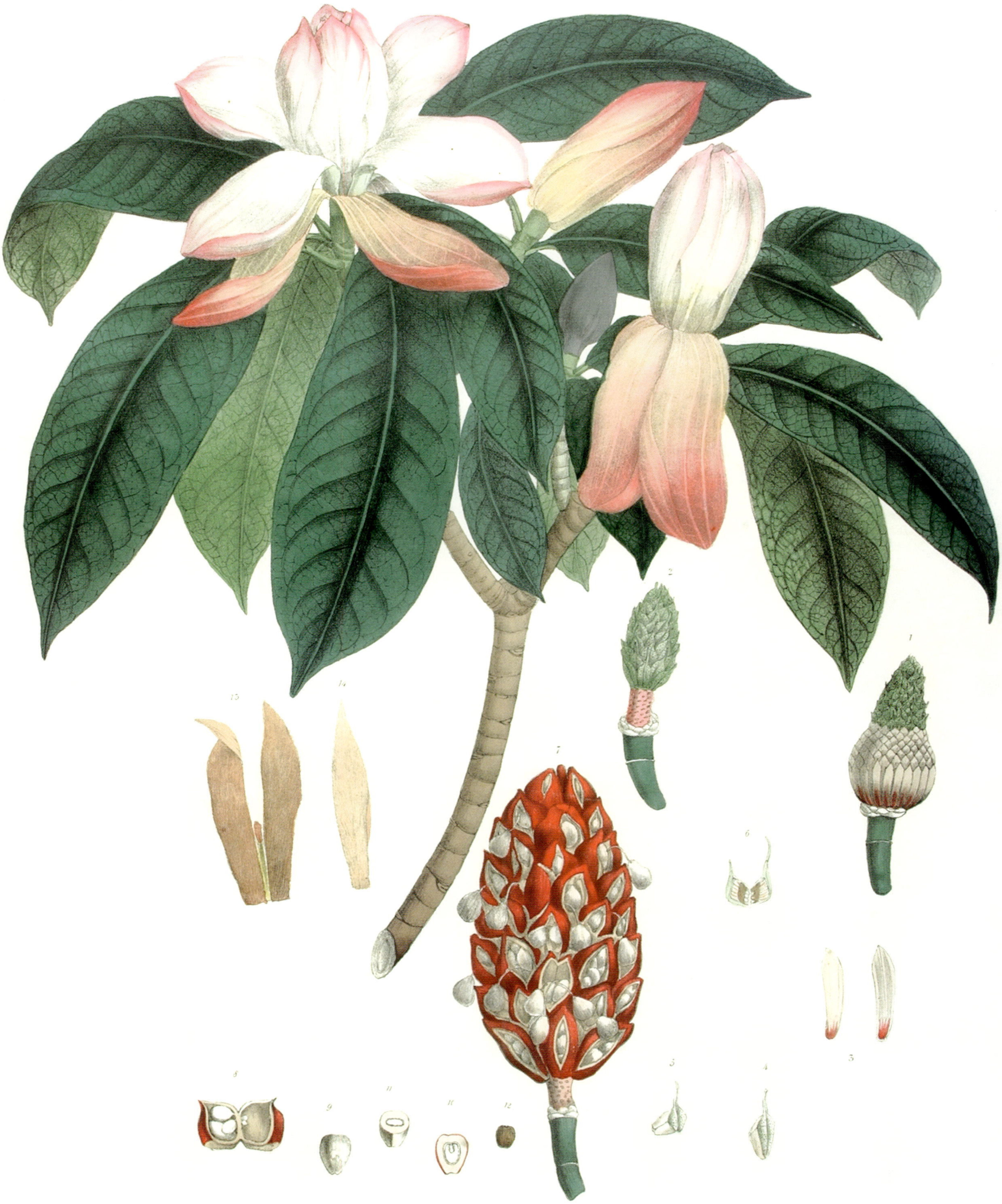

Magnolia insignis

In ever-loving memory of
our dearest father
Barzo Irach Taraporewala
from Soonoo, Irach, and Freya

Leopard Cat, Bengal Cat
Prionailurus bengalensis
Synonym: *Felis bengalensis*

The Leopard Cat was originally described from an individual that swam up to and boarded a ship in Calcutta – hence the name *bengalensis*. It is found in evergreen and mixed forests practically all over India, in Pakistan, and eastwards through Myanmar and southern China into Southeast Asia. It is about the size of a domestic cat but with rather longer legs, and its colour and markings give it the aspect of a miniature leopard. Rounded black ears marked with a white patch on the outer surface distinguish the Leopard Cat and the Rusty-Spotted Cat from other members of the genus. The Leopard Cat is yellowish above, white below, ornamented throughout with black or brownish spots. Both colour and pattern are very variable in the species – the ground colour may show a tendency to grey or rufous and the spots, which usually run in regular lines along the body, may coalesce into uninterrupted longitudinal bands. This beautiful forest cat is an expert tree-climber, and preys upon small birds and animals. It is solitary and nocturnal in habit, and seldom seen. After a gestation period of 65–70 days females give birth to 3 or 4 cubs.

From *JBNHS* Vol. 37, No. 4 (Supplement: *The Wild Animals of the Indian Empire and the Problem of their Preservation*, Part II).

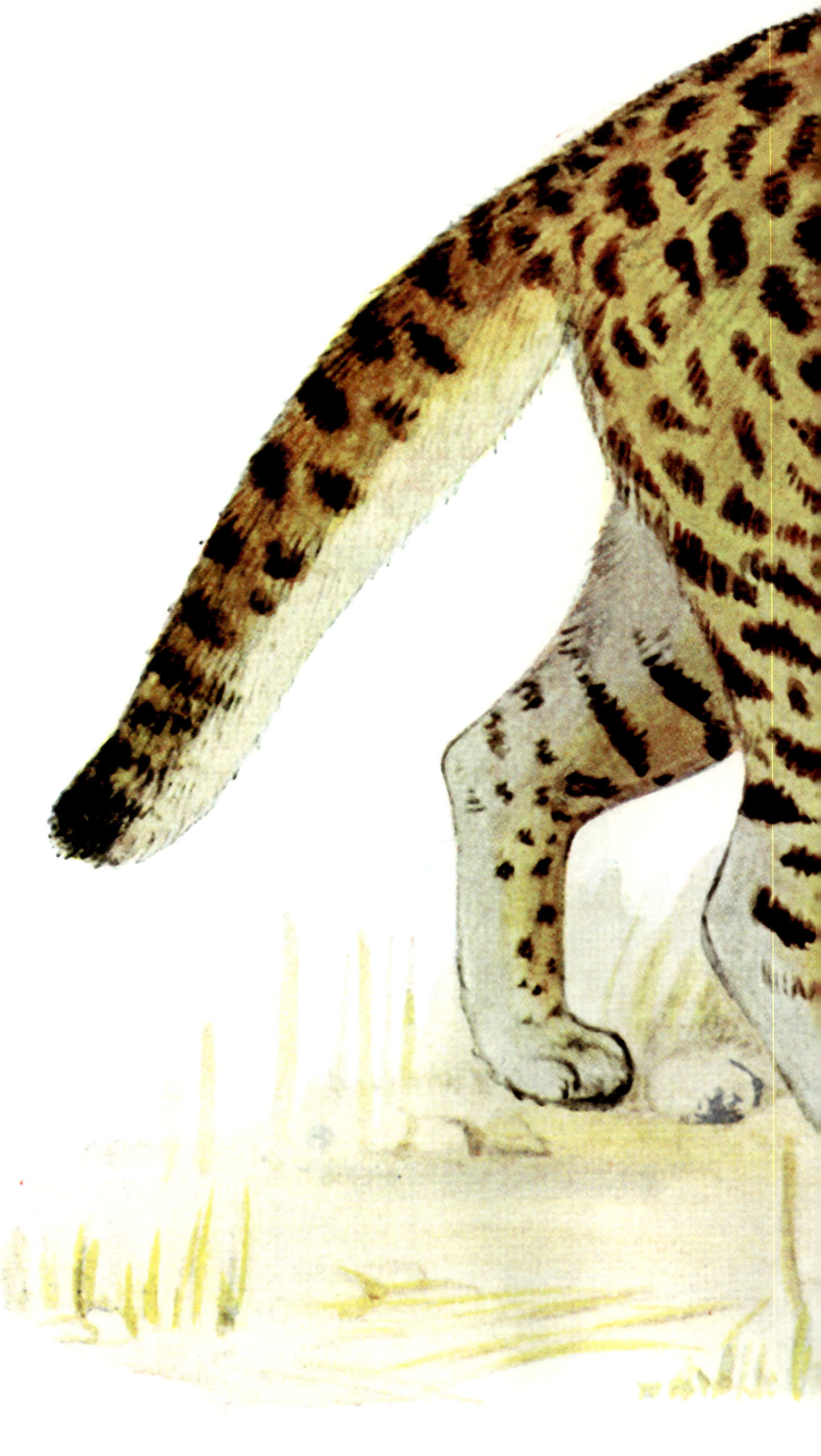

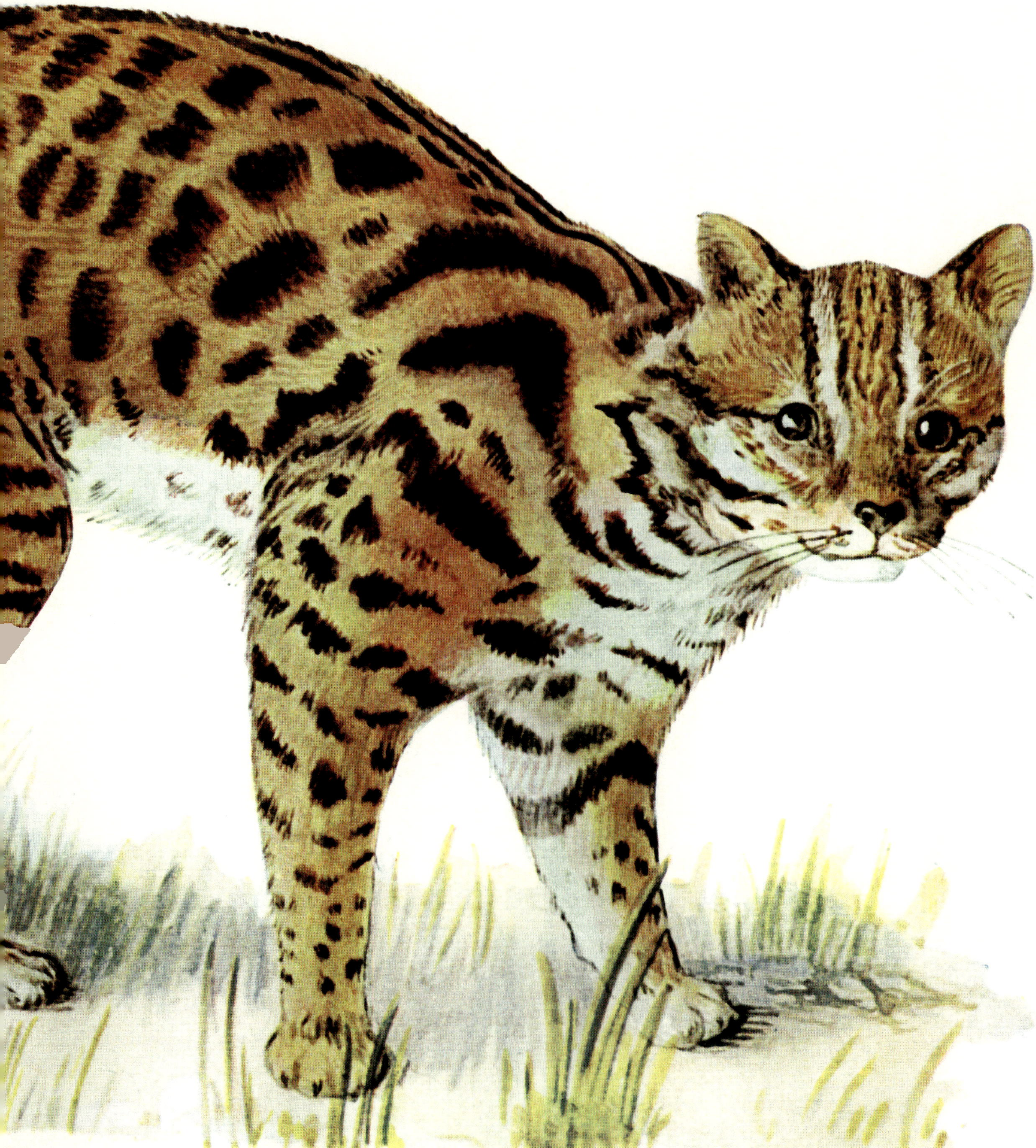

Courtesy HSBC
The Hongkong and Shanghai
Banking Corporation Limited

Leopard Cat (*Felis bengalensis* [*Prionailurus bengalensis*]) in Captivity

"*Felis chaus*" and "*Felis bengalensis*" being proverbially wild and untameable, the following account of a "*Felis bengalensis*" kitten having been tamed may be of interest.

She was caught by a labourer, who said that a second kitten had escaped with its mother, and sent to us by a friend, when about a month old, in August 1919; and was fed every few hours, day and night, with cow's milk diluted with water out of a tea spoon by my wife, who has a wonderful faculty for rearing and taming wild animals. The kitten soon took to the "spoon-feeding", clasping the spoon with both fore paws, and sucking vigorously, and throve well. When old enough, she was fed on raw meat with occasionally a mouse, small bird, or rat. One day when sitting on my wife's lap at breakfast, she seized a bit of cooked meat off the plate, and was allowed to eat it, but it disagreed with her little inside badly, and she nearly died; but my wife's assiduous nursing pulled her through. She used to sleep either at the foot of our bed or on top of the mosquito curtain, generally the latter, and became so tame and gentle that she was allowed full liberty about the house and garden; becoming great friends with our Airedale terrier, and it was very amusing seeing them playing together. When she was about 8 months old we went to stay at a neighbour's estate during his absence on leave, and unfortunately he had left 3 domestic cats in the house, which we hoped might make friends with our kitten, but they refused her friendly advances. And one morning "Kittycat" (as she was named) went out as usual, after sleeping on top of the mosquito curtain all night, and never returned for a week. Then my wife noticed that the cats had taken to sitting on the verandah outside our bedroom window, apparently on the watch: so we had them shut up at night, and at midnight of the 9th day, to our great joy, back came "Kittycat", very hungry and thin, but as tame as ever. In June this year my wife took her up to Kotagiri with her, sitting on her lap most of the way in the train, and the Motor Car! After they had been there some weeks a large stray domestic cat turned up at the house and drove our pet away. However, after two weeks she came back one night, and woke my wife up by shoving her cold little nose against her face, and our children also woke and hurried off joyfully to get her milk and meat. But to everyone's sorrow she was driven away again after a few days by the cat, and never returned, though she was seen several times in the neighbourhood by different people. The last news of her was, that some Badaga boys in the village just below the house had found two lately born Leopard-cat kittens in the scrub between the village and the garden, and had tried to rear them, but that they had died. A great disappointment to my wife, as if they had only told her at once, she would have gone to the place and watched, and probably recovered "Kittycat" and her kittens.

A.M. KINLOCH, F.Z.S.
Palagapandy Estate, Malabar, 10th October, 1920

From *JBNHS* Vol. 27, No. 3.

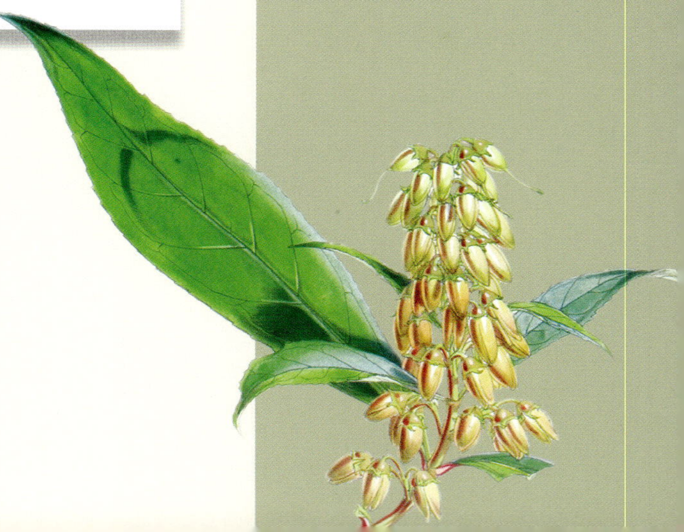

The Hunting Leopard (*Cynaelurus jubatus* [*Acinonyx jubatus*])

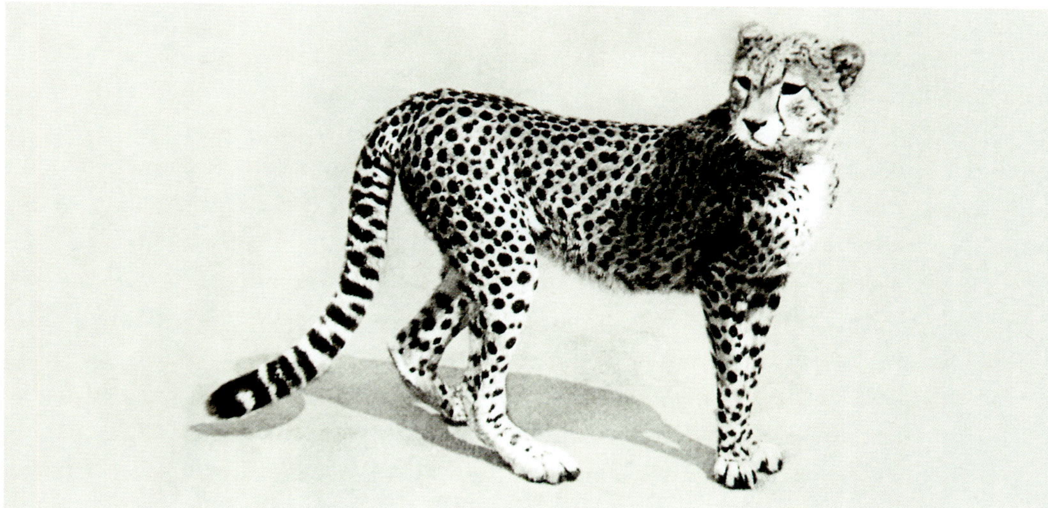

F.W. Bond

I have never had the fortune to see the Hunting Leopard in the feral state. I heard of one in the Buldana District of Berar in 1912. I beat up the animal's quarters, but found only tracks. It was said that its mate had been captured by *pardis*. These animals appeared to be less uncommon in Berar than in many localities. I saw in 1890 the skins of three which had been shot in one beat in the Melghat Forest in the Satpura Hills North of Ellichpur by Mr. Ballantyne of the Forest Department. Captain Winter of the Hyderabad Contingent Artillery saw two or three and shot one when sitting over a kill or a tethered goat at Damangaon near Ellichpur in 1894. Another was shot in 1895 in the Wun District of North Berar by Captain Barnard, 4th Lancers, Hyderabad Contingent, when it came down to drink at a pool of water.

In an article in the *India Sporting Review* for February 1857, there is a reference to Chesney's "Journal of the Euphrates Expedition", in which this species is said to be more numerous in Asia Minor than in Persia and Mesopotamia: its occurrence is also noted in Arabia and in the vicinity of Aleppo. The writer of the article says that the chief supply of these animals in Upper India is from the Jeypur District. According to "Buchanan Hamilton", it is found in most of the hilly parts of India, but is nowhere very numerous except near Hyderabad, Deccan. It is stated by Mr. Ure, Surgeon at Hyderabad, that Hunting Leopards were numerous near that place, and live in holes among the rocks on the hills, or rocks that are near the plains which the antelope frequent.

Sir Samuel Baker, in his "Eight Years' Wanderings in Ceylon", published in 1855, says this animal is common there and "frequently caught at Newera Ellia". The late Sir Montagu Gerard told me he had ridden down and speared this species in Central India.

R.G. BURTON, BRIG.-GENL.
Charlton Kings, England, June 1920

From *JBNHS* Vol. 27, No. 2.

IN LOVING MEMORY OF OUR DEAR SKIPPY
OUR FAITHFUL DOG AND A FAMILY MEMBER
WHO DEFENDED US AND GAVE US PROFOUND LOVE
– ARUNA & CHANDRAKANT PATEL, HOUSTON, TX
DHRUV, KIRAN, DEV, & MAYA AMI, HEMAL, VEENA, & SAIRA

Porcupine's Mode of Attack

In 1907, I was stationed at Fatehgarh in the United Provinces and driving home one evening from the Club with my wife we saw in the waning light our dog "Peggy" a bull-terrier, then about two years old, rush after a porcupine that crossed the road in front of us. The porcupine ran for some distance with the dog in full chase over the maidan when the former suddenly stopped and ran backwards into the dog who gave a yelp and commenced rolling on the ground. I at once jumped out of the trap and got hold of her pulling out a quantity of quills from the head and chest, all of which I kept and took back to the bungalow. One quill we noticed was the butt end and this circumstance together with the uneasiness of the dog for the next few days convinced my wife that she still had a quill in her somewhere. There being no Veterinary Officer in Fatehgarh my wife wrote to the young Surgeon attached to the 2/10th Gurkhas then quartered in the Cantonment, asking him to have a look at the dog. After a long examination he at last located the quill and the next day the dog went up with the small dogboy to the Hospital, and our friend pulled out a half quill of 4⅛" long. The butt end measures 3¾" so that together they make a whole quill of 7⅞ inches long. This gives a guide to the diameter which is ¼ inch. The dog, an extremely intelligent beast, seemed to know that the probing was being done for her good, and never moved a muscle the whole time, but licked the Doctor's hand when it was all over.

I may add that I once shot a big male Panther on the Sendra-Ghat some thirty miles from Ajmere and found on skinning him that his two fore feet were full of porcupine quills, pieces from 1 inch to 2½ inches long being embedded in the pads, and some even up the fore arm. The poor beast must have suffered agonies, but I think that it is often so difficult for a Panther and for a Tiger to obtain a meal, that they go for the first animal they come across which may be with dire results to themselves.

A peculiarity of the Porcupine is that his quills drop out a short time after they have been damaged even in the slightest manner. I presume that the time taken for dropping out depends on the damage done. The quill is then replaced by a new one which probably forces the old one out. I have on many occasions picked up these quills and on examining them invariably found the flaw, mostly made I should imagine by projecting rocks, for *Hystrix cristata* [*Hystrix indica*] is a "cave dweller" although he makes an earth when not in a rocky country.

The natives have the idea that he has the power of throwing quills like darts at his enemies, and probably it is got from finding these old quills lying about.

REGINALD H. HEATH
Chipstead, Surrey, June 1918

From *JBNHS* Vol. 26, No. 1.

Peregrine Falcon,
Shaheen Falcon
Falco peregrinus peregrinator
Synonym: *Falco shaheen*

A powerful, alert, fast-flying broad-shouldered falcon, slaty-grey above with black head, nape, and prominent moustache-like stripes. The throat and breast are white, the rest of the underparts reddish-orange. The wings are pointed. As in other raptors, the female is larger than the male. It is seen solitary or in pairs on cliffs, rock pinnacles, hill forts, minarets, etc. or gliding on its powerful wings, swooping suddenly to catch smaller birds, often in mid-air. In the breeding season pairs indulge in spectacular aerial display, stooping on each other, looping, turning, and twisting all the while in speedy flight. It is resident throughout the hilly portions of India, in the Himalayas from Himachal Pradesh to Bhutan and the southern Assam hills, forested hills of central India, Western and Eastern Ghats, and in Sri Lanka, Pakistan, Bangladesh. During the nesting season (March–May) a nest of twigs, wires, grass, is assembled on a high rocky ledge, in which are laid 2 or 3 pinkish-buff to pinkish-red eggs with brick red or purple markings.

From T.C. Jerdon, *Illustrations of Indian Ornithology*, 1847

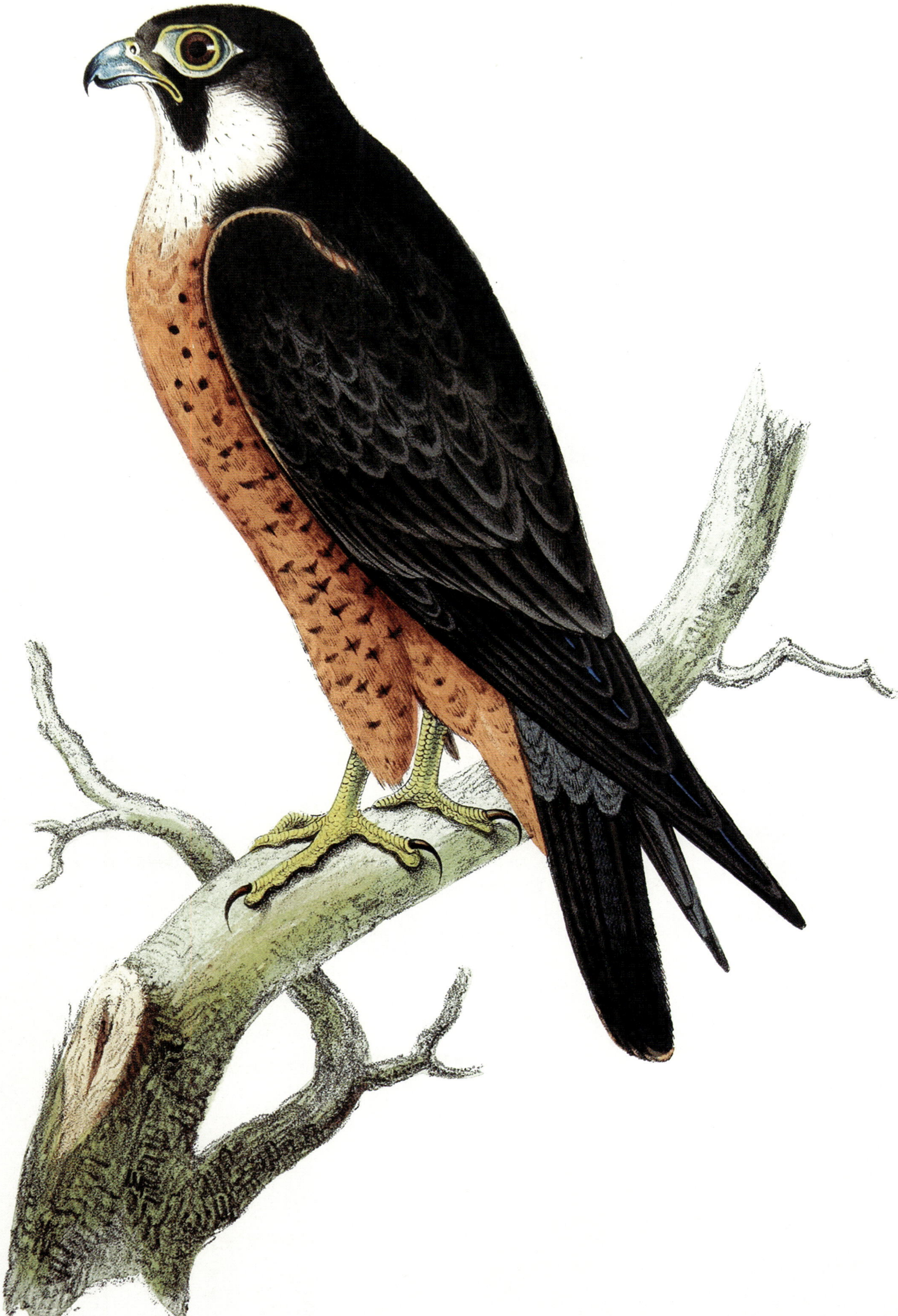

Falco peregrinus peregrinator

Courtesy
Mahindra & Mahindra Ltd.

Recovery after a Bite by a Russell's Viper

On August 6th, 1920, a man was brought to the Indian Station Hospital, Sehore, stating that he had been bitten by a snake about an hour and a half previously. The man said that while carrying a bundle of freshly cut grass on his head, he felt something moving on his arm, he looked and found that the snake had reached his forearm, he hit it with his hand, the snake bit him, and he killed it with a stick.

On examination two small punctures three inches apart were found on the right forearm. The man was quite calm and normal pulse 86. The punctures were scarified and rubbed with crystals of Potassium Permanganate 40 c.cs. Antivenine were injected subcutaneously. The ligature which the men with him had tied above his right elbow was removed.

The snake which the man said had bitten him was shown to Col. Luard who identified it as a Russell's Viper about a month old. The snake was 14 inches in length.

I do not think that the treatment had anything to do with the man's recovery. That was evidently due to his receiving a quantity of the poison too small to produce toxic symptoms.

Since coming under observation the man has shown no abnormal symptoms.

J.B. MOLANY, Capt., I.M.S
Indian Station Hospital, Sehore, 12th August, 1920

From *JBNHS* Vol. 27, No. 3.

Tigers in Trees

In the *Bengal Sporting Magazine* for 1834 it is recorded that a tiger, struck on the back of the head by a bullet on the previous evening, was found quite dead on the lower branch of a pipal tree at least fifteen feet from the ground. The marks of his claws on the bark were so clear that there could be no difficulty in deciding that he had scrambled up cat-fashion. He then ran along the branch, at the extremity of which he lay down across it, his legs on either side being kept in balance by small twigs, there he died. This animal had not attained his full growth. An instance is given in the *Bengal Quarterly Sporting Review* for 1843 of a hunted tigress mounting to a bough twenty and a half feet from the ground to seize a man, the man was wounded, and the tigress then lost her balance and fell.

In 1856 "Teutonius" recorded two such instances in the *India Sporting Review.* In the first case he had news of a tiger sitting in a tree where it was said to be blockaded by villagers. He rode to the place, which was within five hundred yards of a village, and there saw a large

Padam, Wild Himalayan Cherry
Prunus cerasoides
Family: Rosaceae
Synonyms: *Cerasus puddum, Prunus pudum, Prunus majestica*

This is a handsome deciduous tree found in the Himalayan forests from Himachal Pradesh to southwestern China at an altitude of 1,200–2,400 m. It grows up to 30 m in height; its bark is reddish-brown, smooth, and peels off in thin horizontal strips. The leaves are elliptic, pointed with toothed margin, 5–7 cm long. The leafstalk bears one or more reddish glands. The pink or crimson flowers appear in October–November or sometimes January–February, and fade to white; they are 2–4 cm across, with five petals and long stalks, clustered at the ends of branches. The fruit is yellow ripening to red, ovoid, 1.2– 2 cm long, acid in taste, with a single wrinkled stone. The fruit is eaten and the stones are strung into rosaries and necklaces. Gum is obtained from the trunk, and a green dye from the leaves and fruit. The branches of this tree are often used to make walking sticks.

From N. Wallich, *Plantae Asiaticae Rariores*, 1832.

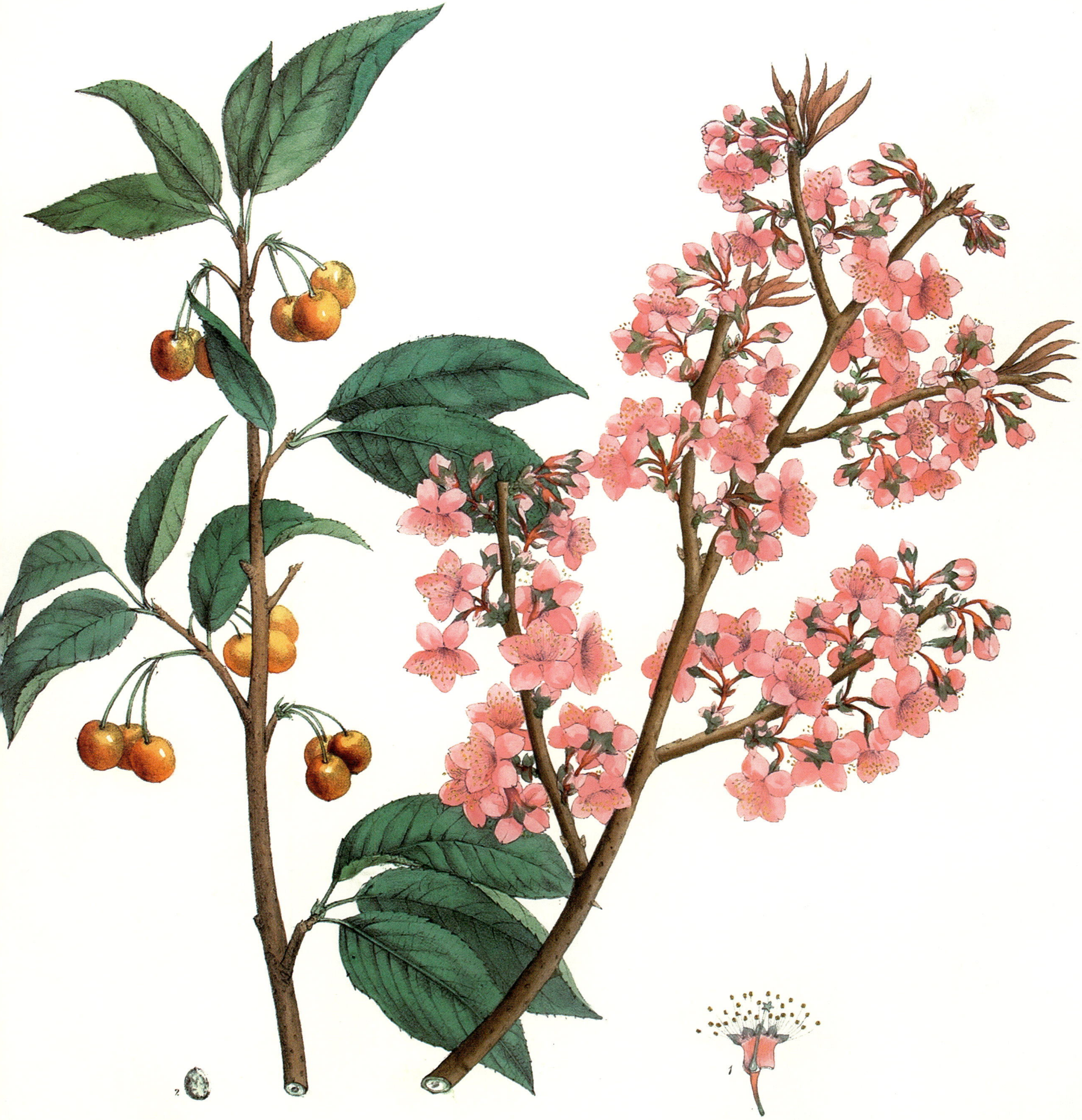

Prunus cerasoides

Courtesy
S.G. Jhaveri Consultancy Pvt. Ltd.

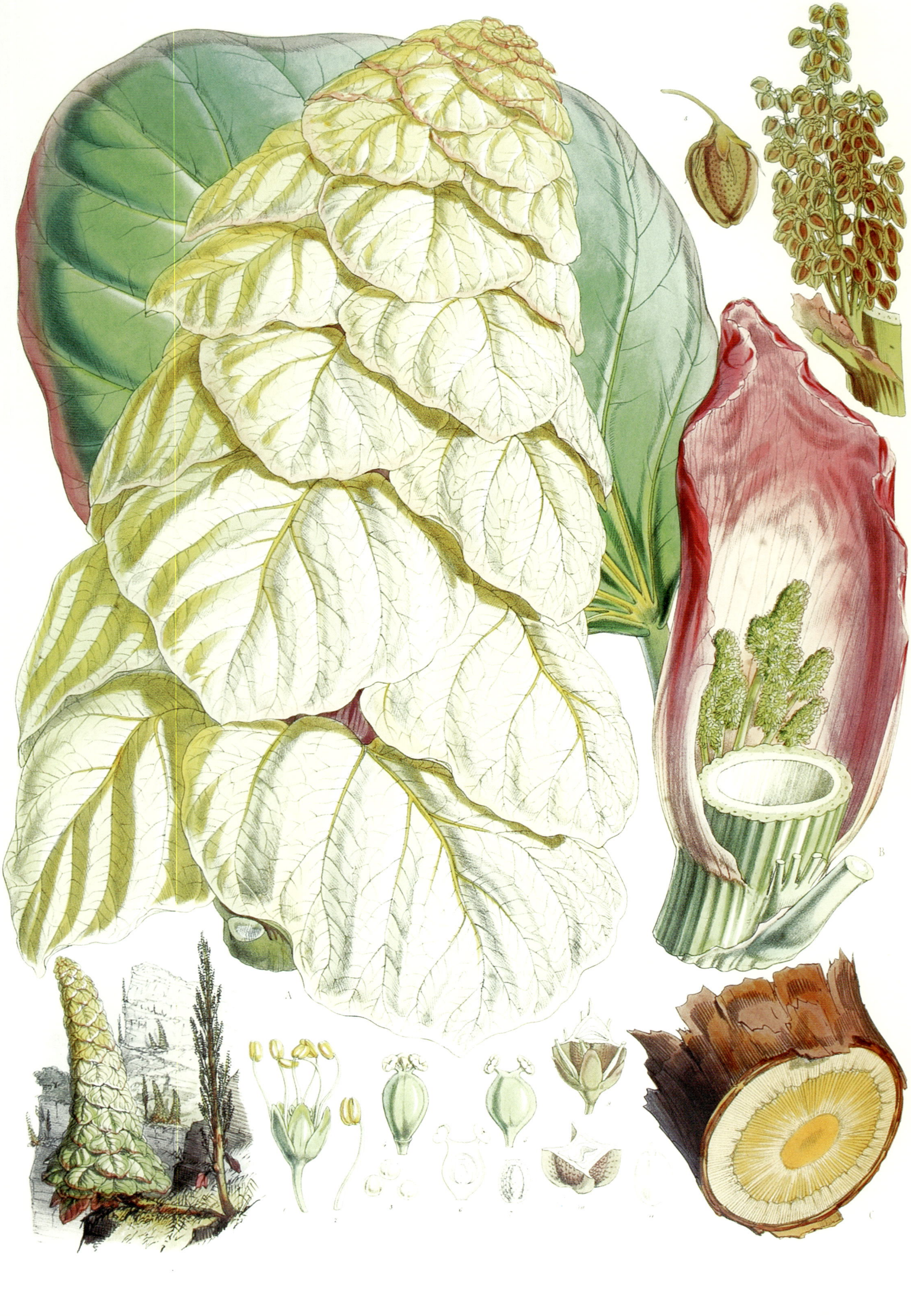

Rheum nobile

In memory of the late Pandit Ravishankar Shukla,
Freedom Fighter, Premier of C.P. & Berar (1936–47),
and Chief Minister of Madhya Pradesh (1947–57)
– Pratibha Pande

Noble Rhubarb,
Sikkim Rhubarb
Rheum nobile
Family: Polygonaceae

Rheum nobile was considered by Hooker the most handsome herbaceous plant in Sikkim, and he described it thus: "On the black rocks the gigantic rhubarb forms pale pyramidal towers a yard high, of inflated reflexed bracts, that conceal the flowers, and overlapping one another like tiles protect them from wind and rain; a whorl of broad green leaves edged with red spreads on the ground at the base of the plant, contrasting in colour with the transparent bracts, which are yellow, margined with pink. It is called 'Pechuca' and the acid stems are eaten both raw and boiled; they are hollow and full of pure water; the root resembles that of the medicinal rhubarb but is spongy and inert; it attains a length of [over a metre] and grows as thick as the arm. The dried leaves afford a substitute for tobacco...." The plant is found in the inner ranges of the Himalaya from eastern Nepal to southeastern Tibet, at an altitude of 3,600–4,500 m. It flowers in May–July, the branched clusters of tiny green flowers being 7.5–10 cm long. The small fruit has three or four wings. On a clear day the plant can be seen from a long distance, being 90 to 120 cm high. George A. Gammie mentions in the *Gazetteer of Sikhim* (Risley 1894), "In the late military expedition to the Tibetan frontier a batch of plants of this rhubarb growing in a sequestered valley were mistaken in the distance, one misty morning, for a surprise party of the Tibetan army."

From J.D. Hooker, *Illustrations of Himalayan Plants*, 1855.

pipal tree round which people were picketed. A spearman was standing almost underneath it. When he approached within a hundred yards of the tree "there appeared, standing on a sturdy branch high aloft in the tree, the tiger, erect and calm and fearless with black, yellow, and white colours in stripes, looking beautiful in high relief". The height of the animal above the ground was found to be twenty-five feet when subsequently measured. On receiving a shot he caught a lower branch with his two arms in falling, hung for a minute, and then dropped dead. The villagers had found the animal in the morning asleep under a mango tree. On being roused he at first tried to hide himself in the drain of a tank, and eventually mounted the tree, which from its size and low stout branches was easy of ascent. They said that once during the day he tried to descend, but that he was driven higher up by their shouts. This was a young tiger about eight feet in length.

The second case, which was very similar, occurred on the 17th May 1856 near the village of Tuchezra. A tiger attacked a buffalo near the village at about daybreak, but was driven off by the herdboy. The villagers then turned out and the tiger got up a pipal tree, where a dozen villagers remained to prevent it getting down. The sportsman approached to within forty yards, mounted on an elephant, and related that as he stood up in the howdah with his rifle levelled at the tiger's chest, it appeared to be a few feet higher than the rifle, and he calculated it to be about nineteen feet. The tiger, which was killed with three shots, proved to be a male measuring nine feet eleven inches. "How he got up the tree I could not well make out, as with the exception of an intervening branch, and many large notches, the trunk, measuring in circumference at least 38 feet, was at least fourteen feet high, and then branched out like a banyan tree, and there was plenty of room where he could stand and lie."

In Volume I, No. 5, of the *Oriental Sporting Magazine* for 1866, it is related that a herd of buffaloes drove a wounded tiger into a rhododendron tree, but no other particulars are given. In Volume IV, No. 41, of the same magazine for 1871, an instance is given of a tigress charging a man in a tree, and springing up eleven feet six inches from the ground, as proved by the claw marks. It is also said that she began tearing the bark off the tree with her teeth at ten feet four inches from the ground.

Mrs. Colin Mackenzie in her book "Life in the Mission Camp and Zenana", says: "Captain J. told me he never knew a tiger up a tree but once; that was at Hingoli. An unfortunate man who thought himself quite safe cried out 'here she comes'! The tigress heard him, went up the tree, pulled the man down, and bit him on the knee so severely that he died in hospital soon after. It was so remarkable an occurrence that the tree was cut down and brought into Hingoli as a curiosity."

But the most remarkable instance of tigers climbing trees was recorded in the *South of India Observer* in December 1870, as follows :–

"More tigers! Now that our monsoon is over, or greatly moderated, our sportsmen have been able to resume operations against the feline race. We learn that two gentlemen killed two tigresses last week, one on the 2nd instant, the other on the 5th. They may be called the Peermund and Aniculmund tigers. Both afforded considerable sport, and one carried on such games as possibly never were heard of before in tiger shikar. It appears that the one alluded to was found in a small sholah, about 200 yards long and 50 or 60 broad. The villagers and dogs had scarcely been put in at the top of the beat when a sudden screaming and skedaddling among our canine friends was heard, which showed that there was some animal of dangerous proclivities inside. This had hardly passed when a huge tigress bounced out at the foot of the sholah, immediately going heels over head in a hidden nullah which she had not seen. This was all so instantaneous that there was barely time for one of our sportsmen, who was seated calmly on the grass within 25 yards of where this happened, to blaze a shot at the tigress as she flashed back into the sholah. The men continued beating down and presently Stripes again showed for a moment at almost the same place, and another hasty shot was fired. She then disappeared for some time, and next showed herself near the top of the sholah and tried to escape in another direction, but the villagers made such a din that they drove her down again. Now comes the curious part of the story. One of the shikaries, watching at the top, spied Mrs. Stripes suddenly appear on the top branches of one of the highest trees in the sholah. He immediately called out, 'the tiger has gone up a tree.'

"This was good news, and one of the sportsmen ran up the side of the sholah and sure enough saw Mrs. Stripes laid out on a branch at least thirty feet from the ground. To knock her off this perch with a brace of bullets took but a moment of time, and down she came with a thundering crash, apparently lifeless. The beaters began to cheer, thinking their work was over; but they say a cat has nine lives and so it proved on this occasion. She presently began growling and snarling at the dogs that had got round her. The gentleman who had been busy with her, as she was nearest his side then crawled into the sholah and presently saw Mrs. Stripes get on her legs and move off; but the brushwood was so thick that it was impossible to see in what direction she was coming, so he skedaddled. He had hardly got out, when to his amazement he saw Stripes shinning up the same tree, just as any ordinary house-cat would. The tigress got back to almost the same part of the tree as before and stood on the fork of two branches upright, looking down, and exposing her great chest. To tumble her off again was a momentary affair, and down she came with a tremendous crash. The sportsmen examined the tree after the battle was over, and it proved to be perpendicular for about 25 feet. There were the claw-marks, showing that she had gone up the same tree twice, once wounded, as seen by the blood. The claw-marks were mere scratches, and how this great animal could climb a perpendicular tree is a perfect marvel. Everyone has seen a house-cat run up a post or tree, and this huge tigress did this with just as much ease. From this experience we might suppose that tigers drop on their prey from branches, and also reconnoitre their game from

Ativisha, Monk's Hood, Wolf's Bone
Aconitum heterophyllum
Family: Ranunculaceae

A perennial herb found in alpine and sub-alpine regions of the western Himalaya at altitude 1,800–4,500 m. The plant is up to 50 cm tall, with an erect stem. The helmet-shaped flowers are 2.5–3 cm, bluish with purple veins, and appear in August–September. This plant is known for its medicinal properties. It is the source of the oldest aconite drug in India, a valuable febrifuge, a bitter tonic, given after malarial and other fevers. It is recorded as having been used in the treatment of diarrhoea, dysentery, and chronic gastric disorders. The nomadic tribes of the Himalayan region collect the roots and store them for medicinal use. The powdered root mixed with honey is given to children suffering from cough, cold, or vomiting. The root powder is also used to alleviate toothache caused by cavities.

From J. Forbes Royle, *Illustrations...*, 1839.

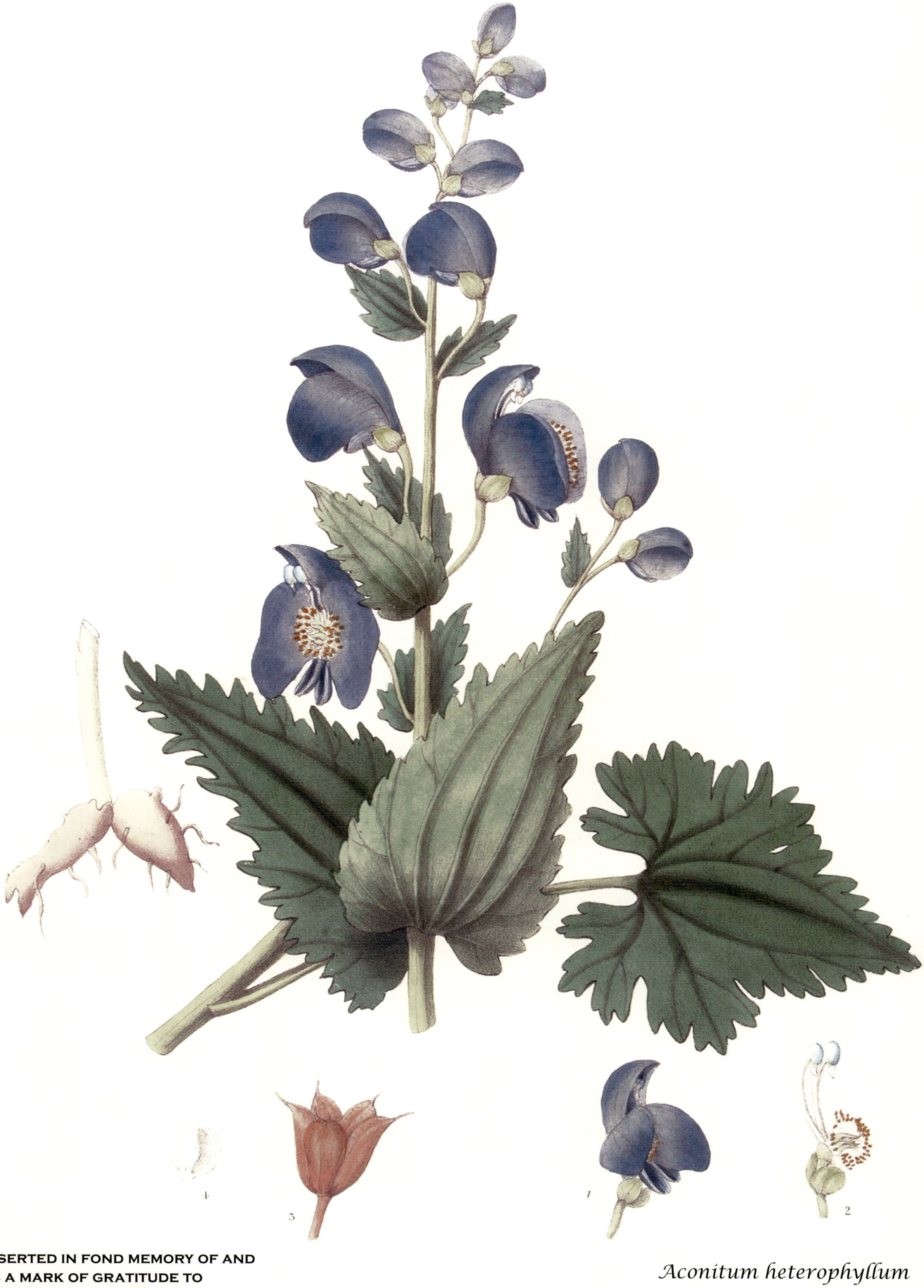

Aconitum heterophyllum

INSERTED IN FOND MEMORY OF AND
AS A MARK OF GRATITUDE TO
THE LATE PROFESSOR P.V. BOLE
– DHUN PANTHAKI

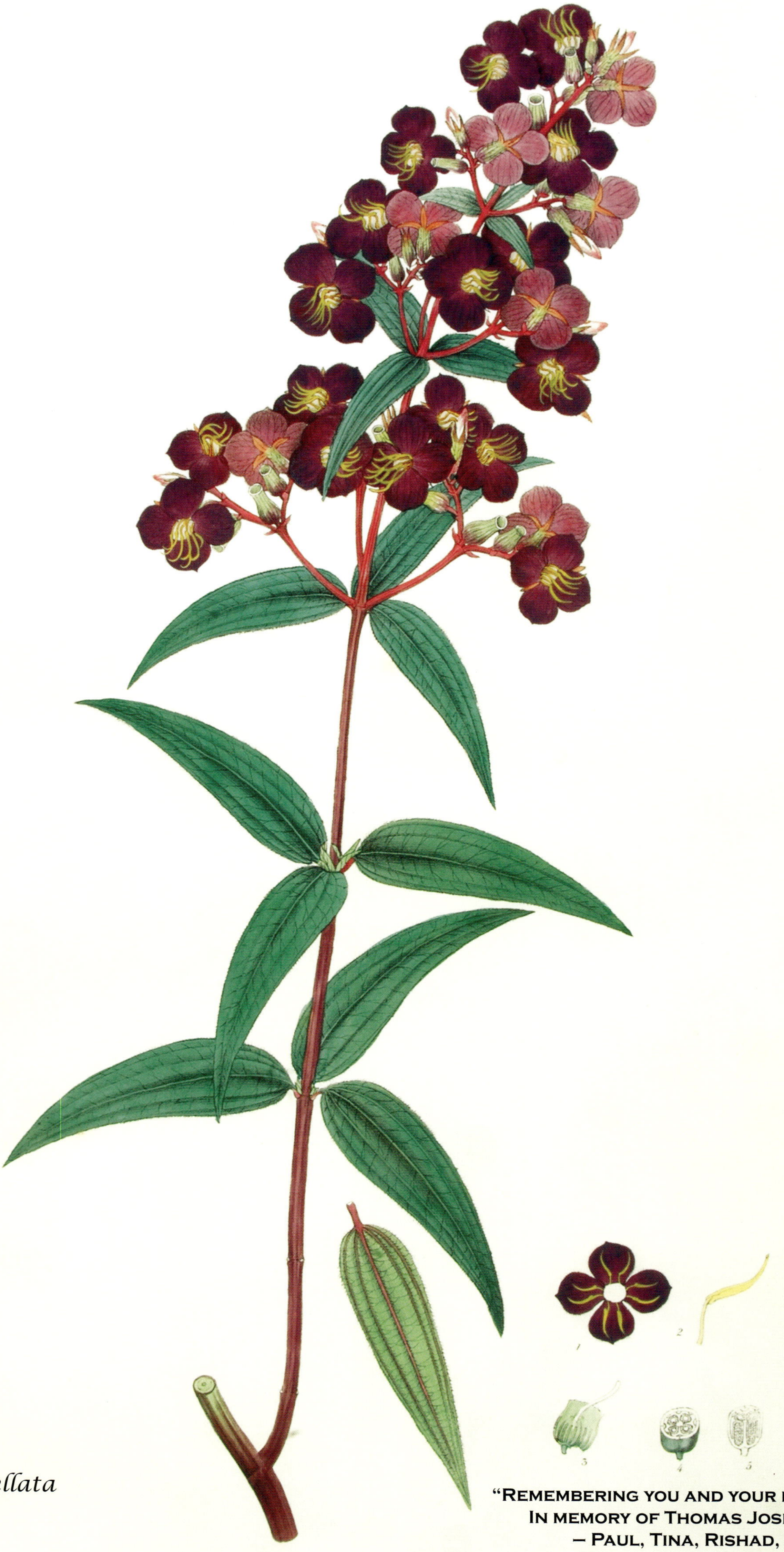

Osbeckia stellata

"REMEMBERING YOU AND YOUR LOVE FOR NATURE"
IN MEMORY OF THOMAS JOSEPH MOZHACHERIL
– PAUL, TINA, RISHAD, & SAHIL ABRAHAM

Osbeckia stellata
Family: Melastomataceae
Synonyms: *O. ternifolia, O. rostrata, Melastoma pulchella*

A hairy shrub, abundant in the Himalaya from Uttarakhand to Darjeeling and Sikkim hills, extending to Khasi hills, Assam, and Nepal, Bhutan, Myanmar, at an altitude of 1,200–2,400 m. The shrub with its tetragonal branches is often seen hanging from rocks and banks. The leaves are opposite, 7–15 cm long, entire, and lanceolate, with five nearly parallel veins and short hairs on both the surfaces. Large pinkish-purple flowers appear in terminal clusters in July–October. Each flower is 5–7 cm across and usually has 4 rounded petals. The ovoid fruits are 1.2–1.8 cm long and hairy. The genus *Osbeckia* was named by Carolus Linnaeus for the Swedish explorer and naturalist Pehr Osbeck (1723–1805).

From N. Wallich, *Plantae Asiaticae Rariores,* 1832.

high trees. To show what a difficult tree this was to climb two young active villagers were offered a rupee each, if they would climb the tree, and they failed to do so."

In a letter to the *South of India Observer,* Lieutenant-Colonel Christie confirmed the story, and wrote that he shot the tiger in the way described. He wrote that "the tree in question will doubtless retain the double set of claw-marks for years to come, so any curious or dubious gentleman may go and look at it near Peermund. The length of the tigress was about 9 feet 6 inches."

R.G. BURTON, BRIG.-GENL.
Charlton Kings, England, June 1920

From *JBNHS* Vol. 27, No. 1.

On the Occurrence of Tigers on the Islands of Bombay and Salsette

A large tiger was shot in the vicinity of the Vehar Lake, Salsette on Tuesday, January 22, 1929. The animal was killed by Mr. J.J. Sutari, to whom I am indebted for the following particulars. Mr. Sutari and a party of friends were out after the usual type of game the Salsette jungles provide, which is mainly wild boar. They took up their positions in the vicinity of the south end of the lake shortly after sunset and waited for something to turn up. Towards 10 p.m. Mr. Sutari's attention was aroused by the sounds of some animal approaching. One can well imagine his astonishment when a tiger walked out of the shadows into the moonlight. The tiger came steadily on, when at a distance of 12 yards, Sutari fired his 12-bore loaded with ball and dropped the animal in his tracks. The tiger in question, a straggler from the main land, probably crossed over by swimming the Thana Creek. An animal doing so would find immediate shelter in the jungles which cover the hilly portions of Salsette.

Tigers appear to have been fairly plentiful in Salsette at the end of the eighteenth century. Hector MacNeil (*Archaelogia,* vol. vii, 1873) tells us that in 1761 "the Governor and most of the gentlemen of Bombay used to go annually on a pleasure party to Salsette to hunt Wild Boar and Royal Tiger both of which were found there in great plenty." Records of the occurrence of tiger in these islands during the nineteenth century are few and far between. In 1806, two tigers were seen near General Macpherson's bungalow at Kurla, while a few days previously two persons were carried off from a village a little further north, it is presumed by the same animals. On February 9, 1822, a tiger on Malabar Hill came down and quenched its thirst at Gowalia Tank and ran off up the hill between the Hermitage and Prospect Lodge. The imprint of its feet were clearly visible the next morning (*Bombay Courrier,* February 10,

Stanley Henry Prater (March 3, 1890 – October 12, 1960) was a long-time affiliate of the Bombay Natural History Society and the Prince of Wales Museum of Western India, Bombay. He was born in the Nilgiris where his father William Prater was a coffee planter. He joined the Bombay Natural History Society in 1907, and developed an intimate knowledge of the mammals of the Indian subcontinent during the Society's Mammal Survey (1911–23). In 1923 he became curator of the BNHS collections and the Museum's Natural History Collection, which positions he held for 25 years. During the last 25 years of his tenure at BNHS he >

1822). The *Bombay Courrier* of December 1829, records the sudden appearance of a tiger at Mazagon, the animal apparently swam across the harbour and landed near the ruined Mazagon fort. It was driven into the compound of Mr. Henshaw's bungalow where it was eventually shot by the guard of the Dockyard and several Arabs. It measured 8' 8". On March 2, 1858, the crew of the steamer *Aden* killed a large tiger which was swimming across to Mazagon from the opposite shore. The animal attempted to board a small boat and was kept off with hand pikes by the lascars. It was eventually dispatched with "six balls through its head". (*Bombay Times,* March 6, 1858). In May of the same year a tiger was killed in Mahim woods by a young Portuguese, while on January 26, 1863, another tiger was killed at Mahim after mauling a Parsi cart-owner and committing other damage. (*Bombay Times,* January 27, 1863). James Douglas (*Bombay and Western India*) writing about tigers in Salsette gives an amusing narrative of a "traveller (was it Silk Buckingham?) in Salsette who was suddenly surprised by his palkee being dropped and the coolies bolting. The palkee was closed, and he soon felt outside the *Jhilmils* something of a fee-faw-fum character. Stripes was wide awake and the coolies, up a tree, were wide awake also. He didn't sleep much that night I tell you." In 1907, a tiger was shot at Pir Pau, Trombay, near Sandow Castle by Mr. Mullan of the Bombay Port Trust. This with the one cited above are the most recent records.

S.H. PRATER, C.M.Z.S.
Bombay Natural History Society, June 22, 1929

From *JBNHS* Vol. 33, No. 4.

How the Female Dugong Carries Her Young

In my paper on the Dugong *(Journ. Bom. Nat. His. Soc.,* Vol. xxxiii, No. 1, p. 96) a reference was made to the manner in which the mother dugong carries her young. Langvel *(Der Zoologische Garten Ann.,* p. 37, 1896) states that the baby dugong supports itself on the back of its mother. According to Petit *(Bull. Mus. d'Hist. Nat.,* Paris, No. 5, 1927), this was also the opinion of the Malagassy fishermen. It was pointed out that even if the baby dugong was able to take refuge under the breast of its mother, it was not apparent how it could be embraced by the comparatively short flippers of the parent. I have been endeavouring to collect further information on this point and have since received a letter from Mr. M.C.C. Bonnington, Divisional Forest Officer, South Andamans, who writes, "The Andamanese harpooned a female dugong on the 8th of March carrying a fully formed young. They inform me that females with young are found chiefly during the south-west monsoon. [Mr. S. Wesche-Dart tells me that he observed a pair of dugongs with a young one in the Andamans in the month

> became executive editor of the *Journal of the Bombay Natural History Society* (*JBNHS*). His major work, *The Book of Indian Animals,* published in 1948 remains a classic even today.

Oxyspora paniculata
Family: Melastomataceae

This shrub with spreading, drooping branches and long pendulous flower clusters is found from central Nepal to Tibet, Sikkim, Darjeeling, and Bhutan, covering the hillsides. According to Wallich, "It is an attractive handsome shrub ... with its long pendant peduncles bearing attractive pink or red flowers". It flowers during August–October. The calyx, petals, and flower stalks are pink-purple. The stamens have curved filaments and the anthers are purple. The leaves are elliptic, acute, 15–25 cm long with five to seven parallel veins and numerous fine cross veins. The fruit is spindle-shaped and eight-ribbed.

From N. Wallich, *Plantae Asiaticae Rariores,* 1832.

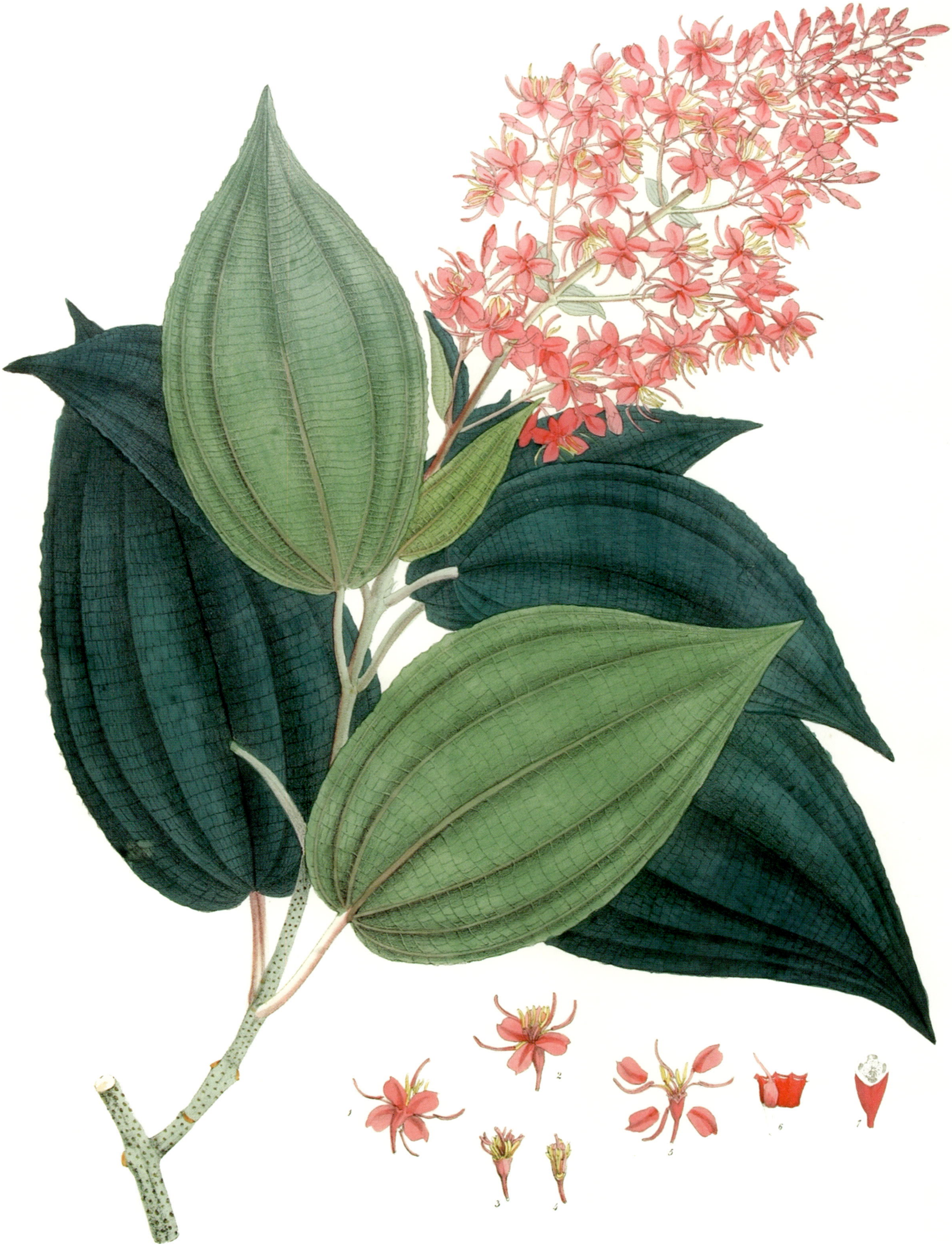

Oxyspora paniculata

Courtesy
Dilipkumar V. Lakhi,
Lakhi Trust

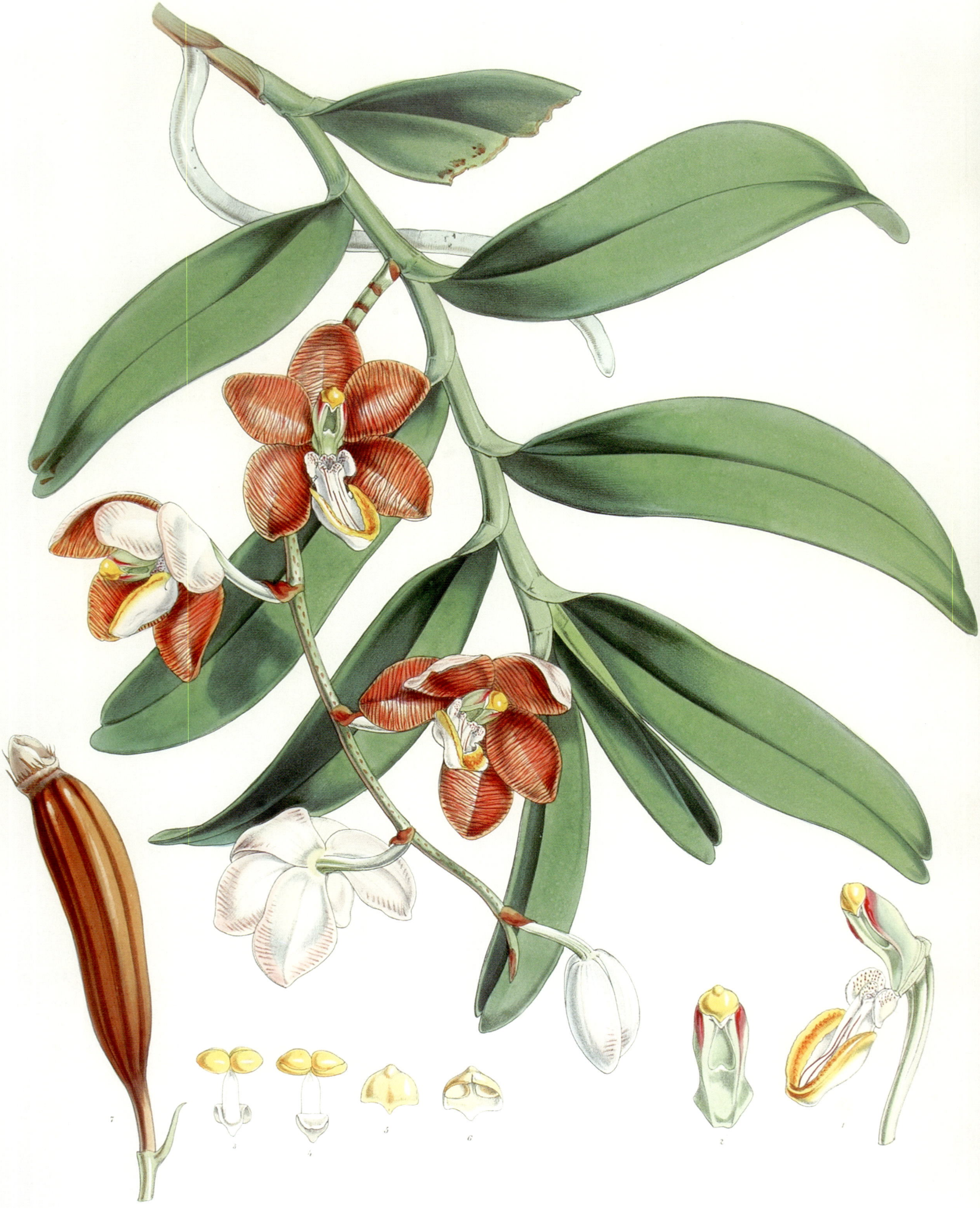

Vanda cathcarti

COURTESY
DINESH SHAH,
ATAM DEEP CHARITABLE TRUST

Vanda cathcarti
Family: Orchidaceae
Synonyms: *Esmeralda cathcartii, Arachnis cathcartii, Arachnanthe cathcartii*

This plant is found in the eastern Himalaya in India, Nepal, and Bhutan, at elevations of 600–2000 m. The round flowers, 8 to 9 cm across, are waxy, strongly fragrant, and very long-lived. Hooker states: "Dr. Lindley says that no more remarkable orchid has been found in Northern India. Though not so showy as the gorgeous *Dendrobia*, among which it grows, it exceeds any other in its singularity and its chaste elegant appearance. A living specimen which I sent to Calcutta flowered in the Botanical Gardens there, but did not survive the voyage to England."

From J.D. Hooker, *Illustrations of Himalayan Plants,* 1855.

of December.] They also tell me that the mother carries its young when travelling firmly held to her breast with her flippers, but that it plays about near the mother when the latter is feeding." Mr. Shunker Narayan Pillay, with whom I have also corresponded, was told by the fishermen of Rameswaram that the baby dugong is held so firmly between the flippers by the mother that three strong men were unable to extricate one from her grasp. Mr. Pillay tells me that between November 1928 and 1929, five dugongs were caught at Rameswaram in the fishermen's nets. The animals, as I have stated previously, are highly esteemed as an article of diet by the Mahomedans of the neighbourhood. Mr. Pillay states that no part is wasted, even the viscera are washed and dried and are eaten, seasoned with salt, pepper, and turmeric. According to Mr. Bonnington, dugongs are still found in the Andamans in herds up to 20 and Col. M.L. Ferrar, Chief Commissioner, Andamans, writing in this connection states that owing to the practical extinction of the Andamanese and the absence of any successors to these *qua* dugong hunters, there should be plenty of dugongs in those waters.

S.H. PRATER, C.M.Z.S.
Bombay Natural History Society, June 6, 1929

From *JBNHS* Vol. 33, No. 4.

Observations on the Indian Elephant

SOLITARY ELEPHANTS

Sanderson in his "*Thirteen Years Among Wild Beasts of India*", an excellent book though it is, makes several mistakes. In regard to elephants he commits two errors which are remarkable in view of the fact that Mr. Sanderson's knowledge about elephants, and his experience in the capture and shikar of Indian elephants has, I suppose, been unequalled. He states in his book that "a really solitary elephant is rarely to be met with." I am certain this is not the case. My experience is that solitary elephants seldom rejoin herds. A solitary tusker will certainly graze with any herd that may happen to be in the same jungle for the time being; but will generally stay behind when the herd moves on to new grazing grounds. I believe that when it becomes "*musth*" a solitary elephant will seek a herd, but directly this sexual state in the elephant has passed off, it will return to its solitary life. The elephants on the Baragurs in North Coimbatore are certainly nearly all solitaries, and I do not suppose that there are ever more than three small herds for the whole length and breadth of that range. A single tusker (rogue) which I shot in March 1928 on the Billigirirangans had been known to be in the same jungle for at least 20 years, leading a solitary existence. I saw him there myself in 1913 or 1914 and when I shot him in 1928, he was approximately in the same place! Solitary elephants are in my opinion not the leaders of nearby herds as supposed by Sanderson. I agree with Sanderson that it is highly improbable that any male elephant is driven from its herd, and I have always expressed this view in regard to both Bison and Elephants. I feel sure that in both cases the bull arrives at an age when it prefers to live a solitary life.

Size of Elephants

Again Sanderson states "here is little doubt that there is not an elephant of 10' at the shoulder." It is extraordinary that Sanderson should have made such an assertion. The two tallest elephants I have ever seen were two single tuskers, shot by Major Brook Purdon R.A.M.C., in 1925, and by me in 1928. In each case, the elephant had only one tusk and both were 10' at the shoulder; in fact the first was just over 10'.

Speed of Elephants

Sanderson's estimate that an elephant can run at a fast shuffle at about 15 miles an hour is no exaggeration, and was supported recently by writers (mostly referring to African Elephants) in the *Field*. Personally I consider from my own experience that an Indian Elephant when "all out" can travel quite 20 miles an hour. I do not think any "mahout" can make an elephant travel as fast as one in full charge, or in full terrified flight. I once saw a pair of elephants cross a grassy flat, and their speed over this stretch of about 300 yards must have been nearly as good as that of a race-horse.

R.C. MORRIS
Honnametti Estate, Attikan P.O., *via* Mysore,
June 10, 1930

From *JBNHS* Vol. 34, No. 3.

The Duration of Life of Tigers

Writing in *The Indian Forester* for June 1932, referring to Mr. Dover's article on the "Longevity of Indian Animals", in which he says:– "Records of extreme longevity show that tigers can live 13 to 19 years and probably more", Mr. A.F. Minchin, I.F.S. quotes a case of a man-eater having lived for about 20 years in the Ganjam District. I give below, his note on the subject:–

A man-eater who had his beat in the Chendragiri forest of Ganjam District was well known in 1913. He did most of his business on a forest road from "Cox's Corner" – named after Mr. Stephen Cox – to the edge of the sal jungle near Tilisingi.

Randolph Camroux Morris, a Scottish planter, was the first to introduce coffee in southern India. His son, Colonel Ralph Camroux (R.C.) Morris (1894–1977), born in India, became an officer in the British Army. After retiring from the army he returned to his father's coffee estate Attikan at Honnametti, near Mysore in the Billigirirangans. He and his father documented the wildlife of the region in the *JBNHS*. Their estate was visited by several people including Sálim Ali, E.P. Gee, and the Maharaja of Mysore. R.C. Morris's father-in-law, Angus Kinloch, of Kotagiri in the Nilgiris, was also a keen naturalist and an expert on snakes. >

Fulvous Whistling-Duck, Large Whistling Teal
Dendrocygna bicolor
Synonym: *Dendrocygna major*

This pale brown or chestnut coloured duck has white uppertail coverts which distinguish it from the Lesser Whistling-Duck. A black line down the back of the neck, a broad neck patch with black streaks, and prominent striped flank plumes are also diagnostic. In flight the dark wings and white rump are seen. It is uncommon, resident from the Deccan north and eastwards, and in Pakistan, Nepal, Bangladesh, and Sri Lanka, where it is seen in jheels and ponds in the plains, in small mixed flocks with the Lesser Whistling-Duck. It is omnivorous but mostly feeds on aquatic plants. It nests in May–June, when it lays 6–8 white eggs on a collection of sticks in a tree fork or hollow. The shrill whistling call gives this bird its name. It was formerly called Tree Duck because of its arboreal habit.

From T.C. Jerdon, *Illustrations of Indian Ornithology*, 1847.

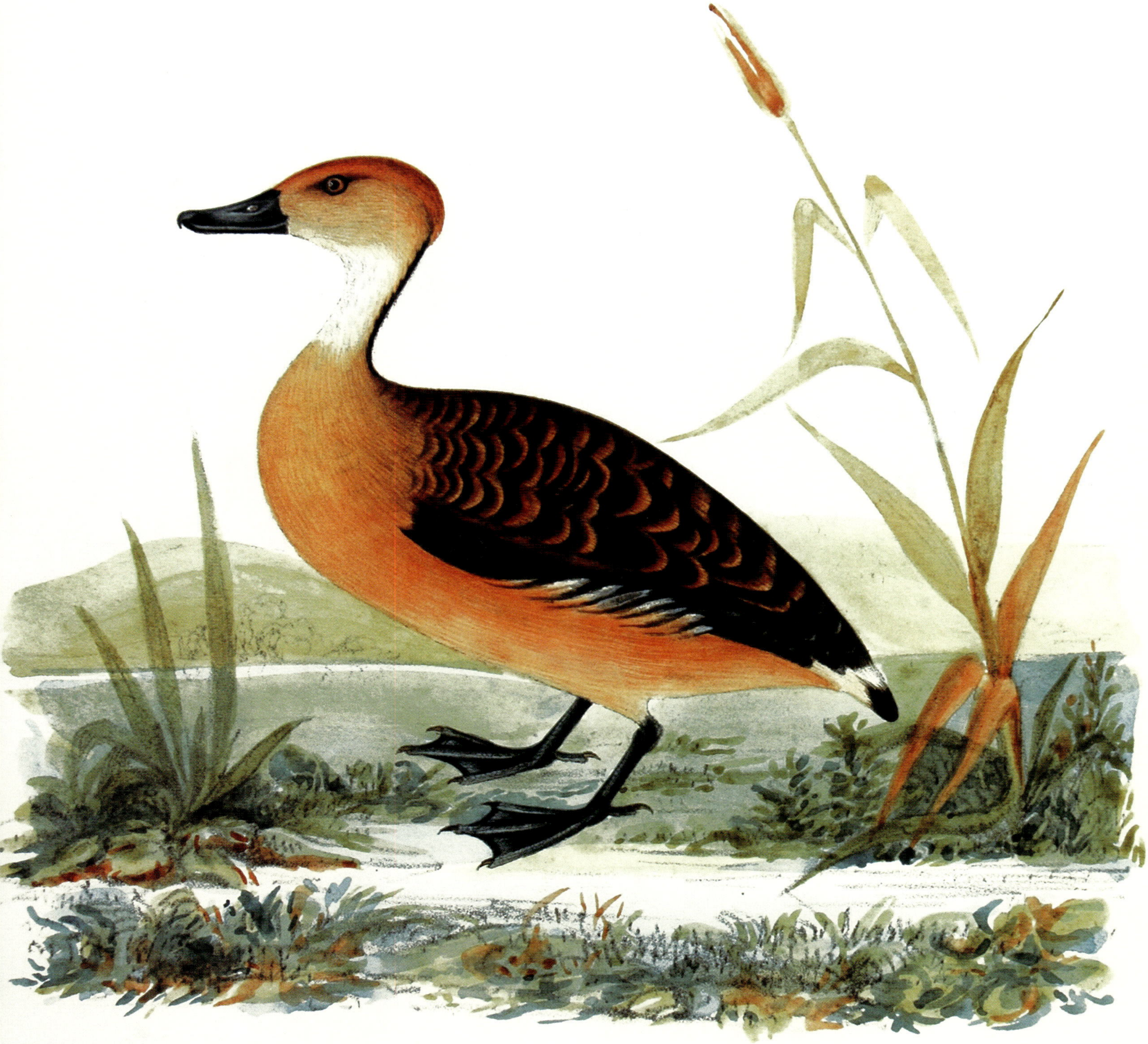

Dendrocygna bicolor

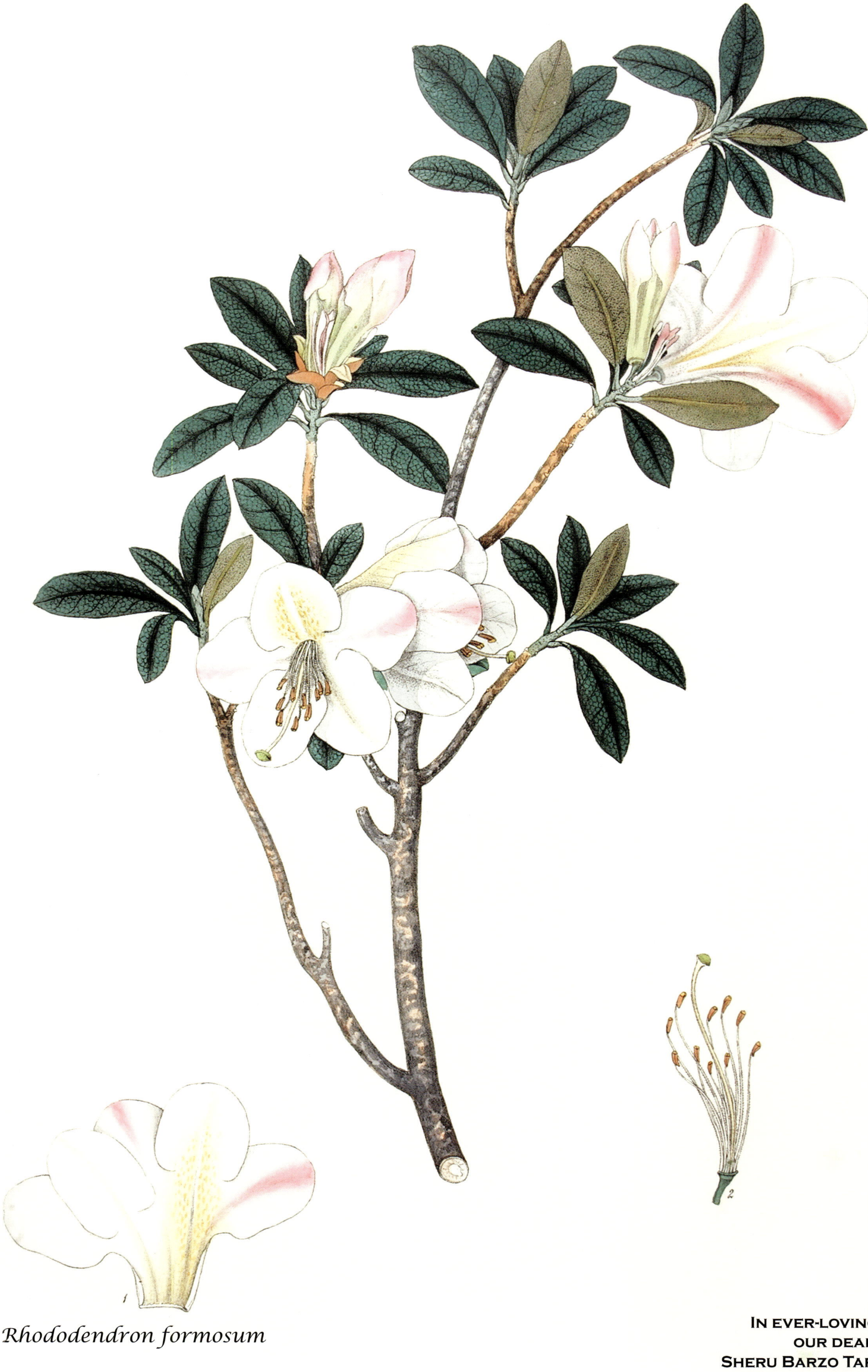

Rhododendron formosum

IN EVER-LOVING MEMORY OF
OUR DEAREST MOTHER
SHERU BARZO TARAPOREWALA
FROM SOONOO, IRACH, AND FREYA

Rhododendron formosum
Family: Ericaceae

This plant is a native of Northeast India (Arunachal Pradesh, Manipur, Meghalaya), found in the Himalaya at elevations up to 700 m. The name *formosum* means beautiful. The species was first published in Nathaniel Wallich's *Plantae Asiaticae Rariores*. According to Wallich the plant was originally collected by a Mr. Smith in 1815. It is a small evergreen shrub, about 1 m tall, with elliptic to linear obovate leaves that have bristly margins. It flowers at different times of the year; the inflorescences have 2 to 6 fragrant white (sometimes flushed pink) flowers, usually with a yellow blotch.

From N. Wallich, *Plantae Asiaticae Rariores*, 1832.

Ganjam tigers are man-eaters by heredity generally, and not "by misfortune". There were nearly always several tigers in the locality referred to; but it is improbable that there has been any confusion about the identity of this particular tiger. He had distinctive mannerisms. For example, after crossing a stream bed the road passed upwards through a cutting. Whilst a train of loaded timber carts were labouring up the incline, out on to the road would stroll our tiger, would sit him down luxuriously and enjoy the ensuing pandemonium. Some magic or telepathy would warn him when to abstain from such performances.

In 1913 he was mature; probably not under four years old. Otherwise he could hardly have appropriated and defended so desirable a beat. In 1915 there were five hundred rupees on his head; and one day he would have come to an end but for ill-timed politeness between myself and another man: my fault it was. I do not know exactly how many people he is supposed to have killed both before and after this occasion. Dozens and dozens.

A man-eater may seem to have been a bad institution. It must be observed though that my friend's victims were frequently women, were never above the rank of Forest Guard, and that, in a forest where the general shooting was poorish, he certainly raised the standard of his tract far above the average. To one of his exploits we owe a memorable Service Order, which runs "K. Lachmana to act as Forest Guard, Fourth Grade, in Tarasingi beat, vice Bamsonia devoured by tiger, with effect from March 3rd, 1915".

The same tiger was there in 1920, when I left the district. In 1925 I toured in Ganjam; and once again in 1929. From what local acquaintances – forest officials and also the Khonds and Ooriyas – had to say, it was still the very same tiger.

In 1929 March, early one hot weather afternoon, I was watching a fire line, along which it seemed the tiger might come back towards his kill. He came unexpectedly early and took me by surprise so that I did not make up my mind to fire quickly enough; but a mental picture of him remained. He had no tawny shades in his coat: there were the dark stripes on a dull silvery grey ground colour. Obviously an old animal. Mr. Ware of the Veterinary service was with me at the time. It was he and I who had had the chance at the same tiger in 1915. After this occasion in 1929 no more was heard of the "Chendragiri Man-eater". There is no story of any one having shot at him; but he ceased to terrorise the neighbourhood.

As far as one can be sure in a matter of this kind, the tiger, which must have been quite four years old in 1913, was the same tiger that was last seen in 1929. That would make him not less than 20 years old at the time of his presumed death.

The above is about equalled by the history of the Bargur Man-eater which was the periodic scare of the Ramapuram-Bargur Ranges (Kollegal Division, Coimbatore District) for 15 or 16 years. This tiger apparently varied its diet with men, cattle and game killing, rather a peculiar case. It was finally shot in 1928 by a poacher, who was out for a stalk in the early morning, met the tiger round a corner, fired, threw down his gun, and bolted for dear life. The tiger was later in the day found lying dead. Needless to say, the man did not claim

> Randolph Morris was gored by a Gaur (*Bos gaurus*) and died of complications. A stone shrine marks the spot where the Gaur attacked him.

The Secretary of the BNHS notes in Vol. 51 of the *Journal* that, at the request of the Kashmir Government, who desired expert advice for rehabilitation of their badly depleted wildlife, Sálim Ali and R.C. Morris visited Kashmir as the Society's representatives in October 1952 and submitted a report recommending suitable measures. R.C. Morris was an early member of the Indian Board for Wildlife, constituted by the Government of India in 1952.

the reward, knowing he would be prosecuted under the Arms Act and for poaching. This tiger must have been quite 20 years old at the time of its death.

RANDOLPH C. MORRIS, F.Z.S.
Honnametti Estate, Attikan P.O., *via* Mysore, S. India,
June 10, 1932

From *JBNHS* Vol. 36, No. 1.

Notes on Monitor Lizards

In the Central Provinces *Varanus monitor* (= *bengalensis*) is the common form and is fairly plentiful though extremely timid and shy. The Common Monitor (*V. monitor* [*V. bengalensis*]) lives in burrows and in cracks in the ground from which it often comes out to bask in the sun during winter, retreating on the slightest alarm. I have also noticed these monitors occupying the roof in less frequented rest houses from the eaves of which they would now and then flop to the ground. During the rains they are more in evidence. I believe that this is their breeding season.

The flesh and skin is much in demand by the local aboriginal tribes who eat the flesh or use it in medicine and use their skins for making their small drums (*khanjeries*) and fiddles (*chikara*). At Nagpur, during the rains, small parties of Gonds regularly hunt these lizards. They take with them a spade, a knife, and a small pack of a few rather under-sized dogs. The *modus operandi* is as follows: –

The hunt takes place immediately after a spell of wet weather. The party, which consists of at least three men, proceeds along the uncultivated borders of fields examining every likely hole or burrow. The presence of a lizard inside a burrow is discovered by its footprints at the entrance to the burrow. If no footprints are present the party moves on to the next burrow. If footprints are visible, one of the party stations himself at the entrance to the burrow and a search is made elsewhere for another entrance to the same burrow; when this is found another man posts himself here while the third begins digging vigorously somewhere near the middle. The lizards on being disturbed make for the entrances and are caught by the men stationed there. On catching a lizard a notch is cut with the knife on the top of the tail, which is now twisted round the neck and a claw of the fore limb is inserted into the notch to prevent the tail from uncoiling. The lizard is now quite unable to move; it is slung on to a stick and carried away. They are kept in this condition till sold when they fetch about twelve annas each.

Rosa macrophylla
Family: Rosaceae

A deciduous shrub or small tree, 1.5–3 m high, usually prickly, found in the Himalaya from Jammu and Kashmir to Sikkim at an altitude of 1,000–3,000 m. The leaves are 5–20 cm long, pinnate with 7 to 11 ovate-elliptic, finally toothed, 7–7.5-cm-long leaflets. Flowers appear in June–July; they are pink or white, 2.5–6 cm across, solitary or in clusters. The fruit is egg-shaped, red, 1.2–5 cm long, glandular, hairy, with single-seeded fruitlets enclosed in the fleshy calyx. The ripe fruit is eaten and becomes very sweet when black and overripe.

From N. Wallich, *Plantae Asiaticae Rariores*, 1832.

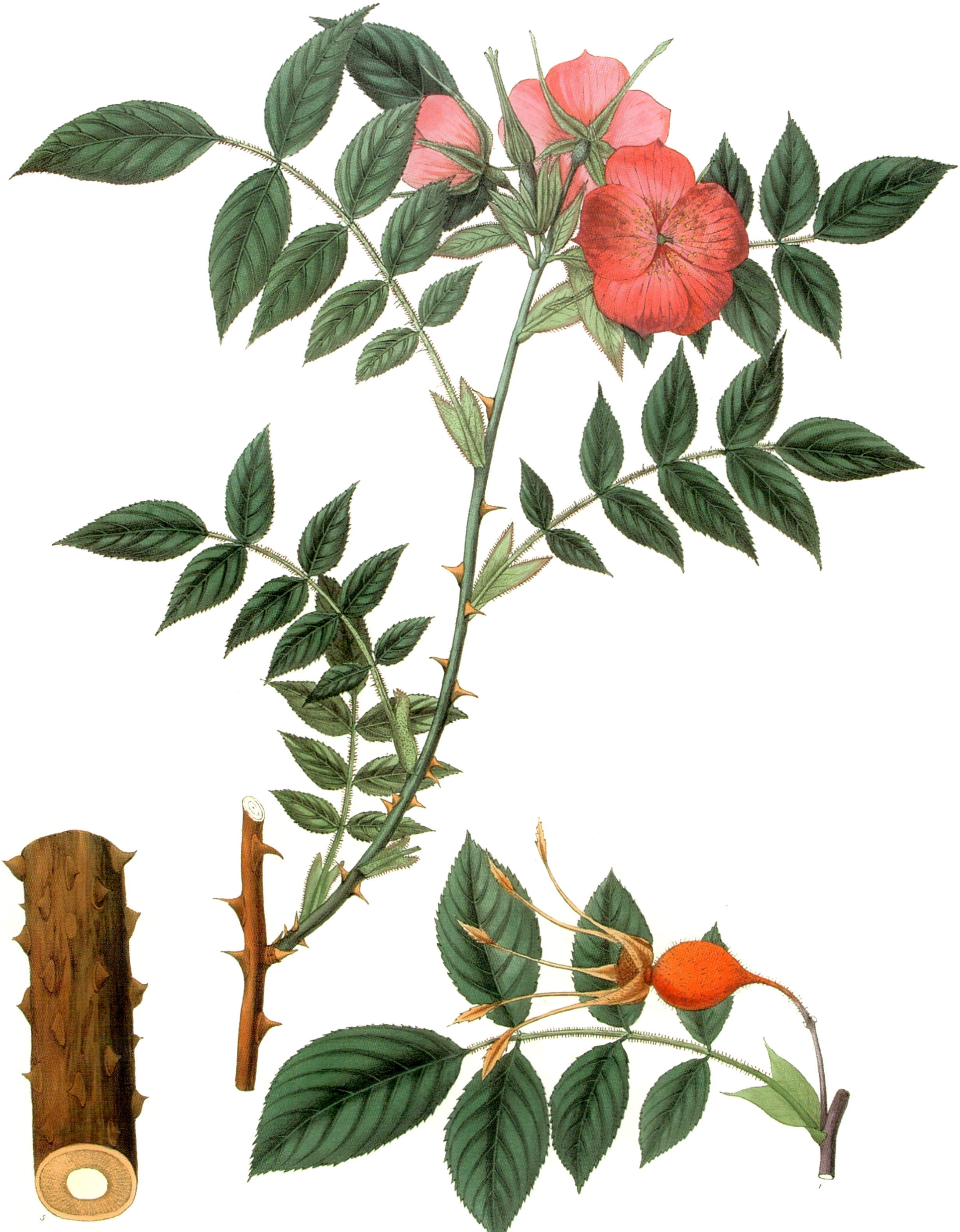

Courtesy
E.A. Kshirsagar,
Honorary Treasurer, BNHS

Rosa macrophylla

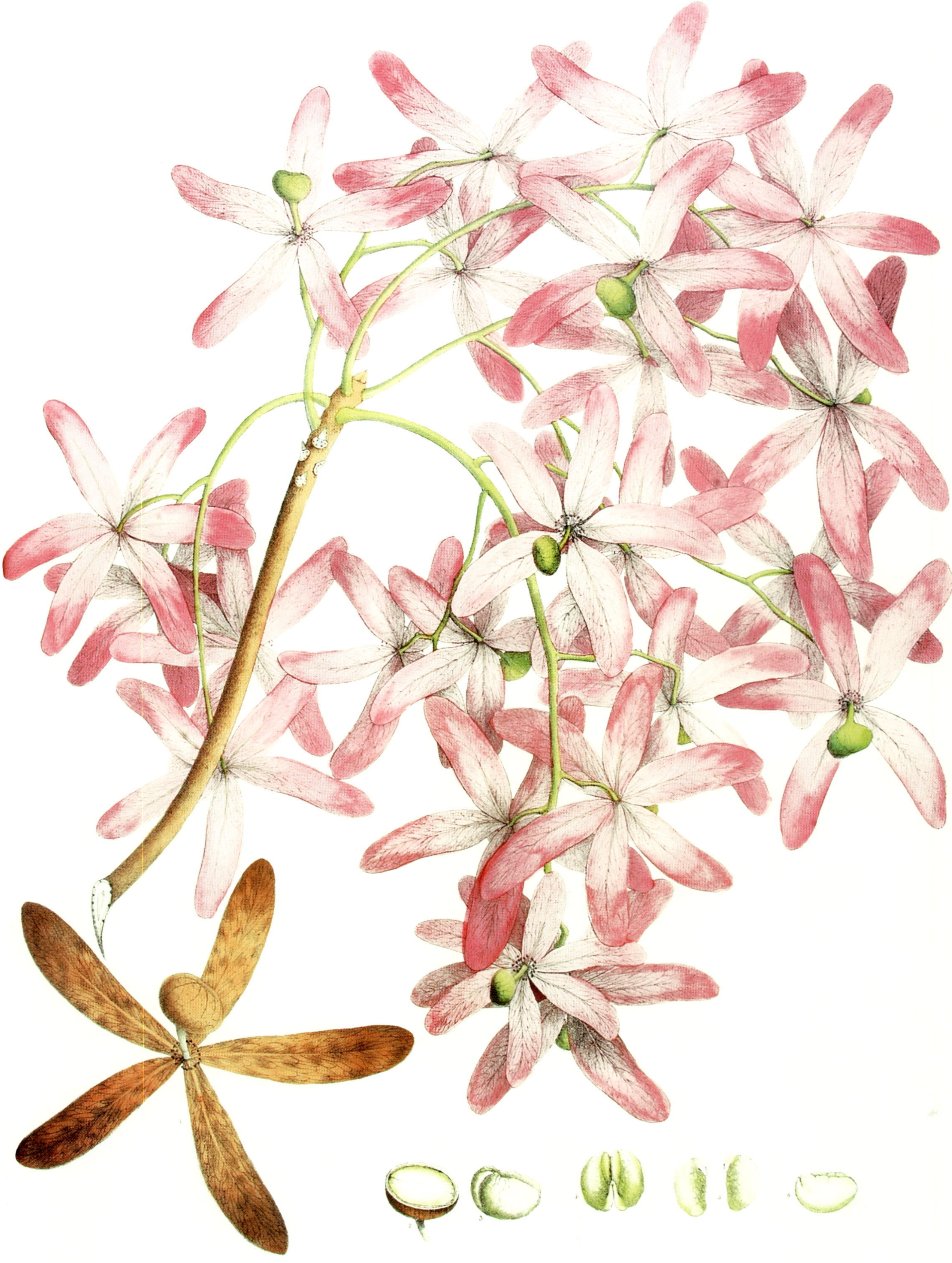

Melanorrhoea usitata

IN MEMORY OF MAHARANI USHAKUMARI OF VANSDA
– MAHARAJA DIGVEERENDRASINHJI I. SOLANKI
OF VANSDA

Burmese Lacquer Tree
Melanorrhoea usitata
Family: Anacardiaceae
Synonym: *Glutu usitata*

A tall deciduous tree found in Manipur, Myanmar, and Thailand. The twigs and leaves are hairy. The tree bears white flowers from January to March. The fruit (illustrated here) is reddish and fleshy. The tree is cultivated for black Burmese lacquer which is gathered from the bark. It thickens rapidly on exposure to air. When applied as a varnish the lacquer provides excellent protection as it is unchanged by acids, alkalis, alcohol, or heat up to 70 °C.
One of the 162 decorative panels depicting flora that form the ceiling of the central hall of the Natural History Museum in London is of *M. usitata*.

From N. Wallich, *Plantae Asiaticae Rariores*, 1832.

The Common Monitor deposits its eggs during the month of September, the number laid is about 21. They are white, oval and soft-shelled like snakes' eggs and measure 45 mm. in length and 26 mm. in breadth.

These lizards thrive well in captivity and can be fed on meat when other vermin is not available but they are dirty creatures, their faeces are voided in a liquid form and are highly odorous, and their cages must be constantly washed. Specimens in captivity have eaten the following:– Rats, mice, shrews, squirrels, frogs, toads, grasshoppers, skinks (*M. maculatus*) fishes, bloodsucker lizards, snakes (*Ptyas mucosus, Eryx conicus, Lycodon aulicus*), small turtles, small birds, and beef. An adult monitor was able to dispose off a full grown *Eryx conicus* and a *Ptyas mucosus* four feet long.

A Red Earth Boa (*Eryx conicus)* when given alive to a monitor was shaken about and bitten first and then the head was held down and it was swallowed tail foremost. This snake was about a foot in length and it did not seem to attack the lizard at all. A living *Lycodon aulicus,* on the other hand, when seized by the monitor bit it vigorously in spite of the mauling. The lizard's body bled profusely from the numerous small wounds it had received from the snake which was game to the last. I have not yet tested these lizards with either scorpions or poisonous snakes. These lizards frequently fall a prey to the python but it takes about a day for a python to crush the life out of even a half-grown monitor.

The Yellow Monitor (*Varanus flavescens*) is the common form in Behar in the Gangetic valley though *V. monitor* is also to be found in these parts.

During the rains the colour of *V. flavescens* is yellow or yellowish brown with broad cross bands of red; these red cross bands are indistinct or wanting at other times of the year.

It will be noticed that in these lizards the outer hind toe is separated by an appreciable distance from the other four toes; this toe is used in apposition with the others in adjusting the tail of the female during coition.

E.A. D'ABREU, F.Z.S.
Central Museum, Nagpur, C.P.,
April 11, 1932

[The Common Monitor was discovered breeding on Island of Salsette by Mr. C. McCann in June. He took eleven eggs. Newly hatched monitors are very common on the island at the commencement of the rains. – Eds.]

From *JBNHS* Vol. 36, No. 1.

Cases of Snake Bite in Ahmednagar District

(These notes were printed in the *Indian Medical Gazette* Vol. lxvii, Feb. 1932).

Since I have been stationed at Ahmednagar, I have come across the following cases of snake bite, six at Ahmednagar itself and two from the district and one from outside, during the last eighteen months. This is not a "snakey" district, as during the last six years an average of only thirty deaths due to snake bite have been reported annually, in a district with a population of nearly one million and an area of 6,630 square miles.

The following are notes of the cases: –

Case 1. – A woman came to the Civil Hospital, Ahmednagar, at 5-30 p.m. on 30th April, 1930, with a killed snake that was dried up. This snake had bitten her in the morning at about 6 a.m. on the right forearm about 2 inches above the wrist. The whole forearm was red and swollen as far as the elbow. She had applied "Oriental Balm" to the arm. There were no after-effects, the swelling gradually going down.

Result: – Recovery.

The snake was a Trinket Snake (*Coluber helena*) – 3 feet 6 inches long – non-poisonous.

Case 2. – Case reported in the *Indian Medical Gazette* of March 1931 of a person bitten in Dhulia, West Khandesh, on 27th July, 1930.

Result: – Recovery.

The snake was *Echis carinata* [*E. carinatus*] or Phoorsa – poisonous, pit-less viper.

Case 3. – A boy of about 8 years, bitten by a Common Krait (*Bungarus cieruleus* [*B. caeruleus*]). Case reported in the *Indian Medical Gazette,* May 1931, p. 266.

Result: – Death.

Case 4. – A man from the city of Ahmednagar came at 10 a.m. on 24th June, 1931 to the Civil Hospital, Ahmednagar, complaining that when he woke up in the morning he found his face swollen and that he found a small snake near his pillow. His face was slightly red and swollen.

The snake he brought was alive and was a baby *Lycodon aulicus* or the Common Wolf Snake. 8 inches long – non-poisonous.

Result: – Recovery.

Case 5. – A cobra bit one of the hospital staff on 8th July, 1931; the case was reported in the *Indian Medical Gazette* of October 1931.

Snake – Binocellate Cobra (*Naia tripudians* [*Naja tripudians*]) – poisonous. Antivenine used.

Result: – Recovery.

Case 6. – A boy, of about 7 years, who had just recovered from cholera was found vomiting and unconscious on his bed in the morning at a place named Kolhar Khurd about

Clock Vine
Thunbergia hawtayneana
Family: Acanthaceae
Synonym: *Meyenia hawtayneana*

This plant is a native of the Nilgiris, and flowers throughout the year. Wallich says: "This beautiful species is widely different from all other Thunbergias. The leaves are of a rigid and leathery texture, of an ovate cordate form, entire, nearly sessile, stem clasping and very glaucous. The flowers are very large and of a most beautiful deep purple colour."

From N. Wallich, *Plantae Asiaticae Rariores,* 1832.

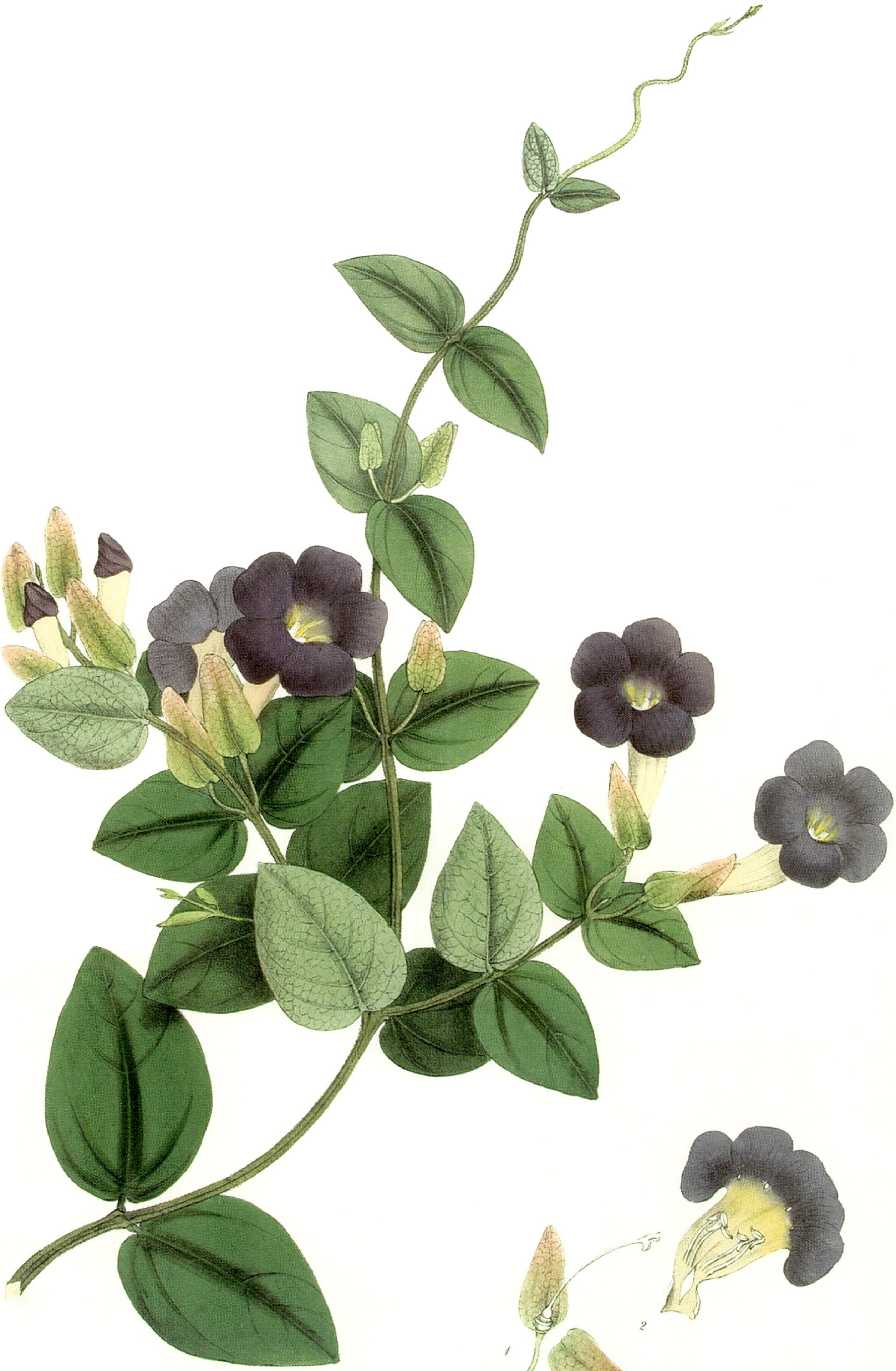

Thunbergia hawtayneana

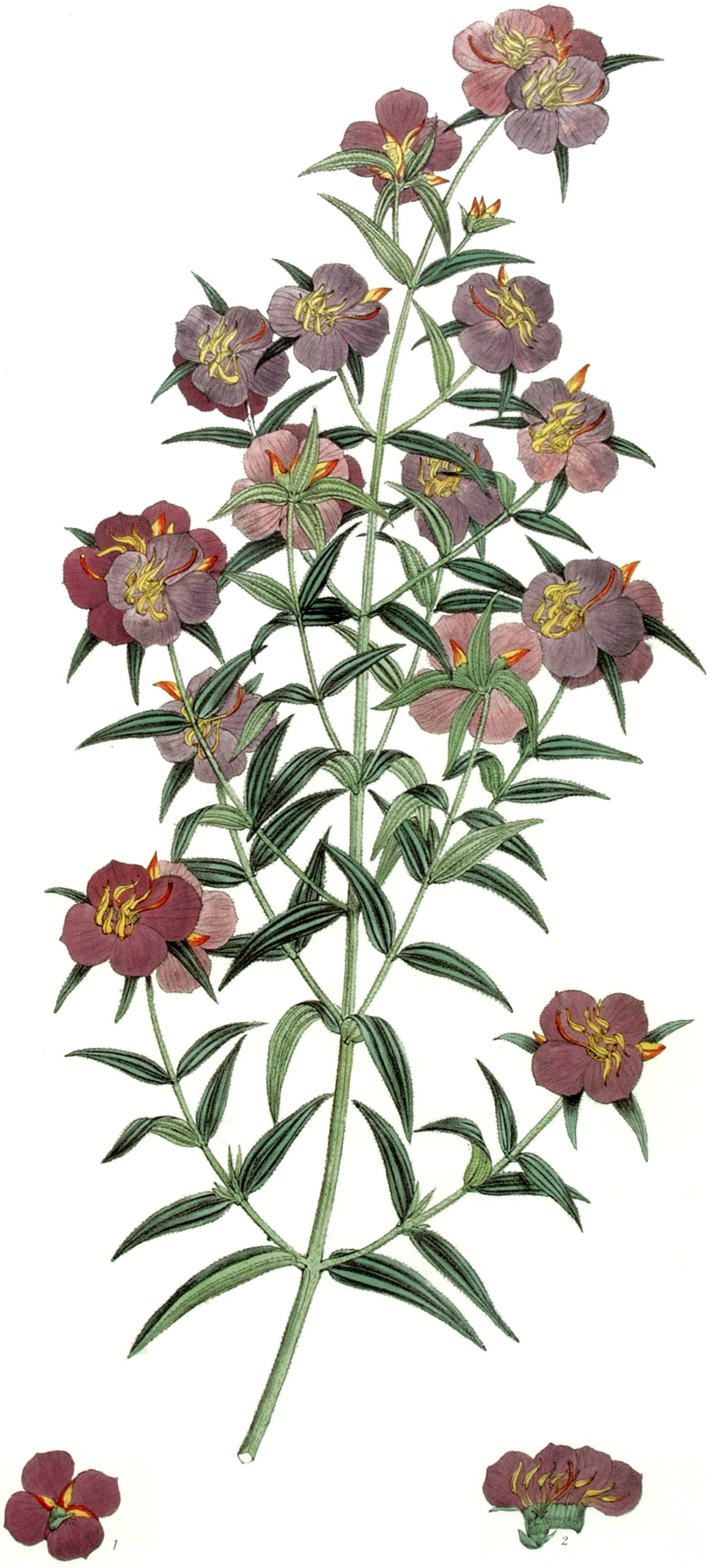

Osbeckia angustifolia

"Remembering you and your love for nature"
In memory of Thomas Abraham Puthupallil
– Paul, Tina, Rishad, & Sahil Abraham

Osbeckia angustifolia
Family: Melastomataceae
Synonym: *O. chinensis*

A small shrub, up to 1 m tall, abundant in the Himalayan Terai from Uttarakhand to Darjeeling, Sikkim, the Khasi hills, Assam, and Bhutan up to altitude 1,400 m. The leaves are subsessile, narrow-oblong with 3–5 parallel veins, hairy on both surfaces, with base round to subcordate and apex acute. In September–October the purple blooms appear, arranged in terminal heads. The fruits are ovoid.

From N. Wallich, *Plantae Asiaticae Rariores*, 1832.

36 miles from here on 14th August, 1931. On removing him from his bed a small live snake was found under him. There were many scratches on the boy's back. The boy was seen by Dr. Shiveshwarkar, Assistant Director of Public Health, who was on tour. He brought the killed snake to me. It was a baby common krait – *Bungarus cieruleus* – 12 inches long – poisonous.

Result:– The boy died at about 12 noon, the same day.

Case 7. – A snake bit a woman below the ankle in the evening of 20th August, 1931, and she came to the hospital at 8 p.m. with the killed snake.

The snake was a Red Earth Boa (*Eryx conicus*) – 16 inches long – non-poisonous. The bite had no effect on the woman.

Result:– Recovery.

Case 8. – Three live snakes were brought to me at the hospital caught in a grain godown on 8th October, 1931. They were all Trinket Snakes (*Coluber helena*) – non-poisonous. After being let out from a tin in which they were kept, they were identified, captured again, and put into the tin. While he was capturing them, one of these bit the catcher on the right index finger on the dorsal surface about ¼ inch behind the nail. There were no puncture holes but there were four scratches, two long ones about ¼ inch long and ¼ inch distant and two small ones on the wrist-side of these two scratches.

The person who was bitten was the same one as in case 5 who was bitten by a cobra before. He was a bit nervous after the bite and complained of slight pain for two hours. Permanganate was applied to the scratches to soothe him. He complained of itching about the eyes, beginning 15 minutes after the bite and lasting for about two hours.

There were no after-effects.

Result:– Recovery.

Case 9. – A snake catcher named Mahaboob Khan lived in a mosque in Bhingar, a suburb within Ahmednagar cantonment limits. He was a reputed snake catcher and used to make money from this job. He came to me on 31st July, 1930, with a big dhaman snake about 8 feet long; he said he had a medicine which he used in cases of snake bite, invariably successfully, and wished me to try it. On talking with him for some time, it was found that he was a practical man of the world and had worked as a stable boy, groom, syce, and a rider at the racing stables in Bombay and Poona. I got the impression from his previous history that he lived more by his common sense, wits, and manual dexterity in catching snakes, than by the efficacy of any snake cures. The medicine, he told me was the powdered fruit of *Michelia champaca* (white variety) – N.O. – *Magnoliacea* – given in goat's milk. Only an occasional tree of this species gives fruit. So the fruit is rare. This Mahaboob was bitten by a snake he caught on 23rd October, 1931, at about 5 p.m. in trying to extract its teeth. He collapsed and became cold, and probably died by 9 p.m. His neighbours came to my bungalow at 12 midnight reporting his condition and said his body was cold. My assistant went there with antivenine at about 1 a.m. but the man had been dead four hours before. The deceased had taken by the mouth his own medicine after having been bitten.

The snake was brought alive to me and it was a full-grown common krait – *Bungarus cieruleus* – poisonous – 3 ft. 5½ ins. long. The punctures on the man were two and were on the dorsal aspect of the left middle finger on either side of the first inter-phalangial joint about 1½ inches distant. As the mouth of the krait was not wide enough for the distance between the fangs on the two sides to be over an inch, it is presumed that the man put the bent joint of his left middle finger in the snake's mouth to keep it open and used his right hand to pull out or break the snake's teeth with tweezers.

Result:– Death.

K.G. GHARPUREY, LT.-COL., I.M.S., Civil Surgeon,
Ahmednagar, Deccan, Bombay Presidency, April 27, 1932

From *JBNHS* Vol. 36, No. 1.

A Carnivorous Bear

We went from Srinagar to Pahlgam for a few days' outing and sport. Pahlgam is a lovely spot in Kashmir, about 63 miles from Srinagar. The main pathway of the ancient pilgrimage to Amarnath begins from there.

It was here, on the 1st. of July, that I made my first acquaintance with the Himalayan Black Bear (*Ursus torquatus* [*Ursus thibetanus*]) – under somewhat strange circumstances, which have led me to write this account. The local shikari brought in news of a bear having killed a calf; and he assured us that the beast would come back to feed on the carcase before sunset. This report we did not credit at first, because the bear is generally known to be herbivorous. Therefore I decided to investigate their report personally, as never have I known, or heard, before of an animal being attacked and killed by a bear for food. However, on visiting the spot, I found the carcase of a calf which had been killed a couple of days previously; and about half of it had been eaten already. So I decided to sit up. A fine, big bear came at about 7 p.m. and was shot; but what interested me was that the animal came to the kill just as carnivora would do. I talked about this experience of mine with the Game Warden of this State, who told me that he had heard from natives about these Himalayan Black Bear occasionally doing some damage among the natives' live stock; but this was the first time he had had reliable first-hand information of a bear returning to a kill.

MADANSINH of Kutch
Srinagar, July 11, 1932

A *Vaccinium salignum*
Family: Ericaceae
Synonym: *Agapetes saligna*
B Himalayan Lantern, Creeping Agapetes
Agapetes serpens
Synonyms: *Pentapterygium serpens, Vaccinium serpens, Thibaudia myrtifolia*

Hooker writes: "In the lower part of the eastern Himalaya, Malay Peninsula, Java and other of the Malayan islands, there is an extensive section to which the two species figured belong. They are epiphytical shrubs, having the lower part of the stem often swelling out into a prostrate trunk, as thick as the human arm or leg, and sending out branching fibrous roots that attach it to the limb of the tree upon which it grows. These trunks are soft and spongy internally and are reservoirs of moisture and nutriment; they send out a few slender, generally pendulous branches which bear often gorgeous flowers. ... they are wholly unknown in the peninsula of India, and in Ceylon, where some of the terrestrial shrubby species grow. The leaves of the present species [*Vaccinium salignum*] are used as a substitute of tea by the natives of Sikkim. It was discovered in Bhotan by Dr. Griffith. ... [*Agapetes serpens*] was discovered by Griffith in Bhutan and found abundantly by Dr. Thomson and myself, inhabiting the trunks of lofty trees at various elevations between [1,000 and 2,300 m]." Himalayan Lantern is a beautiful shrub with arching stems and tiny hanging lantern-like flowers. The leaf blade is dark green above and pale green on the underside. The flowers hang down in fascicles of 1–3; the berry is globose and 5-winged.

From J.D. Hooker, *Illustrations of Himalayan Plants*, 1855.

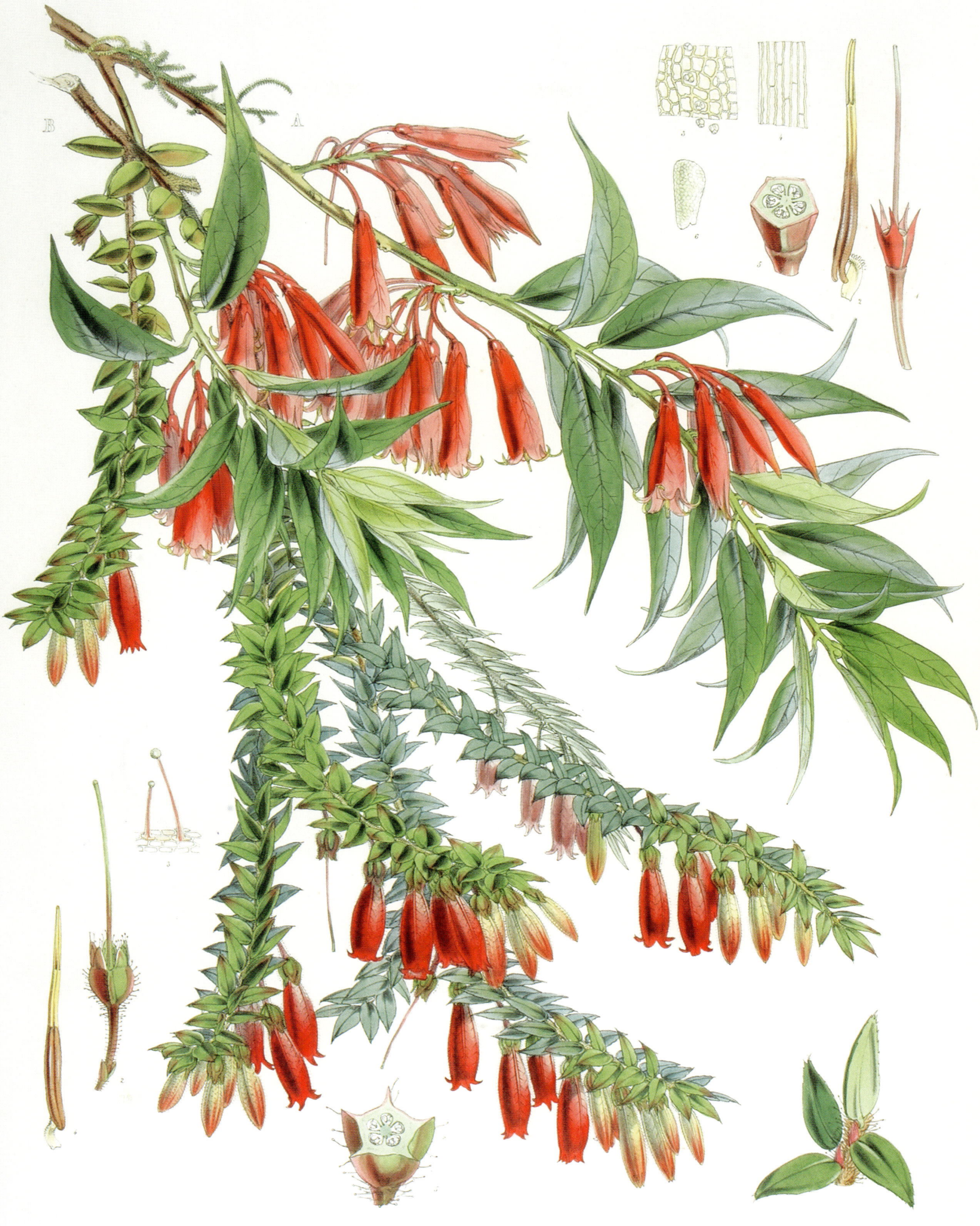

Vaccinium salignum,
Agapetes serpens

COURTESY
VENUS JEWEL

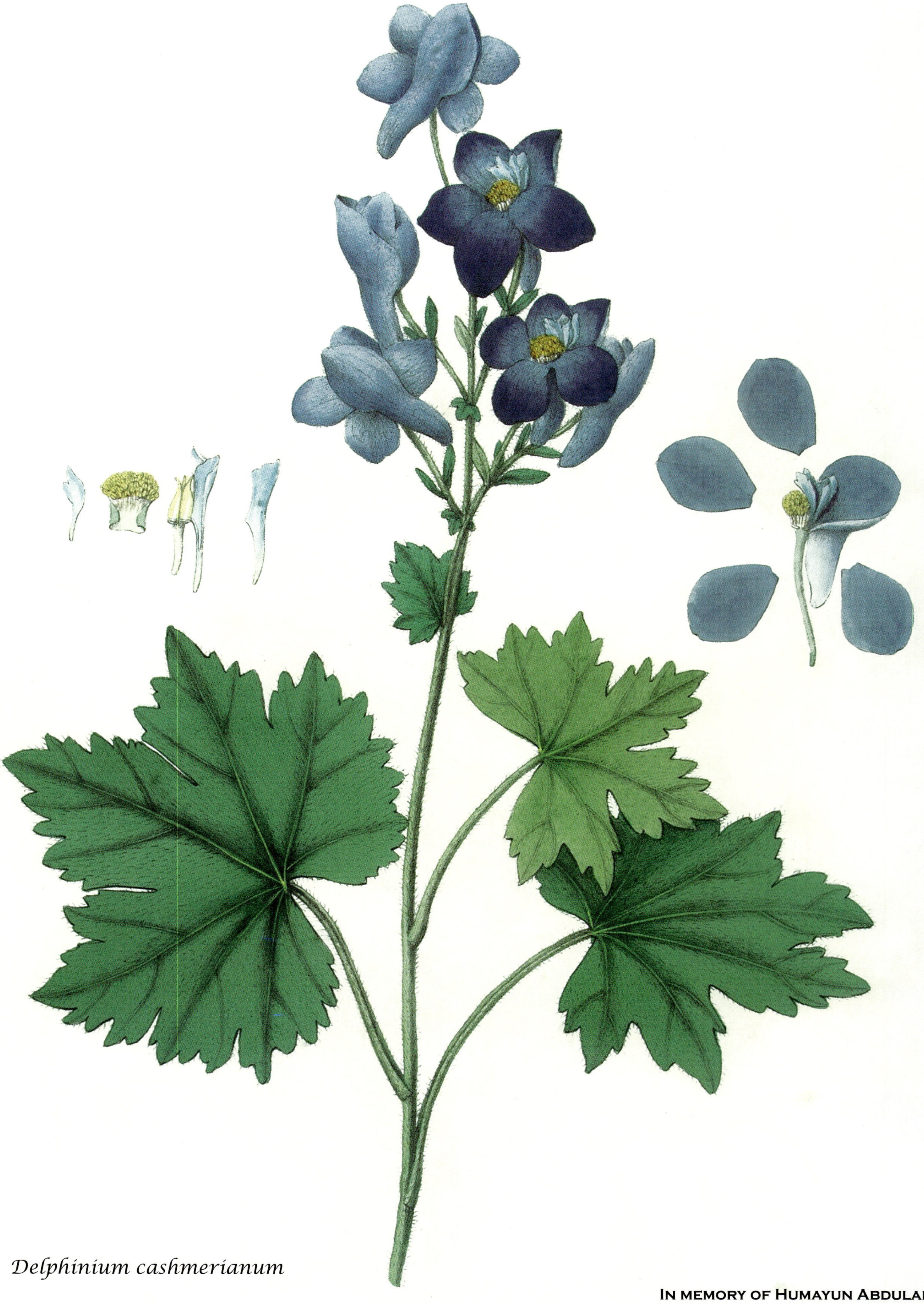

Delphinium cashmerianum

IN MEMORY OF HUMAYUN ABDULALI,
NATURALIST & ORNITHOLOGIST PAR EXCELLENCE

Delphinium cashmerianum
Family: Ranunculaceae

This plant is found in the western Himalaya from Kashmir to Uttarakhand and western Tibet, at 2,700–4,500 m on damp stony and rocky slopes. The leaves have 3–7 toothed, pointed lobes. The flowers appear in August and are bluish purple; their long stalks arise from different points of the axis but grow in such a way that the larger flowers are nearly at the same level. All parts of the plant are highly toxic, especially so in young plants. A parasiticide is obtained from the leaves, which is used locally to destroy ticks on animals.

From J. Forbes Royle, *Illustrations...*, 1839.

[The food of the Himalayan Black Bear (*Selenarctos* [*Ursus*] *thibetanus*) consists mainly of roots and fruits in quest of which this bear is frequently found in fields and orchards. Like most bears it is fond of honey and raids beehives. It has been known to feed on swarming locusts. At the same time the Himalayan Black Bear is the most carnivorous of Indian Bears. Not only does it kill cows, sheep, goats, and deer but it also feeds on carrion. In a note on the habits of these Bears (*Journ. B.N.H.S.*, vol. xvii, p. 833) Mr. C.H. Donald comments on this bear's clumsy methods of killing. He came upon a bear making a kill – the victim was a cow. He shot the bear and put the cow out of its misery. The animal had one leg broken, two huge pieces of flesh taken out of her flanks while her stomach and hind quarters had been horribly lacerated by the bear's teeth and claws. Of 11 fresh kills made by bears and seen by him – one of them, a bullock, was dragged up and placed in the fork of a tree! – Mr. Donald states that except in the case of a fine big buffalo, which was mauled about the neck and shoulders, all the victims were done to death in the same crude manner. It is evident that this bear, from the horrible job he makes of it, is a prentice hand at killing animals. The Brown Bear (*Ursus arctos isabellinus*) may develop carnivorous habits in the absence of its regular food. This consists mainly of herbs, fruits, and roots. This bear restricts its depradations to goats and sheep which it kills with one blow of its powerful paw. The Brown Bear does not attack cattle but has no hesitation in making a meal of a dead cow or buffalo. The Indian Sloth Bear feeds mainly on fruits and insects – but will occasionally feed on dead animals. Mr. Hasted records an instance of one which made a meal of a snake, which the writer took to be a Russell's Viper. – Eds.]

From *JBNHS* Vol. 36, No. 1.

Termite Collecting in South India for Food

It will be interesting to learn that "white ants" have got an edible value. The queen ants were formerly used for medical purposes but now people use them both medically as well as for food. In Madura, Nagercoil, and other places people make big collections of these insects during the rainy months when "swarming" takes place.

There are two popular methods by which these termites are captured. In some places during the swarming season men dig moderately large pits by the side of the termite nests and fill them with water. Near these pits lighted torches are fixed. The insects attracted by the light crowd over the pits and eventually fall into them. The fallen insects are soon caught and filled in empty bags. This is the light mechanism.

In Madura the method is quite different. In these parts children seem to take more interest in termite collection than men. During rainy months they search for the termite holes and finding one, they blow hardly into it making a shrill sound. The winged termites disturbed by the sound, swarm out and in their attempt to fly they are easily captured in cloth-nets.

The wings and the legs of the termites so collected are removed and the rest roasted with a small quantity of salt and pepper added to it. The roasted white ants are widely used as a delicious article of food. In Nagercoil and such localities it is a common sight to see poor men and women sitting by the road-side and selling these roasted insects as they sell roasted ground-nuts or beans.

T.V. SUBRAHMANIAM, B.A.
Government Museum, Madras,
January 28, 1934

[We recall here a note on Red Ants as food (A.M. Long, vol. xiii, p. 356). The Murries of Bastar, the southernmost Native State in the Central Provinces use red ants as a regular article of food. Whole nests of these insects are taken, torn apart, and shaken into a cloth. The occupants mature and immature are beaten into a pulpy mass with a stone and then parcelled into small packets for sale. You may eat the bolus raw after seasoning it with salt, turmeric, and chillies or you may cook it with rice flour, chillies etc. into a thick paste. Regular consumption is said to give great power of resistance against fatigue and the sun's heat. The jungle people of Canara have the same liking for red ants as food and deal with them in much the same way as the Murries do. (Aitken, vol. iv, p. 153.) The squashed insect gives out an acrid smell and when many are being pounded up the "fumes" which arise make the eyes smart. Aitken suggests that the sensation produced when eating them "must be like a torch light procession going down the throat". Apart from food value it is good to know that the acrid aroma of pulverised red ant is a substitute for *smelling salts.* Tamils in Ceylon use it (A.L. Butler, vol. x, p. 330). The modus operandi is to go to a teeming nest in a bush, seize it with both hands and rub ants and nest together violently between the palms, then take a few good penetrating sniffs of the ammonia-like fumes which arise from the crushed and bruised insects. This, we are told, will instantly relieve a severe cold in the head. The termites which form the subject of the above note are no connections of ants. As a family they are considered relatives of the Cockroaches. Cockroaches as an article of food also provide a long and interesting subject. – EDS.]

From *JBNHS* Vol. 37, No. 2.

String Lily
Crinum herbertianum
Family: Amaryllidaceae
Synonyms: *C. zylanicum, C. latifolium* var. *zylanicum*

Crinum is derived from the Greek *crinon*, meaning lily. This perennial herb has a tunicate subglobose bulb with a short, stout neck. The leaves are basal with thick blades, 3–5 cm long, with prominent whitish rib on the underside. The plant flowers in April–July with umbels of sessile, fragrant, showy, red and white striped blooms. Father H. Santapau writes in *Flora of Khandala*: "This is one of the most showy and attractive of Khandala plants. Both in flower and in leaf this plant deserves a place in gardens. Gregarious on steep slopes at the foot of the Beharn's plateau; not seen elsewhere in the district. For several years I have been cultivating this plant in Bombay from two bulbs originally brought down from Khandala."

From N. Wallich, *Plantae Asiaticae Rariores,* 1832.

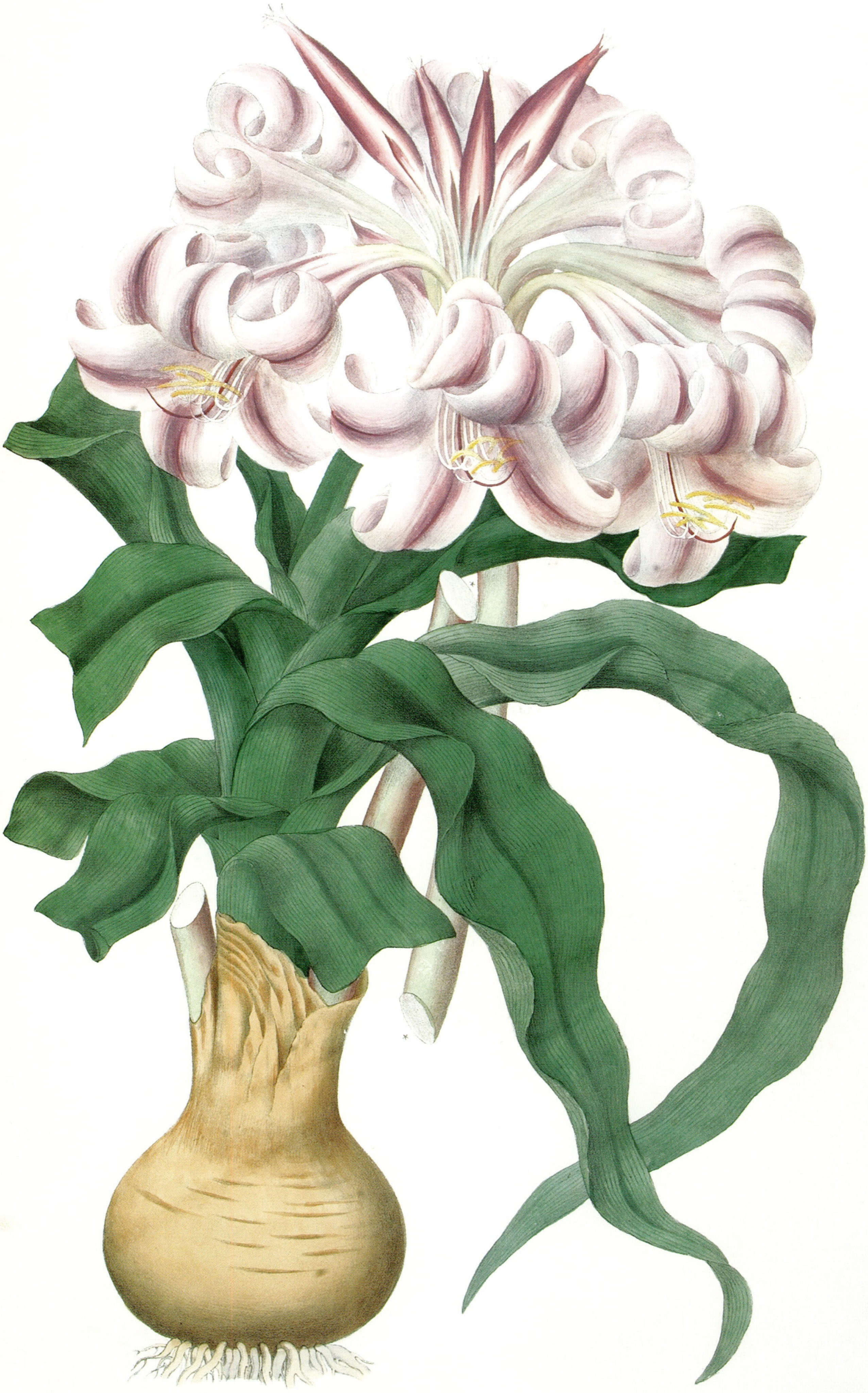

Crinum herbertianum

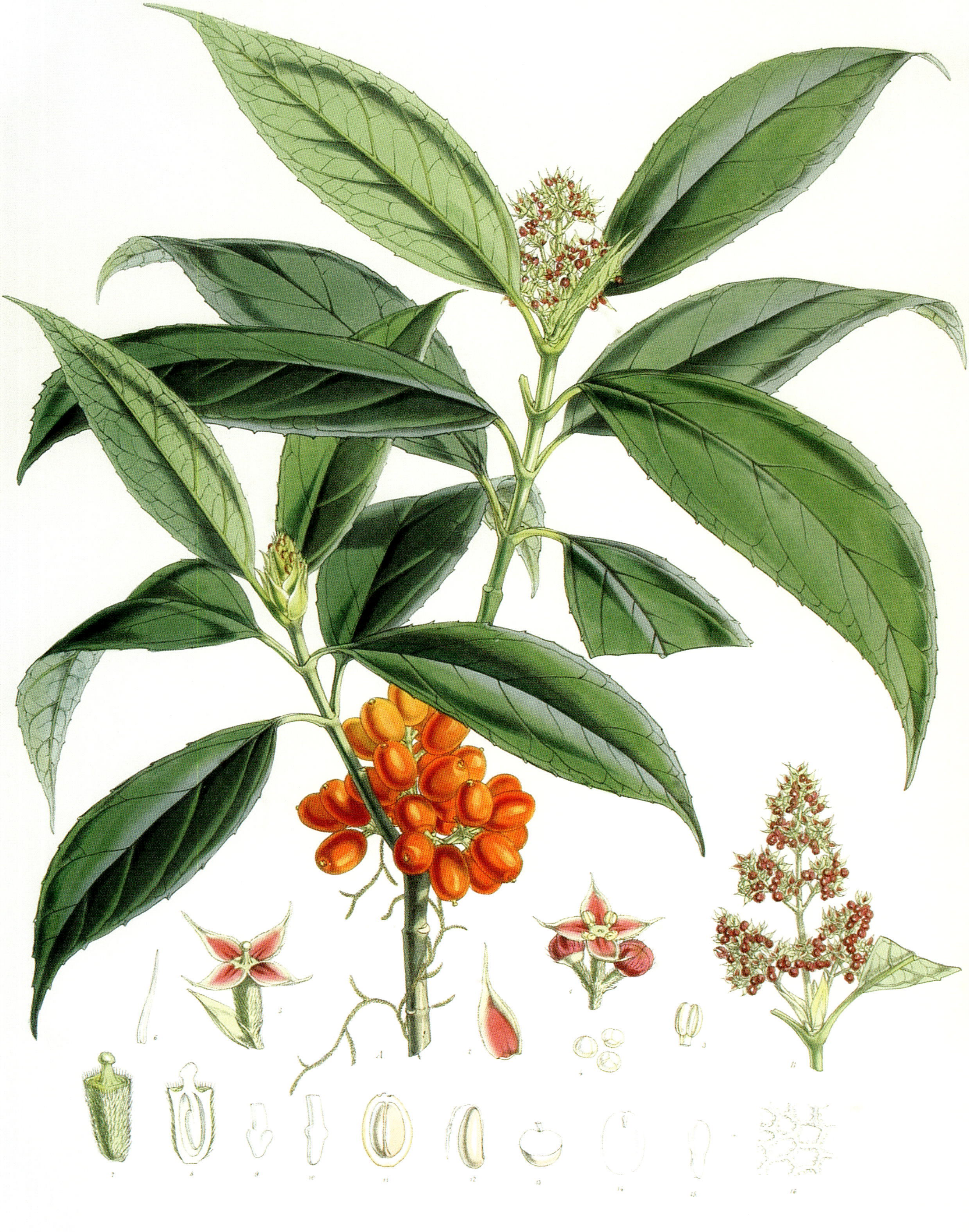

Aucuba himalaica

COURTESY
UNITED PHOSPORUS LTD.

Aucuba himalaica
Family: Garryaceae

An evergreen shrub or small tree, 8–10 m tall, with glossy, thinly leathery leaves. The leaves are opposite, lanceolate, with a few large teeth on the margins. The flowers are small, 4–8 mm across with four purplish brown petals: they are produced in clusters of 10–30 in a loose cyme. The fruit is a red spherical berry of 1 cm diameter. Hooker describes the plant as follows: "In all important characters the Japan and Himalayan plants resemble one another very closely indeed, the only differences which I have been able to detect being that the Himalayan species has considerably longer and narrower leaves with longer, narrower points, and long acuminate points to the petals. These characters together with the immense geographical distance between the native location of the two have induced Dr. Thomson and myself to consider the present as distinct. *Aucuba himalaica* is one of the many striking cases of botanical affinity between the temperate flora of the Himalaya, and especially Eastern Himalaya, China and Japan and which affinity is not shared by the flora of Europe."

From J.D. Hooker, *Illustrations of Himalayan Plants*, 1855.

A Panther with Twenty Claws

On closely examining a panther that I had recently shot I discovered something unique about the claws. On the hind feet it had two extra claws corresponding to the Dew claws in front. Not a malformation, for these claws were fully developed and retractile. I had never met with such an unusual case, nor have my friends, and it must be very rare in feline animals.

It is probably the experience of all big game sportsmen that no two animals are identical, but it would be most interesting to know whether such cases of twenty claws instead of the usual eighteen, are rare enough to be freaks.

I wouldn't say that my panther beats are always productive of interest to naturalists, but even if I shoot another hundred animals, I think I shall find no less interest in my sport than on previous and less exceptional occasions.

VIRBHADRASINH, H.H. Maharana of Lunawada
Lunawada, 14th February, 1946

[There are at least three records in the *Journal* of panthers having twenty toes. In vol. xxx, p. 909 a photograph appeared illustrating such an abnormality. – Eds.]

From *JBNHS* Vol. 46, No. 2.

Tiger Kills Lioness in Mysore Zoo

A tragedy took place recently in the Chamrajendra Zoological Gardens in Mysore City, South India.

Tigers and lions are kept in a large cage in the Zoo, each species in separate compartments. There is an open enclosure with a deep trench all round, and tigers and lions are let out into this enclosure alternately.

Recently a large tiger was let out into the enclosure. The keeper, when cleaning the cage in which the lioness was, admitted her into her adjoining cage without noticing that the door giving access to the enclosure was open. So the lioness went into the open enclosure into which the tiger had already been admitted.

The tiger at once attacked the lioness. A terrific battle ensued and the lioness put up a brave fight for her life. But she was no match for the tiger and everything the keepers and others could do was futile, so the poor brave lioness was soon killed.

Even a full-grown lion has been known to be no match for a full-grown tiger, the lioness therefore had even less chance, and the poor creature lost her life through the carelessness of a keeper.

CHARLES THEOBALD, F.Z.S.
Mysore, 23rd January, 1946

From *JBNHS* Vol. 46, No. 2

How Far Can a Tiger Swim?

The tigers inhabiting the Sunderbans in Southern Bengal have been known to swim pretty long distances.

I present certain facts which give some idea as to their capacity to swim long distances:– The nearest to Sunderbans one could reach from within the district of Midnapur (my home) was at Khedgeree in Gontai Subdivision 18 miles across the river Hooghly, to what is now known as Frasergunj, which is at present devoid of any jungle. But even uptil 1922, while the deforestation was in progress but not complete, tigers crossed this 18-mile wide river with currents and were killed by the villagers at Khedgeree. From 1907 deforestation started at Frasergunj and the Revenue Department started to lease out lands to cultivators. The plan originated from the brains of the late Sir Andrew Fraser, after whom the place is now called. Before 1907 the place was densely wooded and a favourite haunt of tigers. This part of Sunderbans was then known as Mahis Khola Khas-Mahal forests. In the year 1900 a tiger crossed this river and came on to the shore at a place called Horkhali in Tamluk Subdivision in the District of Midnapur and was killed at a village called Deulpata about 14 miles from Horkhali. Now, if the tiger started swimming from Mahis Khola, where the jungle finished, then it would have swam as many as 35 miles before it reached Horkhali. But there was another possibility. The tiger might have walked through thickly populated villages under cover of darkness for 31 miles just on to the opposite bank of Horkhali and then swum across the Hooghly, which is about 6 miles, wide at this point. Having achieved this it might again have walked 14 miles to Deulpata, where it was found early in the morning and killed by one of the nephews of late Hon'ble Mr. Nilmoni Mandal, the local Zamindar.

Either of the above feats would seem, at the outset, to be impossible for a tiger, but I have related the bare truth.

D.P. GARGA, Kumar of Mahishadal
Hampton Court, Hazaribagh (Bihar), 19th October, 1947

Pongamia atropurpurea
Family: Fabaceae
(alt. Papilionaceae)
Synonyms: *Callerya atropurpurea, Milletia atropurpurea*

This evergreen tree grows up to 20 m tall, and is found in Myanmar, Thailand, Vietnam, Cambodia, Laos, Indonesia, and Malaysia. The bark is smooth, the leaves compound, alternate, and spiral, with 3 to 5 pairs of elliptic, opposite leaflets. The purple flowers are arranged in terminal inflorescences.

According to Wallich: "This is a very large tree very common about Moulmein, abundant in the Tenassarim provinces. The crown of the tree is thickly covered by deep purple flowers. Burmese employ the wood for beams and rafters. They eat the tender leaves."

From N. Wallich, *Plantae Asiaticae Rariores*, 1832.

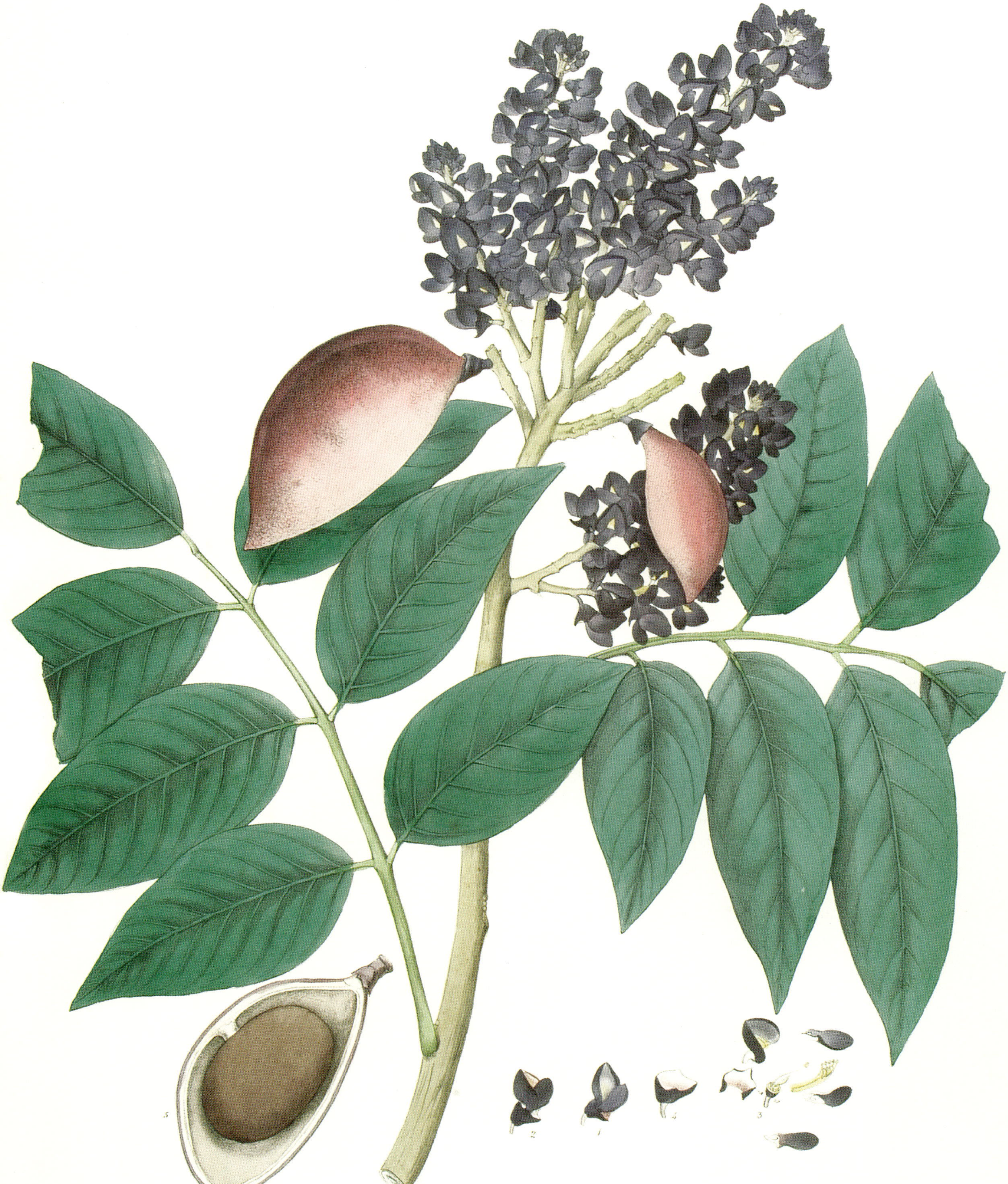

Pongamia atropurpurea

COURTESY
MAIZE PRODUCTS

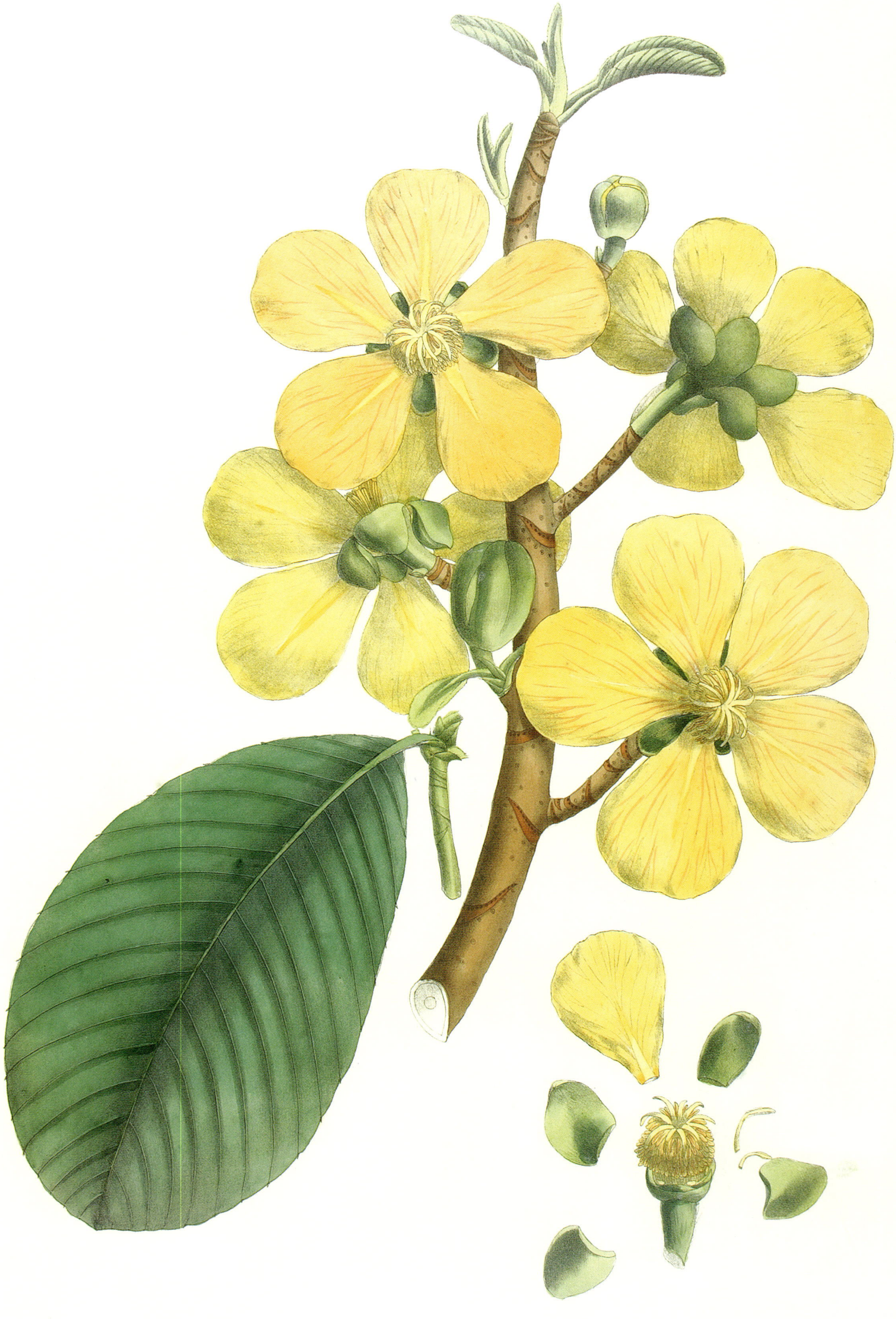

Dillenia ornata

COURTESY
AMBUJA CEMENT

Burma Simpoh
Dillenia ornata
Family: Dilleniaceae
Synonyms: *D. obovata, D. harmandi, Colbertia obovata, Wormis obovata*

A small bushy tree up to 12 m high, found in Myanmar, Indonesia, Malaysia, Thailand, and the Philippines. The twigs, leaves, and flower buds bear soft hairs. The leaf blade is obovate, blunt or slightly tipped, rounded; the edge finely toothed, the underside velvety. The wide yellow flowers appear singly on slender hairy stalks, near the ends of the twigs when the tree is leafless and new leaves have just begun to appear. The fruits are dull yellow and rounded; the seeds dark brown.
One of the 162 decorative panels on the ceiling of the central hall of the Natural History Museum, London, illustrates a flowering branch of this tree.

From N. Wallich, *Plantae Asiaticae Rariores*, 1832.

[Tigers have been shot on Bombay and Salsette Islands, which they must have reached by swimming across the Thana Creek of Bombay Harbour. A list of records ending with one at the Vihar Lake in 1929 is published in *J.B.N.H.S.* Vol. xxxiii, p. 973.* – Eds.]

From *JBNHS* Vol. 47, No. 3.

Altitudinal Limit of the Indian Elephant

While crossing the Bompu La (9,600') in early September I was surprised to find fresh traces of a number of elephants on the very summit, and to learn that one had actually been encountered on the path a few days before. The Bompu La is in the Se-La Subagency of the North-East Frontier of India lying some 30 miles east of the eastern boundary of Bhutan. The range whose highest peaks are over 10,000' rises straight from the plains and the southern base is covered in evergreen forest. Above 9,000' this becomes a dense growth of Ringal Bamboo under a cover of Rhododendrons and other evergreens and it is apparently this bamboo which attracts the elephants during the summer months only, since December to February the Pass is regularly under snow.

I should be very interested to learn the record altitude at which the Indian elephant has been found.

F.N. BETTS, Political Officer, Subansiri Area
N. Lakimpur, Assam, 23rd October, 1947

[Prater in *The Book of Indian Animals* (p. 186) says, "In Burma they wander at all seasons of the year into bamboo forests at a height of 10,000 ft., while in Sikkim their tracks have been seen in the snow 12,000 ft. above sea-level."– Eds.]

From *JBNHS* Vol. 47, No. 3.

Peculiar Behaviour of the Darter (*Anhinga melanogaster*)

In October, 1939, I was fortunate enough to visit the famous Keoladeo Ghana in Bharatpur and among many other curious and wonderful sights, we noticed that Darters which were fairly common on the babool trees overhanging water, would, when approached, let

* Reproduced in this volume on page 155.

themselves drop plumb into the water with closed wings, as if shot dead, knocking against the branches as they fell. They would also fly out of the trees and then suddenly flop into the water and disappear below without any preliminary hovering in the air or alighting on surface as is usual among other such birds.

I was reminded of this note when I recently saw a pair of Darters perched on a Jamun overhanging water at the Powai Lake near Bombay, drop in similar fashion. This therefore appears to be a widespread habit though I cannot recollect having seen it published before.

HUMAYUN ABDULALI
Messrs. Faiz & Co., 75, Abdul Rehman St., Bombay 3,
October 30th, 1947

[This habit is recorded on p. 387 of *The Book of Indian Birds* (4th edition). – Eds.]

From *JBNHS* Vol. 47, No. 3.

Acacia Thorn in the Stomach of a Toad
(*Bufo melanostictus* [*Duttaphrynus melanostictus*])

Among a number of toads supplied for dissection in the class, a fully grown female toad (*Bufo melanostictus* Schneid.) measuring about 95 mm., distended, with ripe eggs was found to have swallowed a biradiate Babul thorn (*Acacia arabica* Willd. [*Acacia nilotica*]). The two arms of the thorn were at right angles to each other within the stomach and measured 22 mm., and 20 mm. One of the pointed ends of the thorn rested freely against the middle region of the muscles overlying the parasphenoid in the buccal cavity and the other arm of the thorn had penetrated through the oesophagus right into the left urinary bladder. Due to this, the oesophagus had become greatly dilated so as to present little difference between the oesophagus and the stomach in size. The wound caused by the penetration of this thorn appeared to be completely healed up. On opening the stomach, I noticed bits of food such as cuticles and muscles of insects. This indicated that the toad was apparently without any discomfort and probably was leading a normal life.

BASANTA KUMAR BEHURA
Maharaja's College, Parlakimedi, Ganjam, Orissa,
18th November, 1947

From *JBNHS* Vol. 47, No. 3.

White-cheeked Barbet, Small Green Barbet, Green Barbet
Megalaima viridis
Synonym: *Bucco viridis*

This small grass-green bird has dark brown crown and nape, a blackish band running backwards from the eye, and the distinctive broad white cheek stripe. The chin and throat are whitish; the breast whitish streaked with brown. Abdomen and vent are pale green. It is a common resident in the evergreen and moist deciduous forests of southern India, also frequenting gardens and groves in the neighbourhood of towns and villages. It is absent in Sri Lanka. Fig trees like Banyan, Pipal, and Umber are favourites with these birds, who also eat other fruits, berries (especially coffee berries), and insects. They are highly vocal, particularly when feeding and breeding. They nest from December to June, in a hole in a dead or rotten branch of a tree, 2–10 m from the ground. They lay 2–4 eggs in a clutch.

From T.C. Jerdon, *Illustrations of Indian Ornithology*, 1847.

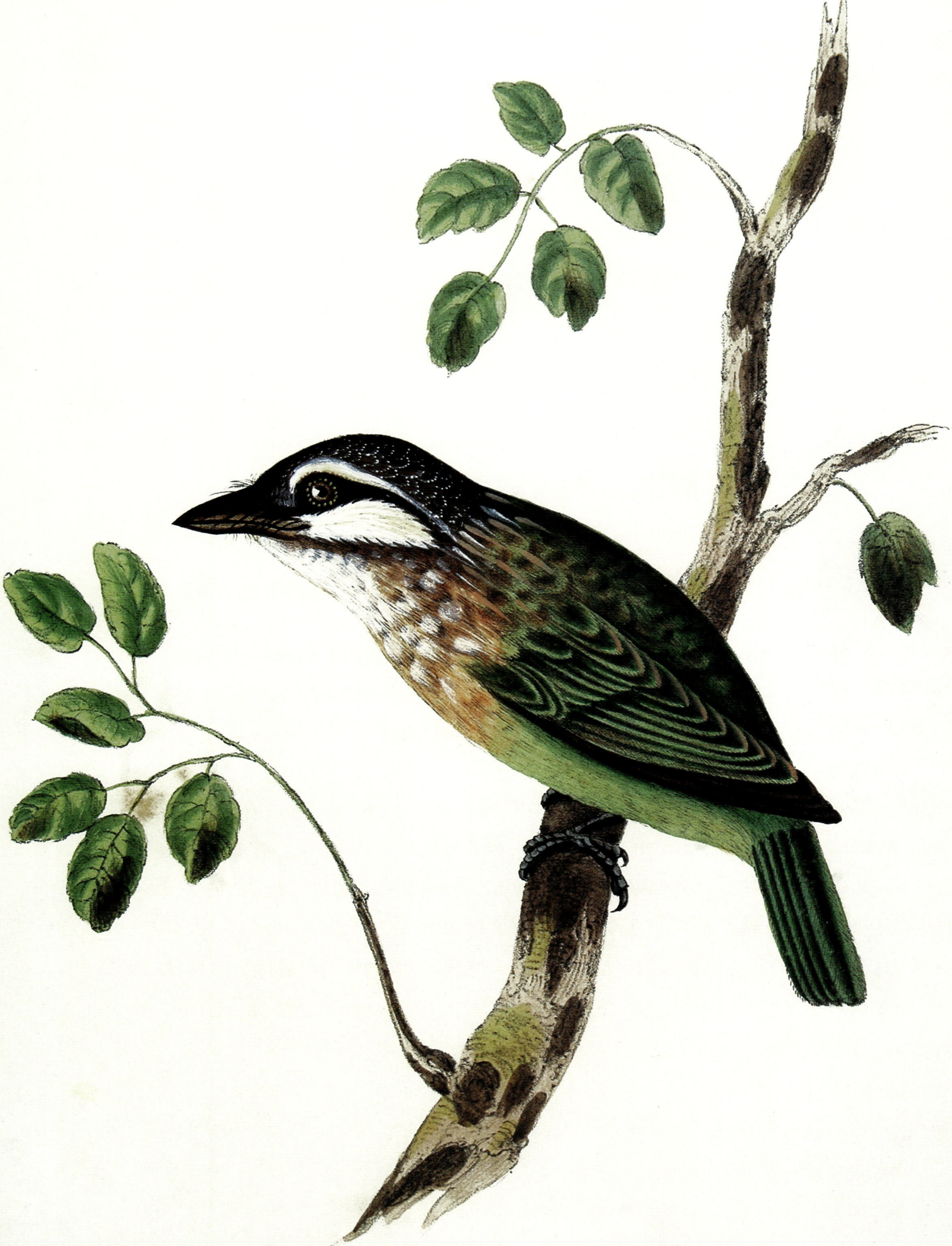

In memory of Mr. J.S. Serrao,
keen naturalist and former Librarian,
Bombay Natural History Society

Megalaima viridis

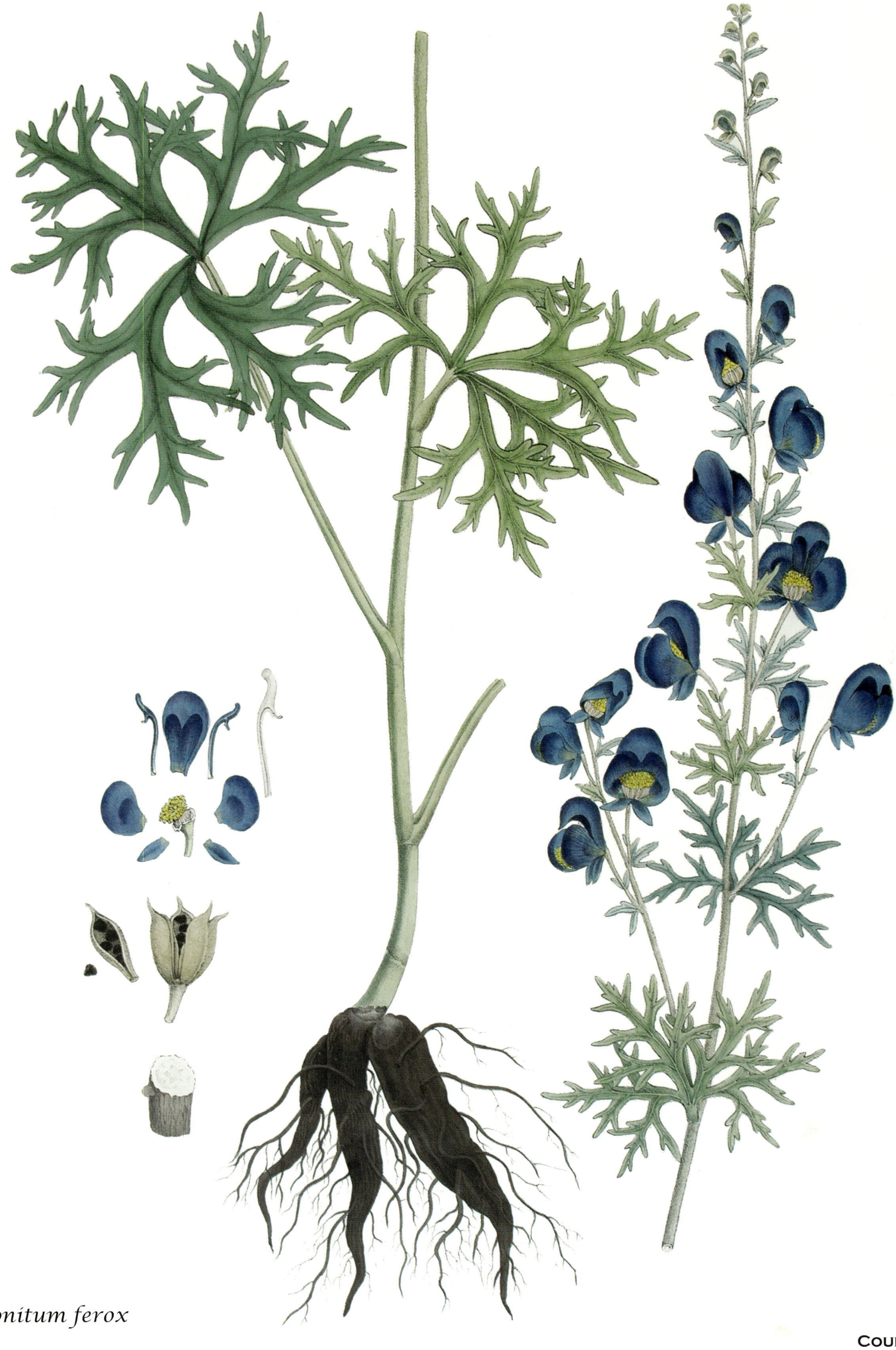

Aconitum ferox

Courtesy
Ambuja Cement

Himalayan Monkshood, Indian Aconite
Aconitum ferox
Family: Ranunculaceae
Synonym: *A. virrorum*

This plant is found in the temperate subalpine Himalaya from Uttarakhand to Sikkim, Nepal, and Bhutan, at 2,100–3,600 m. According to Wallich, it is "a deciduous perennial which grows to 1 metre high. The tall erect stem is crowned by racemes of large, eye-catching blue, purple, white, yellow, or pink zygomorphic flowers with numerous stems. There are 2 to 10 petals. The upper two petals are large and placed under the hood of the calyx and are supported by long stalks. *Aconitum ferox* supplies the Indian poison." The plant flowers in August–October and the flowers are commonly large, and pale dirty blue colour. The tuberous root has anti-diabetic, anti-pyretic, and diuretic properties. If taken frequently or in an overdose it is poisonous (the Sanskrit name of this plant is Ati-visha, strong poison; Hindi: Vish, Mitha zahar, Bishnak; Punjabi: Moura-bikh; Tamil: Visha-navi; Telugu: Ati-vassa). An ointment made from the root is useful in the treatment of rheumatism and neuralgia. Balfour writes in his *Cyclopaedia of India* (1885), "A preparation of the root is much used in all the hilly districts of Northern India to poison arrows for the destruction of wild beasts; .... It has been used on several occasions to poison wells and tanks as means of defence against the invasion of territories .... The Gurkhas say that they could so infect all the waters with the dreadful root that no enemy can advance into their mountain fastnesses."

From N. Wallich, *Plantae Asiaticae Rariores*, 1832.

Animals "Shamming" Death

In Vol. 46, No. 2, Mr. Dunbar Brander, the well-known naturalist, says in his letter entitled "Shamming Death", that animals have no knowledge of death. Question of time is infinity to them. But my idea is not that. If they do not realise what death is, they certainly preconceive a danger. I do not mind if you term it "Instinct". But how can instinct tell goats that one is going to be slaughtered and the other is not when both are led along in the same fashion? The one which is not going to be slaughtered will in all probability follow without any sign of resistance, but the other requires to be hauled. He gives out successive cries and shows resentment. I know cases and also I say from my experience that when chickens are tamed they can ordinarily be caught without any difficulty, but before slaughter the same will invariably fly and run about and refuse to be caught.

2. With regard to death-cry of animals Mr. Dunbar Brander rules out excessive fear being the cause of it and questions whether it is instinctive. I would rather rule out both and am inclined to suggest – it is the outcry of agony and excessive pain that precedes a violent death. If some one says that when an animal gets a sudden shock all its nerves are benumbed and it does not feel the death-agony, I would then attribute this cry to unexpectedness of the stroke or hit rather than call it instinctive. The hit startles the animal to such an extent that it gives vent to a succession of cries which are never uttered under ordinary circumstances.

D.P. GARGA, Kumar of Mahishadal
Phul-Bag Palace, Mahishadal, Midnapur Dt.,
June 6, 1947

[Mr. A.A. Dunbar Brander, to whom the above note was submitted, comments as follows:– "Animals possess very strong instincts and very acute perceptions with regard to *danger* and unless special precautions are taken, it would be apparent to the animal that some design with respect to itself was contemplated. But can this consciousness of *danger* be extended so as to include consciousness of *death* itself? Personally I believe not, and all my experience of animals is that they are not conscious of death, in other words of time; to them their life appears infinite. Be it fully understood that I merely express an opinion, I have no certain knowledge on the subject, and my original letter was written with the object of inducing a correspondence which might throw more light on the subject.

"That the death cry is due to fear, or excessive pain or agony I am in a position to deny factually. Much more painful wounds than an actual death wound, are frequently endured by animals in silence. The death cry is never uttered save immediately on the point of dissolution: it is intimately connected with death itself. I recollect smiting a noble stag through the heart with a .577 bullet. He was quite unaware of my presence, had no fear, anticipated no danger. He stood up on his hind legs extending himself to his fullest extent, pawed the air with his fore legs and letting out a heart-rending death cry, fell dead. As most of your readers are probably aware, the usual procedure under such circumstances, is for the stag to gallop blindly forward for 50 yards or so and then fall dead. But what I have related above happens now and again.

"Does the desire to preserve its life necessarily imply that the animal is conscious of its ultimate death? The discussion on this point is in the realm of philosophy rather than that of natural history."– Eds.]

From *JBNHS* Vol. 47, No. 3.

Land Leeches

I have had 30 years experience of leech-infested jungles in the South Indian hill ranges, including the Malanaad of Coorg and Mysore, the Nilgiris and Nilgiri-Wynaad, the Nelliampathi and Anamallai Hills, and the High Wavy Mountains in the Madurai District.

Some 20 years ago, in the Anamallais, I did some small experiments on leeches with the help of our estate doctor. We dissected some, as far as we were able to do so with limited facilities, with the object of finding out what they ate. In none of our specimens did we find anything but humus and leaf mould. At the same time we tried a number of experiments in test tubes to try and find out what common reagents were lethal to them.

The result of these experiments was most interesting. Any acid reagent had little or no effect on them, in fact they appeared to survive quite happily in contact with quite strong solutions of hydrochloric acid.

On the other hand they reacted very quickly to anything alkaline. A small spot of free lime was quite sufficient to kill a leech very quickly. Very dilute caustic soda was equally lethal. Such things as tobacco decoctions, and coarse soap emulsions, were also very effective in killing leeches besides acting as repellents.

The killing power of alkalis was what I expected to find. The effectiveness of coarse soap is probably due to the residue of soda ash in the soap, as the better the soap the less effective it is. In our jungles soils are generally of high acidity, derived either from the high humus cover in heavy rain forest, or from the laterite of more open forest, and it was logical to expect that leeches would be happy living in an acid medium. I also tried malariol on them which seemed to have little or no effect.

Ever since these experiments I have used coarse soap rubbed into ankle puttees as a protective. It is the finest protective I know. It will remain effective on 6 in. monsoon days,

Meconopsis nepalensis
Family: Papaveraceae

This plant with golden-yellow flowers grows on the temperate Himalayan slopes of Darjeeling, Sikkim, and Nepal, at 3,000–3,800 m. Young parts are clothed with soft hairs, and the oblong, lanceolate leaves are hirsute. In the words of Hooker, "This superb plant when seen from a distance, resembles a small yellow Hollyhock. It was first discovered by Dr. Wallich's collectors in Nepal and I found it in the damp interior valleys of Sikkim, growing amidst a rank and luxuriant herbage on the skirts of Silver-fir forests, at [3,300–3,600 m] above sea level. The accompanying figure is taken from a sketch of my own, of a specimen that was [c. 1.5 m] high. The whole plant like its congeners, abounds in a bright, chrome yellow acrid juice, and is considered highly poisonous."

From J.D. Hooker, *Illustrations of Himalayan Plants*, 1855.

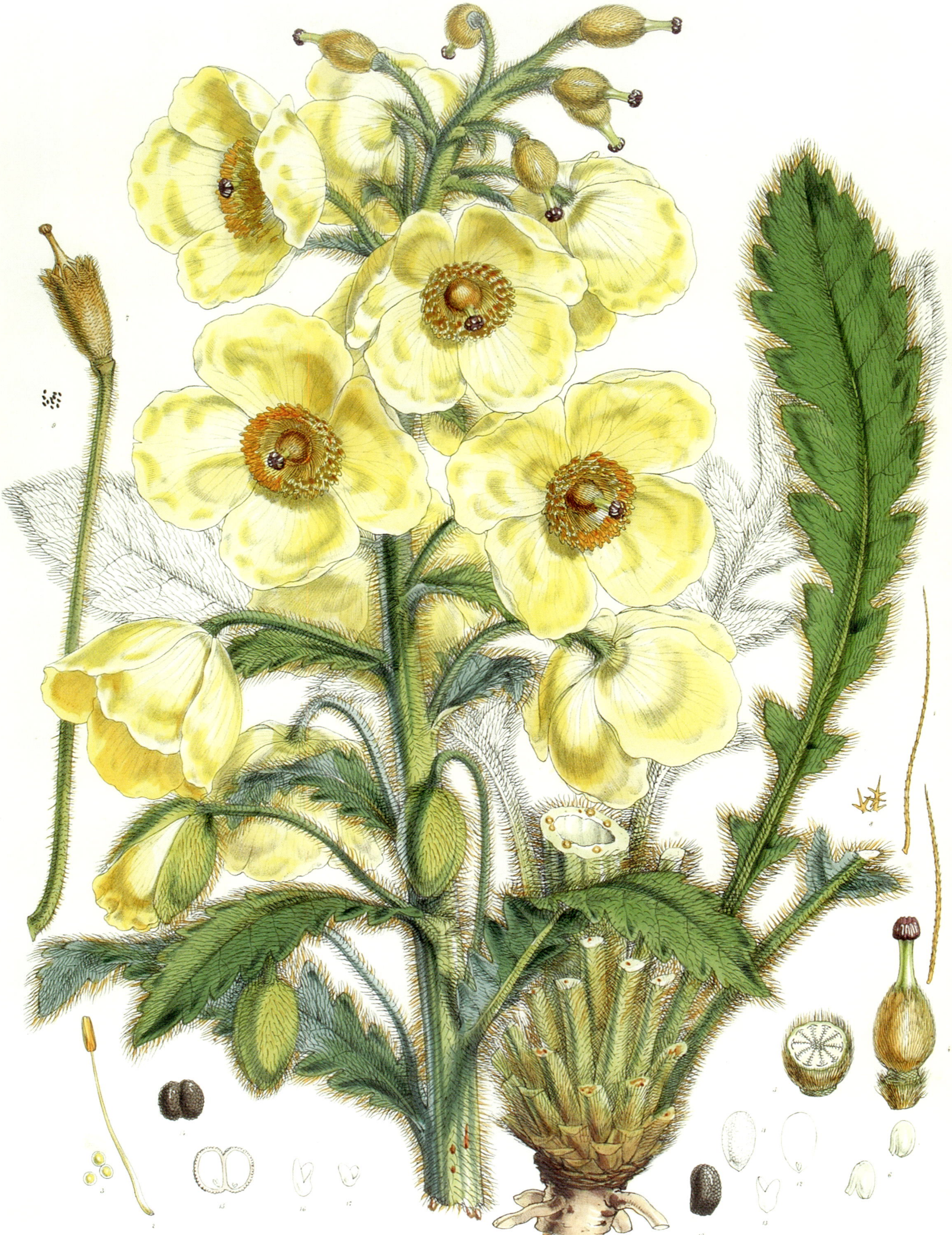

Meconopsis nepalensis

Courtesy
Pheroza & Jamshyd Godrej

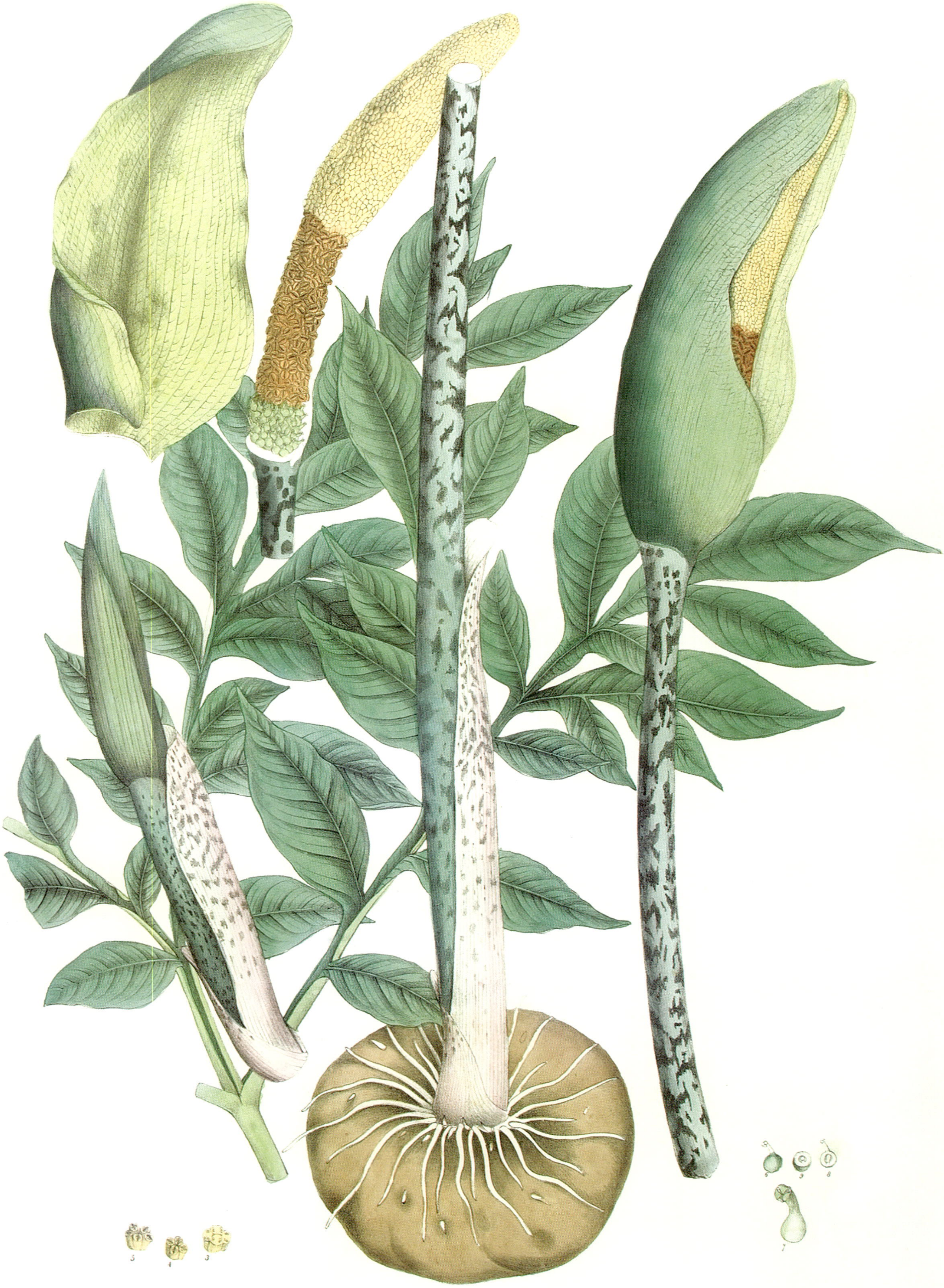

Thomsonia nepalensis

In memory of Dosa & Kekoo Naoroji
– Rishad Naoroji

Thomsonia nepalensis
Family: Araceae
Synonym: *Amorphophallus nepalensis*

This plant is named after Dr. Thomson, a medical officer in the Bengal Army, an eminent scientific botanist and a traveller; author of *Travels in the Northwest Himalayas*, and joint author with Hooker of *Flora Indica*. It is found from Nepal (previously the species name was spelt *napalensis*) to Sikkim, between 600 and 1,800 m. The inflorescence appears before the leaves, in April–May, emitting a strong peculiar smell. The spathe is large and green or pale yellow, boat-shaped, 30–45 cm; the spadix exposed, very stout with a rounded apex covered with tubercles, green changing to yellow, 13–25 cm long and 2–3 cm broad. The male flowers occur in the upper zone of the spadix while the female flowers are found in the lower zone. The leaf is divided into oblong-lanceolate leaflets. The petiole is stout and long, up to 40 cm. The tuber is 10–15 cm across. This seems to be the poisonous plant described by Dr. Kirtikar in *JBNHS* Vol. 7 under the name of *Pythonium wallichii*, and according to him very common in the forests of Thane, Maharashtra.

From N. Wallich, *Plantae Asiaticae Rariores*, 1832.

and stand up to occasional crossing of streams. If it is well rubbed into the puttees the rain keeps a constant dribble of soapy water flowing over the boots for a considerable period. Protection is almost complete in the worst leech jungle.

Many planters in South India use a leech proof stocking of fine woven cotton which is good until the leech climbs over the top. Silk stockings or socks also seem to defeat their efforts at penetration.

A curious fact about leeches is that elephant, bison, sambar, Nilgiri Thar [tahr], wild dogs, panther, and tiger live quite happily in sholas and forest which are so leech infested that the local jungle races will not enter them in the monsoon. I have shot many wild animals in such forest, and none of them have ever shown signs of having been worried by leeches. On the other hand tame dogs, cats, horses, and domestic cattle will all be attacked freely and viciously. Domestic cattle – particularly those with English blood – will suffer severely in condition if they graze continuously in leech-infested areas. I feel sure that wild animals living in these jungles would be unable to live there if they did not have some natural protection. It is a fact that *Bos gaurus* is the only ox which is known to sweat. My guess is that the natural protection of all wild animals living in leech-infested jungles is an alkaline skin secretion.

I have never seen a leech climb a tree, and I have never had a leech drop on me from a tree. In very bad jungle they can be picked up off undergrowth shrubs, but even so they are rarely above knee height. From that height they may easily escape notice until they have worked up to the back of the neck.

It is a well-known fact that in bad jungle it pays to walk first in the column, and that those following always collect more leeches than those in front. I suspect that the leeches feel vibrations – probably heart pulsations – and start moving as soon as they do so. When I was a young assistant we used to have leech races on the club billiard table. The owners put their leeches at one end and then pricked their fingers and held them against the cushion at the other end. Most of the leeches would finish the course without trouble and surprisingly quickly.

As far as South India is concerned it is not true that heavy burns destroy the leech population.

I saw a good deal of the Anamallai jungle felled and opened in tea. It was magnificent evergreen ironwood rain forest at from 3,000–4,000 ft. elevation – often from 100–120 ft. high. The system then was to fell the jungle in August, and put a fire into it in January. A good burn – which meant a very fierce burn – was desired, and it was often impossible to walk over the burnt area for four or five days after the burn. In spite of this very fierce burn, many areas of tea in the Anamallais are still infested with leeches. It is possible that they may have re-invaded the cleared areas from swamps and unfelled jungle on the boundaries, but the fact remains that in many places they are still plentiful under the tea and in short grazing grass.

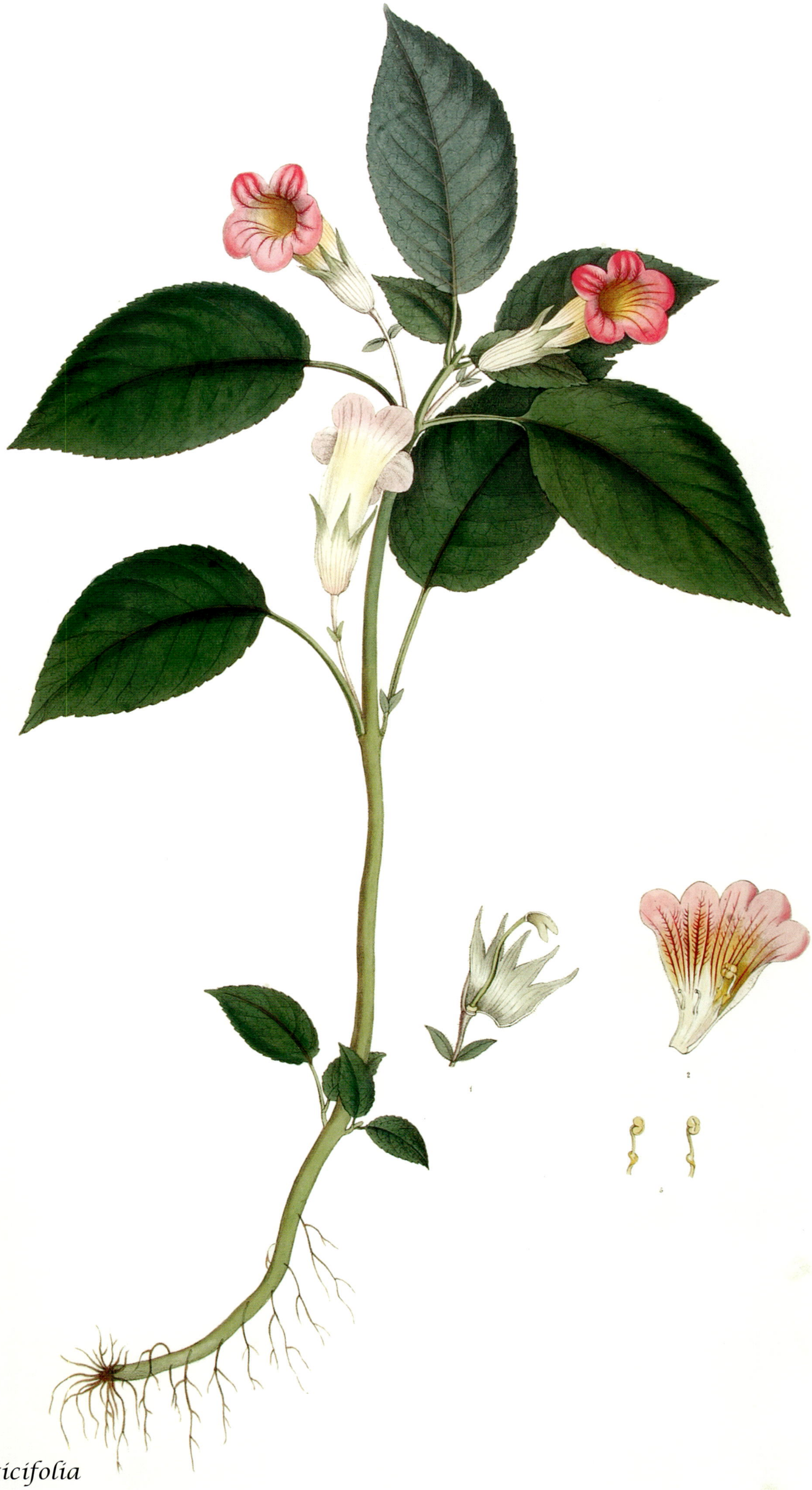

Chirita urticifolia

COURTESY
VIJAY SALES,
MUMBAI

Nettle-leaved Chirita
Chirita urticifolia
Family: Gesneriaceae
Synonym: *C. grandiflora*

This herb, with unbranched stem, about 40 cm high, is found in Arunachal Pradesh, and throughout Nepal, Bhutan, western China, and northern Myanmar. The leaves are stalked, opposite, 2.5–16 cm long, 1.7–16 cm wide, elliptic, serrate, hispid above and on the veins beneath. The red-purple flowers appear in August–September. They are streaked with yellow, funnel-shaped, and drooping, one or several borne on hairy stalks. This plant thrives well on moist ground, under trees.

From N. Wallich, *Plantae Asiaticae Rariores*, 1832.

In the grass hills above the Anamallais – between the Anamallais and the High Range – at elevations from 6–8,000 ft., leeches are common in the short open tussocky, grass land, most of which is burnt every dry weather and has probably never been under heavy forest. They can be a nuisance when fishing for trout in the Kornelar river and can be picked up there even in the middle of the dry weather. Incidentally a sure way of making leeches drop off is to stand or wade in running water.

It is very curious that leeches are entirely absent from all the sholas and grass land on the top of the Nilgiri Plateau. The sholas would appear to be ideal for them. The elevation, undergrowth, rainfall, and acidity of the soil do not differ in any obvious respect from that of the High Range and the Anamallais grass hills where they abound. Leeches are found below, and on the edges of the plateau, and there is no obvious reason why they should not have colonised the plateau land.

I have tried sifting the top soil under heavy rain forest in the dry weather, and have never found any traces of adult leeches. Within a matter of 48 hours after the fall of the first really heavy shower they will appear in thousands – many of them apparently full grown.

Leech bites do not usually go septic unless one is run down – they may cause annoyance by refusing to stop bleeding for quite long periods, and they nearly always irritate if they are on a small blood vessel and not a relatively large one. In bad leech jungle loss of blood may be considerable but once one has overcome repulsiveness they do not worry one unduly.

What happens to a leech after it is fully gorged with a blood meal, and why they should want a blood meal at all, are unexplained mysteries. Their normal function would seem to be aeration and movement of the soil similar to that of earthworms.

J.L.H. WILLIAMS
Tea Estates India Ltd., Post Box No. 13,
Coimbatore, March 15, 1954.

From *JBNHS* Vol. 52, No. 3.

On Nature's Trail: Biographical Notes on Some Pioneering Naturalists in India

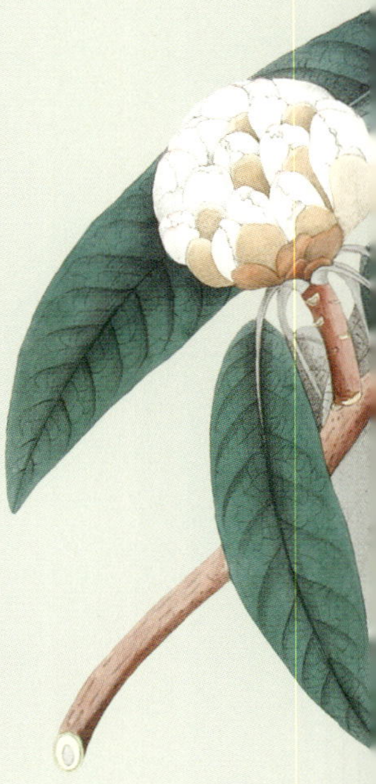

It is not surprising that the study and practice of medicine, which in most cases must have stemmed from a more general inclination for the biological sciences, often went hand-in-hand with a keen interest in natural history. Many of the most distinguished naturalists of the 18th and 19th centuries belonged to the medical profession.

James Forbes was a British artist and writer. He was born to a Scottish family in London in 1749. In 1765 he was appointed writer for the East India Company and he spent nearly 20 years in India. He wrote and made sketches on all aspects of life in India, its fauna and flora, and architecture. He was one of the first Europeans to draw the Taj Mahal, in 1781. His notes and sketches were put together and published as *Oriental Memoirs* in four volumes, 1812–13. The engravings for this work, based on Forbes' drawings, were prepared by a British botanical artist named William Hooker (no relation of William Jackson Hooker and Charles Hooker). The four volumes are a valuable document of the natural history and culture of late-18th-century India. Forbes died in England in 1819.

William Roxburgh was born at Craigie in Ayshire, Scotland on June 29, 1751. He studied medicine and botany at the University of Edinburgh. At the age of 17 he joined an East India Company ship as surgeon's mate. He joined the Madras Medical Service in 1776 as Assistant Surgeon, and was promoted to Surgeon in four years.

In 1781, considering his qualifications in botany, the East India Company appointed him Superintendent of the Samalkot Botanic Garden in the Northern Circars of Madras Presidency, now in Andhra Pradesh. He employed local artists to make botanical drawings and collected 700 illustrations by 1790, when he became Naturalist to the Madras Government. In 1793 the Government of Bengal asked him to take charge of the Calcutta Botanic Garden. Here Roxburgh continued to work assiduously, cataloguing the plants and having drawings made of them. A complete catalogue of the plants in the Calcutta Botanic Garden, titled *Hortus Bengalensis*, was published in 1814. As a member of the Asiatic Society of Bengal he published several papers in their Journal.

Roxburgh was also a pioneering meteorologist. He collected detailed data on the weather over many years, on the basis of which he published his theories on the role of climate change and years of famine.

William Roxburgh returned to Edinburgh where he died on April 10, 1815. His work contributed significantly to the study of botany in India. His major work, *Flora Indica, or Descriptions of Indian Plants* was published posthumously in two volumes, in 1820 and 1824.

Nathaniel Wallich was born in Copenhagen on January 28, 1786, as Nathan ben Wulff. His father was Wulff ben Wallich, a merchant from Altona near Hamburg who had settled in Denmark. Nathaniel obtained his diploma from the Royal Academy of Surgeons in Copenhagen in 1806, and the following year sailed for India where he had been appointed Surgeon at the Danish settlement of Serampore. Soon after, the British East India Company took over the settlement, and Wallich was imprisoned, to be released on parole in 1809 because of his scholarly achievements. From 1814 Wallich joined the Company's service, and was soon appointed as assistant to William Roxburgh in Calcutta. From 1817 until 1846 he served as the Superintendent of the Botanic Garden in Calcutta.

Wallich passionately pursued his interest in India's flora, undertaking expeditions to Nepal, western India, and lower Burma. He readily helped any plant enthusiast who stopped in Calcutta on their way to the Himalaya on collecting expeditions. He became a member of the Asiatic Society of Bengal, and proposed the forming of an Oriental Museum by the Society, which is today the Indian Museum. He was appointed the Honorary Curator, and then the Superintendent of this museum which grew rapidly under his guidance. Most of the contributors to the museum were Europeans, with the exception of Babu Ramkamal Sen, later the first Indian Secretary of the Asiatic Society. Wallich himself contributed numerous botanical specimens to the museum. He was responsible for packing specimens to be shipped to England, and his innovative methods led to his having one of the best records for keeping plant material alive.

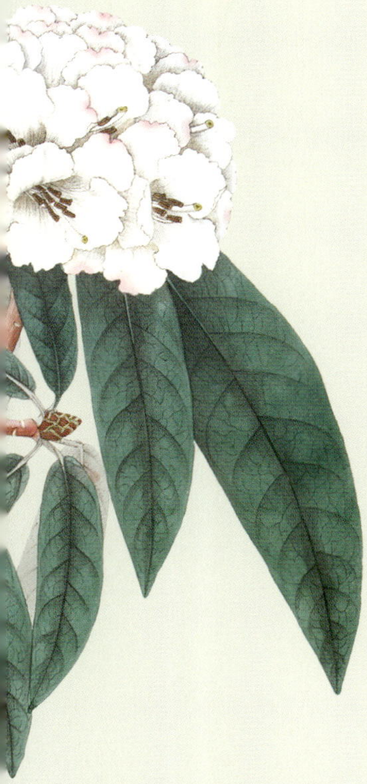

Wallich prepared a catalogue of more than 20,000 plant specimens, and published two important works – *Tentamen Florae Nepalensis Illustratae* (2 volumes, 1824–26) and *Plantae Asiaticae Rariores* (3 volumes, 1830–32). The latter included drawings by artists employed by the Calcutta Botanic Garden – 146 by Gorachand, 109 by Vishnupersaud, and one by Rungiah. The rest of the plates were by John Clark, and three by William Griffith. Wallich encouraged Indian artists to sign their works, although this was not the practice.

During 1837–38 Wallich was Professor of Botany in Calcutta Medical College. He was also an honorary doctor at the University of Copenhagen and a member of the Royal Danish Academy of Sciences and Letters. He retired to London in 1847, and on his death on April 28, 1854 he was buried in Kensal Green Cemetery. His main herbarium is at the Kew Gardens, with duplicate specimens in Copenhagen's Botanical Museum and a full set of his collections in the Central National Herbarium of the Botanical Survey of India in Calcutta (Kolkata).

Wallich's phenomenal contribution to Indian botany was recognized in the naming of several species of plants for him, many of them Himalayan species, e.g. *Pinus wallichiana* (Blue Pine), *Taxus wallichiana* (Himalayan Yew), and *Ulmus wallichiana* (Himalayan or Kashmir Elm). The beautiful Cheer Pheasant (*Catreus wallichi*) is also named for him. The author abbreviation Wall., cited at the end of a botanical name, indicates Wallich.

John Forbes Royle, a British botanist and teacher of materia medica, was born in Kanpur (then called Cawnpore), India in 1799. Like Roxburgh before him, he entered the service of the East India Company as Assistant Surgeon. He devoted a lot of his time to studying botany and geology, making large collections from the Himalaya during the ten years he held the post of Superintendent of the Company's botanical gardens at Saharanpur. He also investigated medicinal plants in India and their use among local peoples, recording his findings in an essay, "On the Antiquity of Hindoo Medicine" (1837).

In 1837, Royle was appointed Professor of Materia Medica in King's College, London, a post he held for nearly 20 years. During this period he conducted a special department of correspondence related to vegetable products at East India House in London, and also created there a museum of technical products from India. In the Great Exhibition of 1851, Royle superintended the Indian section.

Royle's major works are *Illustrations of the Botany and Other Branches of Natural History of the Himalaya Mountains and of the Flora of Cashmere* (2 volumes, 1839). He published several papers in scientific journals, and essays including "An Essay on the Productive Resources of India" (1840), "On the Culture and Commerce of Cotton in India and Elsewhere" (1851), and "The Fibrous Plants of India fitted for Cordage" (1855).

John Forbes Royle died at Acton near London on January 2, 1858.

Thomas Caverhill Jerdon was born in Durham county, UK on October 12, 1811, and studied at Edinburgh University. He went to India in 1936, as an Assistant Surgeon in the British East India Company.

Jerdon started collecting birds soon after his arrival in India. He was posted in the Deccan and the Eastern Ghats, and, after his marriage to Flora MacLeod, in the Nilgiris (Ooty) and then in Nellore and in Tellicherry where he was Civil Surgeon until 1851. Flora had an interest in botanical art, particularly in orchids.

Jerdon sent his specimens of birds collected during his early travels to William Jardine in Scotland for identification, but by the time they reached there they had been infested with moths. After this Jerdon relied on his own identifications. His *Catalogue of the Birds of the Indian Peninsula*, published in 1839–40, included 420 species, almost twice as many as had been published in the early 1830s by Colonel W.H. Sykes (1790–1872).

Jerdon had a broader interest in natural history, and his studies included descriptions of plants, insects, amphibians, reptiles, and mammals, as well as birds. In 1852 he was promoted to Surgeon and posted in the Central Provinces, where he served during the 1857 Uprising. In 1858 he was made Surgeon Major in the Madras Regiment. Before joining his new post he went on leave to Darjeeling. On his way there he met the Viceroy Lord Canning and proposed his scheme of preparing a series of manuals of the vertebrates of India. Jerdon was subsequently transferred to the Government of India and placed on special duty for this purpose. By 1868 when he retired to England, the manuals on mammals and birds were published while the one on reptiles was in press. These manuals were the forerunner of *The Fauna of British India* series that was prepared under the editorship of W.T. Blanford.

Jerdon's most important work was *The Birds of India* published in 2 volumes in 1862–64. Other books included *The Game Birds and Wildfowl of India* (1864) and *The Mammals of India* (1867). About an early work of his, *Illustrations of Indian Ornithology*, which featured illustrations by Indian artists, he later wrote: "I published a selection of fifty coloured lithographs, chiefly of unfigured birds of Southern India; and the excellence and faithfulness of the drawings (the originals of all of which were painted by natives, and half the number also lithographed and coloured at Madras) has been universally allowed."

T.C. Jerdon died at Norwood on June 12, 1872. His pioneering work in the field of Indian ornithology has led to his being considered among the founders of Indian ornithology.

Sir Joseph Dalton Hooker, OM, GCSI, CB, MD, FRS, Charles Darwin's closest friend, was one of the greatest 19th-century British explorers, and one of the founders of geographical botany.

Born in Suffolk on June 30, 1817, Joseph Hooker was the second son of Maria Sarah Turner and Sir William Jackson Hooker, a famous botanist. From the age of seven, Joseph attended his father's lectures at Glasgow University. His interest was kindled in plant distribution and in the voyages of explorers like Captain James Cook. He studied medicine in Glasgow, and after completing his MD in 1839, he joined Captain James Clark Ross's expedition to the South Pole as Assistant Surgeon on board ship.

In November 1847 Hooker left for India where he would go on a three-year Himalayan expedition. From his base in Darjeeling, where he stayed with naturalist Brian Houghton Hodgson, he travelled through Sikkim, Nepal, and Tibet, and later eastwards to the Khasi hills in Assam. He was briefly taken prisoner by the Diwan of Sikkim, when he lost some of his specimens which he later replaced. The first European to collect plants in the Himalaya, Hooker dedicated his *Himalayan Journals, or, Notes of a Naturalist in Bengal, the Sikkim and Nepal Himalayas, Khasia Mountains Etc.* (published by the Calcutta Trigonometrical Survey Office in 1854) to Charles Darwin.

Hooker authored several major works on the flora gathered during his Antarctic voyage, in New Zealand and Tasmania, and along the Niger. With George Bentham he co-authored the famous *Handbook of the British Flora* (1858) and *Genera Plantarum* (1862–83). He also continued his father's project *Icones Plantarum* (Illustrations of Plants), producing volumes 11 to 19. His travels in India added to this prolific output: *Rhododendrons of the Sikkim-Himalaya* (1849–51), *Illustrations of Himalayan Plants* (1855), *Flora Indica* (with Thomas Thomson, 1855), *A Century of Indian Orchids* (1859), *The Flora of British India* (1875–97), *Handbook to the Ceylon Flora* (1898–1900), and *An Epitome to the British Indian Species of Impatiens* (1904–06).

In 1851 Hooker returned to England where in 1865 he succeeded his father as Director of the Royal Botanical Gardens, Kew, serving for 20 years. He later went on expeditions to Palestine (1860), Morocco (1871), and the western United States (1877). Having been elected a Fellow of the Royal Society at the age of 30, he was its President from 1873 to 1877. He received three of its medals – the Royal Medal in 1854, the Copley Medal in 1887, and the Darwin Medal in 1892. The Linnean Society awarded him the Wallace-Darwin Medal. For his magnum opus, *Flora of British India*, he was made a Knight Grand Commander of the Star of India. On completing 90 years of age he was awarded the Order of Merit.

Joseph Hooker died at the age of 94, on December 10, 1911. He was buried alongside his father in the churchyard of St Anne's on Kew Green, close to the Gardens he had looked after for so many years of his life.

Kanhoba R. Kirtikar was born in Bombay on May 24, 1849. After the usual college education he joined the Grant Medical College in 1871. Three years later he left for England to compete for the Indian Medical Service. He returned to India in 1877 and was placed on general duty in Bombay. During the Second Anglo-Afghan War he was on field service from 1878 to 1880. For his gallant behaviour at the battle of Maiwand, Kirtikar was appointed Civil Surgeon of Thana in 1881. The following year saw him in a great variety of offices: Fellow of the Bombay University, Syndic in Medicine, Professor of Anatomy, Botany, and Materia Medica at the Grant Medical College. In addition he held a number of medical appointments. In 1902 he became Brigade Surgeon-Lieutenant-Colonel. In 1904 he retired after completing the 55th year of his age and 27 years of useful and distinguished service.

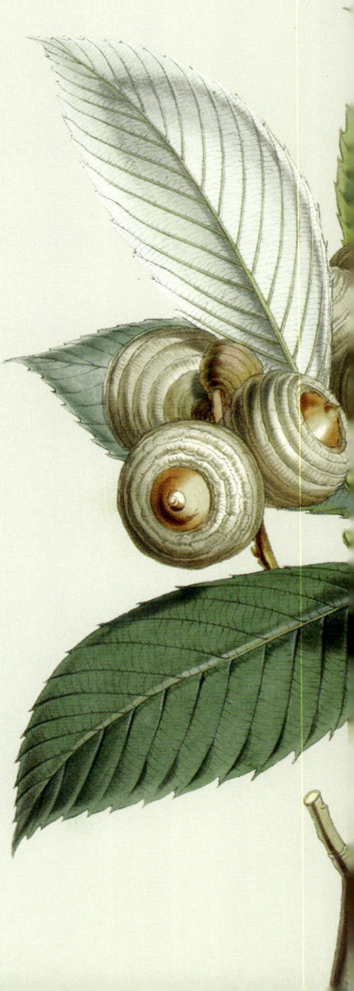

Lt. Col. Kirtikar's interests were many and varied – social, literary, and scientific. In one subject especially he showed not only a keen interest but also a marked talent, viz., Botany. After his retirement from public service he chose Andheri, a suburb of Bombay, as his residence for the rest of his life. Content amidst his books, chiefly botanic, he delighted in showing visitors his valuable volumes, his microscopes, his collections of dried plants, his watercolours of Algae and Fungi, and many other things that interest only an enthusiast. All this was the result of his spare hours (which had been few, for he had been a busy man); he had kept his eyes open, he had read a good deal, he had seen much in many lands, he had taken notes on many botanical subjects and jotted them down in books and on slips of paper that were scattered all over his library. There is no department in Botany, except perhaps physiology, which he did not cultivate. His many contributions to the *JBNHS* were written at a time when professional duties claimed all his energy, and it is astonishing that he was able to do so much. He died in 1917, and his book on the *Medicinal Plants of India* was published posthumously by his friend Major B.D. Basu to whom he had entrusted the work.

Edward Charles Stuart Baker, CIE, OBE, FZS, FLS (b. 1864) was a British ornithologist. In 1883, after his education at Stratford-upon-Avon, he followed his father into the Indian Police Service. He first served in Bengal but in August 1886 was transferred to Gauhati (Guwahati) and the remainder of his service was in Assam, except for a few months in Dacca (Dhaka) in East Bengal.

He joined the Society in 1898 and his first paper, "On the genus *Chloropsis*" appeared in the sixth volume of the *Journal.* Many other contributions appeared from his pen, but it will be for his papers on Indian Ducks and Game Birds that he will always be remembered. The series of articles were so popular that W.S. Millard, Honorary Secretary of the BNHS, decided to publish them in book form. *Indian Ducks and Their Allies* was an immediate success and soon out of print. Apart from the financial success, the book was the means of drawing attention to the activities of the Society, thereby increasing the membership. Stuart Baker was therefore encouraged to write a further series of papers on Game Birds, starting with the Snipe, Bustards, and Sandgrouse, followed by the true game birds, and finally on Rails and Waders.

Stuart Baker collected and studied birds' eggs, making a special study of the eggs and habits of different Indian cuckoos and their fosterers. In 1942 he published *Cuckoo Problems,* the result of years of careful observation and collecting.

In 1911, Stuart Baker returned to England where he assumed charge of the Port of London Police. He devoted much of his spare time to ornithology and in 1913 was elected Secretary and Treasurer of the British Ornithologists' Union, an office held for ten years.

In 1913 Stuart Baker began publishing reviews on Indian birds, beginning with the Silver Pheasants and Drongos. Early in 1920 he was recommended to the editor of *The Fauna of British India* for the job of preparing the new edition of the volumes on Birds. He first prepared a "Handlist of the Birds" which was published in *JBNHS.* For the next eight or nine years he was busy with the revision and by working long hours he was able to bring out the first volume in 1922, astonishingly quick for a work of this nature. The first edition by Oates and Blanford had four volumes; these were expanded to six in the second edition, with two additional volumes on synonymy.

Stuart Baker was a great shikari and wrote interesting accounts of his experiences with big game in *JBNHS* as well as in *Field* and *Asian.* He was a good shot and was twice tossed by a gaur, trampled on by a rhinoceros, and even lost his left arm in an encounter with a man-eating leopard near Silchar. He was following the animal which had taken refuge in a villager's hut, when it charged, knocked the rifle out of his hand, and seized him, but he managed to save himself by thrusting his left arm down the animal's throat. His shikaris came to his aid and dispatched the leopard. Despite the loss of an arm he continued to shoot both with rifle and with shot-gun. With his single hand he would dexterously place a small egg in a box of sand and neatly inscribe it.

An accomplished tennis player and horseman, in 1932 Stuart Baker was awarded the C.I.E. for his services in India, and after WWII was made a Companion of the Order of the British Empire. He died at his home in Upper Norwood, London, on April 16, 1944, age 80.

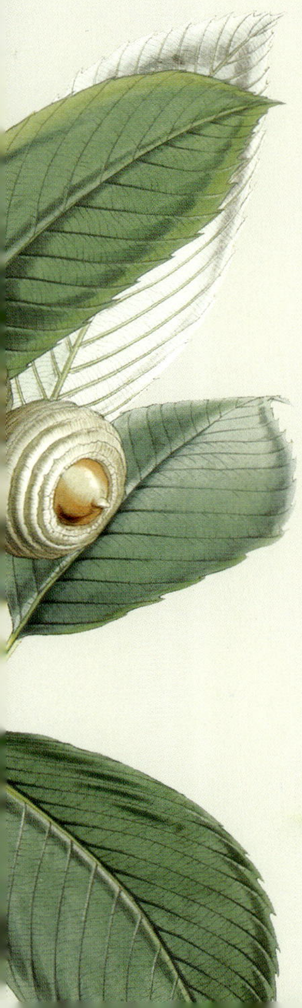

Compiled by the Editors from various sources, primarily Wikipedia online. The biographical details for Lt. Col. K.R. Kirtikar and E.C. Stuart Baker are taken from issues of *JBNHS.*

Bibliography

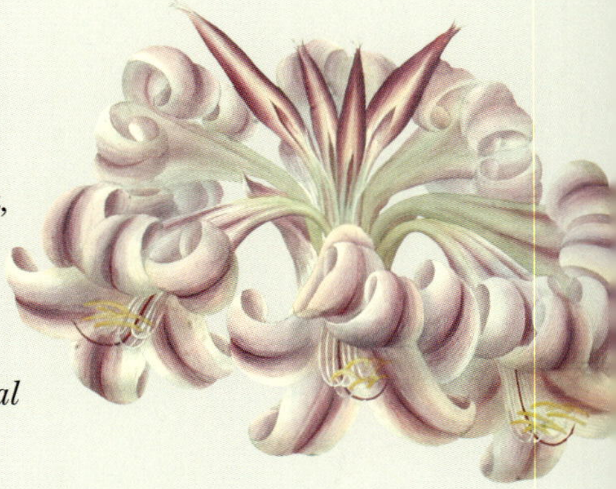

Aitken, E.H. *The Common Birds of Bombay*, 1894.

Ali, Sálim. *Birds of Kerala*, 3rd edition, revised by R. Sugathan, 1999.

Ali, Sálim. *The Book of Indian Birds*, 13th edition, 2002.

Ali, Sálim and S. Dillon Ripley. *Handbook of Birds of India and Pakistan*, 10 vols., 1968–74.

Ali, Sálim and S. Dillon Ripley. *A Pictorial Guide to the Birds of the Indian Subcontinent*, 1983.

Archer, Mildred. *Company Paintings of the British Period*, 1992.

Atkinson, Captain George F. *Curry and Rice, The Ingredients of Social Life at "Our" Station in India*, 1859.

Badhwar, R.L and Sr. R.R. Fernandez. *Edible Wild Plants of the Himalayas*, 1964.

Baker, E.C. Stuart. *The Fauna of British India. Birds*, 8 vols., 2nd edition, 1922–30.

Baker, Lt.-Col. H.R. and C.M. Inglis. *The Birds of Southern India, Including Madras, Malabar, Travancore, Cochin, Coorg, and Mysore*, 1930.

Balfour, Surgeon General Edward. *The Cyclopaedia of India and of Eastern and Southern Asia*, 3 vols., 3rd edition, 1885.

Barnes, Lieut. Edwin H. *Handbook of the Birds of the Bombay Presidency*, 1885.

Biswas, Dr Kaliprada. *Common Medicinal Plants of Darjeeling and the Sikkim Himalayas*, 1956.

Biswas, Dr Kaliprada. *Plants of Darjeeling and Sikkim Himalayas*, Vol. I, 1966.

Blanford, W.T. *The Fauna of British India, Including Ceylon and Burma (Mammalia)*, 1888–91.

Blatter, Ethelbert. *Beautiful Flowers of Kashmir*, 2 vols., 1927, 1928.

Bose, T.K, S.K. Bhattacharjee, P. Das, U.C. Basak. *Orchids of India*, 2nd revised and enlarged edition, 1999.

Commissariat, Professor Manekshah S. *History of Gujarat*, Vol. II, 1957.

Coventry, B.O. *Wild Flowers of Kashmir*, Series I–III, 1923–30.

Dunbar-Brander, A.A. *Wild Animals in Central India*, 3rd impression, 1931.

Eardley-Wilmot, Sir S. *The Life of a Tiger, and the Life of an Elephant*, illustrated by Iris Eardley-Wilmot, 1933.

Finn, Frank. *Sterndale's Mammalia of India*, 1929.

Forbes, James. *Oriental Memoirs*, 4 vols., 1812–13.

Forsyth, Captain J. *The Highlands of Central India*, 1889.

Gee, E.P. *The Wild Life of India*, 1964.

Ghose, B.N. *Beautiful Indian Orchids*, 1959.

Gould, John. *A Century of Birds from the Himalaya Mountains* (with lithographs by Elizabeth Gould), 1832.

Gould, John. *The Birds of Asia*, 6 vols., 1850–73.

Greenway, James. C. *Extinct and Vanishing Birds of the World*, 1958.

Grimmett, Richard, Carol Inskipp, and Tim Inskipp. *Birds of the Indian Subcontinent*, 1998.

Grindlay, Captain Robert Melville. *Scenery, Costumes and Architecture Chiefly on the Western Side of India*, 1826.

H.A.L., the Old Shekarrey, *Sport in Many Lands*, 1877.

Henderson, M.R. *Malayan Wild Flowers*, 1955.

Hooker, J.D. *Illustrations of Himalayan Plants, chiefly selected from the drawings made for the late J.F. Cathcart, Esq. of the Bengal Civil Service, the descriptions and analyses by J.D. Hooker, the plates executed by W.H. Fitch*, 1855.

Hooker, J.D. *The Flora of British India*, 7 vols., 1875–97.

Issacs, Mozelle. *The Commoner Flowering Plants of Western India*, 1927.

Jepson, Stanley. *Big Game Encounters*, 1936.

Jerdon, T.C. *Illustrations of Indian Ornithology containing fifty figures of new, unfigured and interesting species of birds, chiefly from the south of India*, 1847.

Jerdon, T.C. *The Mammals of India*, 1862.

Jerdon, T.C. *The Birds of India*, 2 vols., 1862–64.

Kanjilal, U.N, P.C. Kanjilal, A. Das, and C. Purkayastha. *Flora of Assam*, 2 vols., 1936.

Kehimkar, Isaac. *Common Indian Wild Flowers*, 2000.

Kipling, Rudyard. *Just So Stories*, 1902.

Knowles, George Hogan. *Terrors of the Jungle*, 1932.

Kothari, Ashok S. *A Celebration of Indian Trees*, 2007.

Kothari, Ashok S. and Boman F. Chhapgar. *Sálim Ali's India*, 1996.

Kothari, Ashok S. and Boman F. Chhapgar. *Treasures of Indian Wildlife*, 2005.

Kunjilal, Rai Bahadur Upendranath. *Forest Flora of Chakrata, Dehra Dun and Saharanpur forest divisions of United Provinces*, 3rd edition, revised by Basant Lal Gupta, 1928.

Lydekker, R. *The Game Animals of India, Burma, Malaya and Tibet*, 2nd edition, revised by J.G. Dollman, 1925.

Lydekker, Richard, Sir Harry Johnston, and Prof. J.R. Ainsworth-Davis. *Harmondsworth Natural History: A Complete Survey of the Animal Kingdom*, 3 vols., 1910.

Mundy, Captain Godfrey Charles. *Pen and Pencil Sketches, being the Journal of a Tour in India*, 2 vols., 1832.

Murray, James A. *Indian Birds or the Avifauna of British India*, 2 vols., 1887, 1890.

Polunin, Oleg and Adam Stainton. *Flowers of the Himalaya*, 1984.

Pradhan, Uday C. and Sonam T. Lachungpa. *Sikkim-Himalayan Rhododendrons*, 1990.

Rasmussen, Pamela C. with John C. Anderton. *Birds of South Asia, The Ripley Guide*, 2 vols., 2005.

Risley, H.H. *The Gazetteer of Sikhim*, 1894.

Royle, J. Forbes. *Illustrations of the Botany and other Branches of Natural History of the Himalaya Mountains and of the Flora of Cashmere*, 2 vols., 1839.

Sanderson, G.P. *Thirteen Years Among the Wild Beasts of India*, 1870.

Stainton, Adam. *Flowers of the Himalaya: A Supplement*, 1988.

Toovey, D.J. and Jacqueline W., eds. *Tigers of the Raj: Pages from the Shikar Diaries – 1894–1949 of Colonel Burton, Sportsman and Conservationist*, 1989.

Troup, R.S. *The Silviculture of Indian Trees*, 3 vols., 1921.

Wallich, Nathaniel, *Plantae Asiaticae Rariores, Descriptions and Figures of Unpublished East Indian Plants*, 3 vols., 1830–32. (A combined volume was published in 1832).

Whistler, Hugh. *Popular Handbook of Indian Birds*, 4th edition, revised by Norman B. Kinnear, 1949.

White, J. Claude. *Sikhim & Bhutan, Twenty-one years on the North-East Frontier 1887–1908*, 1909, reprint 1971.

Williamson, Thomas. *Oriental Field Sports*, 2 vols., 2nd edition, 1819 (first published 1808).

Various issues of the *Journal of the Bombay Natural History Society (JBNHS), Bengal Sporting Magazine, Journal of the Bengal Natural History Society*, and *Journal of the Darjeeling Natural History Society*.

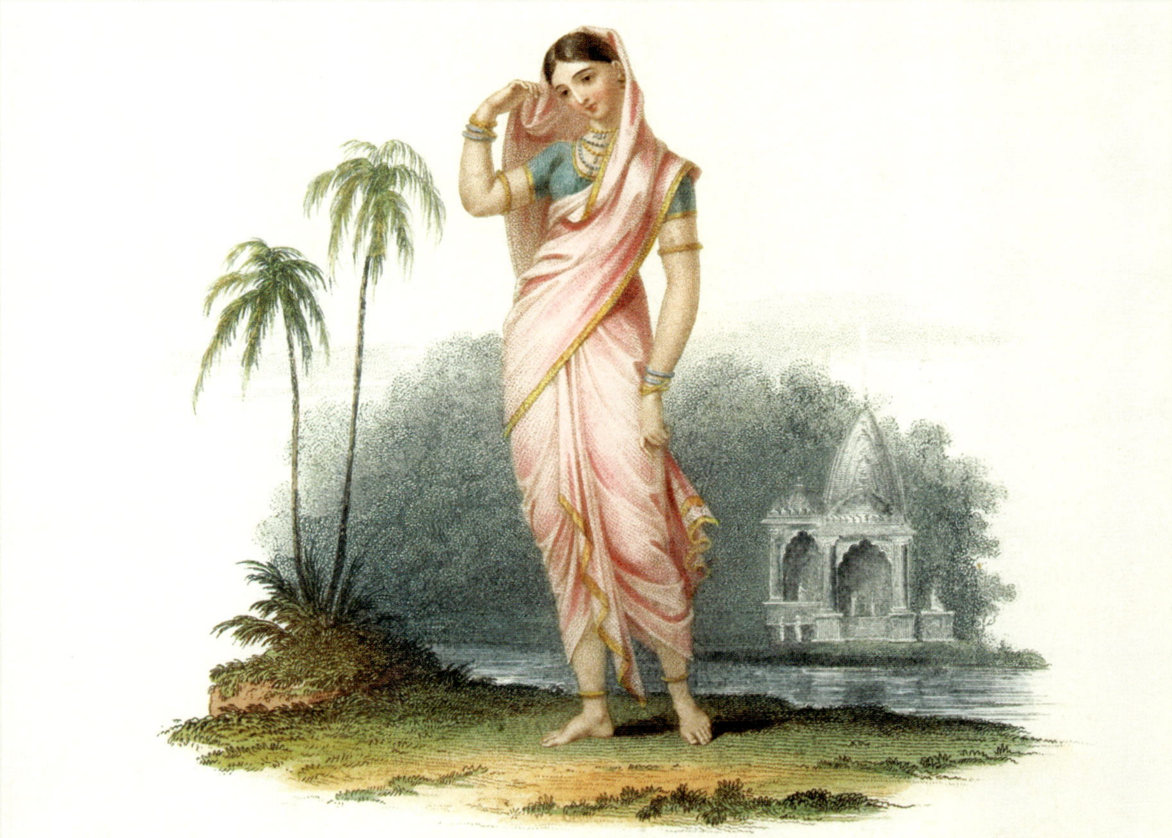

Illustration from the title page of Captain Grindlay's *Scenery, Costumes and Architecture*, 1826.

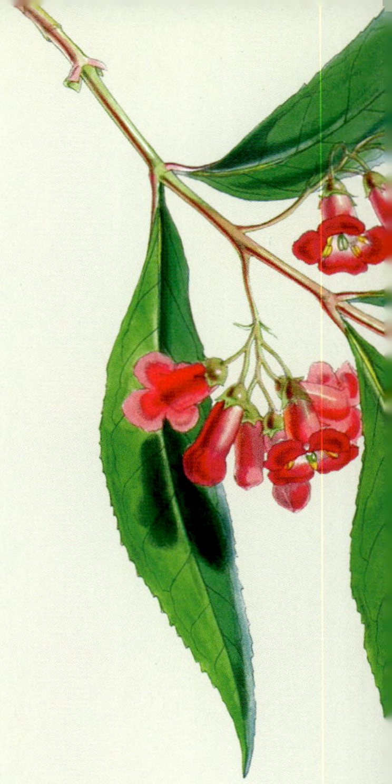

Sponsors

AMBUJA CEMENT
Elegant Business Park, 'D' Block,
MIDC Cross Road 'B',
Off Andheri-Kurla Road,
Andheri (E), Mumbai 400059
(Pages 182, 186)

ARUNA & CHANDRAKANT PATEL
Holiday Inn Express Hotel & Suites,
12439, Northwest Freeway,
Houston, Texas, TX 77092, USA
(Page 145)

ASIAN STAR COMPANY LIMITED
Arvind Shah, Managing Director,
114, Mittal Court-C,
Nariman Point, Mumbai 400021
(Page 47)

ATAM DEEP CHARITABLE TRUST
Dinesh Shah, Managing Trustee,
c/o Asian Star Company Limited,
114, Mittal Court-C,
Nariman Point, Mumbai 400021
(Page 158)

B.G. DESHMUKH
Acacia II, Plot No. 26,
Magarpatta Township,
Hadapsar, Pune 411028
(Pages 88–89, 130)

BALWANT G. BAPAT
4/91 Vijay Nagar Society,
S.N. Marg,
Andheri (E), Mumbai 400069
(Page 134)

BANK OF INDIA
Star House, C-5, 'G' Block,
Bandra-Kurla Complex,
Bandra (E), Mumbai 400051
(Pages 103, 177)

BHANSALI & CO.
Jitendra Bhansali, Partner,
640-646, Panchratna,
Mama Parmanand Marg,
Opera House, Mumbai 400004
(Page 64)

CENTRAL BANK OF INDIA
S. Sridhar, Chairman & Managing Director,
Chander Mukhi,
Nariman Point, Mumbai 400021
(Pages 86, 138)

DHUN PANTHAKI
Wadia Building, 1st floor,
S. Gandhi Marg,
Mumbai 400002
(Pages 80, 85, 118–19, 129, 133, 153)

DILIPKUMAR V. LAKHI
Managing Trustee, Lakhi Trust,
102/115, Prasad Chambers,
Opera House, Mumbai 400004
(Page 157)

DIMEXON DIAMONDS LIMITED
Paresh Mehta, Director,
704, Raheja Chambers,
213 Nariman Point, Mumbai 400021
(Page 35)

EKNATH A. KSHIRSAGAR
Tarangini, Twin Towers Road,
Prabhadevi, Mumbai 400025
(Page 165)

FIRDOS S. CAMBATTA
Trustee, The Serenity Trust,
"Cambatta Central",
Bharat Bobbins Ltd. Premises,
Rakhial Road,
Ahmedabad 380023
(Page 79)

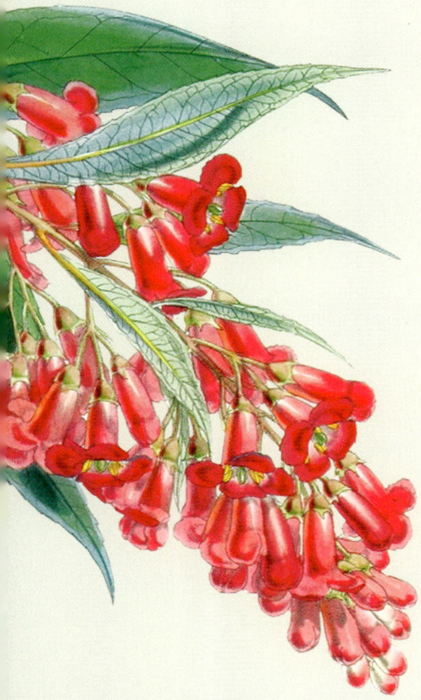

GODAVARI BIOREFINERIES LTD.
Samir S. Somaiya, Joint Managing Director,
Somaiya Bhavan,
45/47, Mahatma Gandhi Road,
Fort, Mumbai 400001
(Pages 21, 22)

HEENA BUILDERS AND DEVELOPERS
Deepak H. Shah, Managing Director,
Vinayak Co-op. Housing Society,
Pond Gauthan,
Baji Prabhu Deshpande Marg,
Vile Parle (W), Mumbai 400056
(Back Endpaper)

HEENA GROUP OF COMPANIES
Deepak H. Shah, Managing Director,
Vinayak Co-op. Housing Society,
Pond Gauthan,
Baji Prabhu Deshpande Marg,
Vile Parle (W), Mumbai 400056
(Back Cover)

HSBC
The Hongkong and Shanghai
Banking Corporation Limited,
52/60 Mahatma Gandhi Road,
Mumbai 400001
(Pages 40–41, 72–73, 124–25, 142–43)

INDUSIND BANK LIMITED
701 Solitaire Corporate Park,
167 Guru Hargovindji Marg,
Andheri (E), Mumbai 400093
(Front Cover, Pages 25, 91, 92)

JEWELEX INDIA PVT. LTD.
Piyush Kothari, Chairman,
605 Aman Chambers,
113 Queen's Road,
Mumbai 400004
(Page 169)

KANSAI NEROLAC PAINTS LIMITED
D.M. Kothari, Vice Chairman,
Nerolac House,
Ganpatrao Kadam Marg,
Lower Parel, Mumbai 400013
(Pages 28, 43)

KANTILAL CHHOTALAL
Ashish K. Mehta, Partner,
B-3, Hermes House,
Opera House, Mumbai 400004
(Page 96)

LONDON STAR DIAMOND CO. (INDIA)
PVT. LTD.
Kamlesh Jhaveri, Chairman,
1610/11 Prasad Chambers,
Opera House, Mumbai 400004
(Page 15)

M. SURESH & CO. PVT. LTD.
Suresh Pethani, Partner,
15-16th Floor, Jain Tower,
17, Mathew Road,
Opera House, Mumbai 400004
(Page 68)

MAHARAJA DIGVEERENDRASINHJI
INDRASINHJI SOLANKI
Digvir Nivas Palace,
Vansda 396580,
District Navsari, Gujarat
(Page 166)

MAHENDRA BROTHERS EXPORTS
PVT. LTD.
Jitendra C. Parikh, Director,
611, Panchratna,
Mama Parmanand Marg,
Mumbai 400004
(Page 52)

MAHINDRA & MAHINDRA LTD.
Keshub Mahindra, Chairman,
Gateway Building,
Apollo Bunder, Mumbai 400001
(Pages 117, 147, 161)

MAIZE PRODUCTS
Priyam B. Mehta, Managing Director,
P.O. Kathwada,
Ahmedabad 382430, Gujarat
(Page 181)

MAMTA & ZUBIN MEHTA
Dallas Diamonds,
5580 L.B.J. Freeway, # 645,
Dallas, TX 75240, USA
(Page 87)

MANGALA & ASHOK ANAND CHANDAVARKAR
Gilder Villa Society,
17, Master Vinayak Cross Road,
Bandra (W), Mumbai 400050
(Page 113)

MONICA BOSE
165/B Dr. Ambedkar Road,
Dadar, Mumbai 400014
(Front Endpaper)

NALKUR SRIPAD RAO
Bungalow No. 1,
Union Park, Khar (W),
Mumbai 400052
(Page 137)

PARIKH FOUNDATION
Kavin C. Parikh, Chairman,
611, Panchratna,
Mama Parmanand Marg,
Opera House, Mumbai 400004
(Facing Page 1)

PAUL ABRAHAM
91B, Maker Towers,
Cuffe Parade, Mumbai 400005
(Pages 154, 170)

PHEROZA & JAMSHYD GODREJ
"The Trees", 1st Floor,
40-D, Ridge Road,
Malabar Hill, Mumbai 400006
(Page 189)

PRATIBHA PANDE
D-I, 112 Rabindra Nagar,
Amrita Shergil Marg,
New Delhi 110003
(Page 150)

PURSHOTTAMBHAI N. PADHIAR
Vadi Road,
Deesa 385535, District Banaskantha,
Gujarat
(Page 66)

RAGHUNATH NADKARNI
42, Beach Towers, P. Balu Marg,
Prabhadevi, Mumbai 400025
(Page 30)

RAIKA & NAVROZE GODREJ
"The Trees", 1st Floor,
40-D, Ridge Road,
Malabar Hill, Mumbai 400006
(Page 36)

RAJEEV DUBEY
1701, Mahindra Heights,
96, Tardeo Road, Mumbai 400034
(Page 51)

RAJSHEKHAR PARIKH
305, North Woodland Street,
New Jersey, NJ 07631, USA
(Pages 56–57)

RISHAD NAOROJI
Godrej Bhavan, Home Street,
Fort, Mumbai 400001
(Page 190)

RIVKA ISRAEL
Marg Publications, Mumbai
(Page 24)

ROHIT J. PARIKH
63E, 9th Street Apt. 7M,
New York, NY 10003, USA
(Page 67)

ROSY BLUE (INDIA) PVT. LTD.
Arun Mehta, Chairman,
1608–09, Prasad Chambers,
Opera House, Mumbai 400004
(Pages 27, 108)

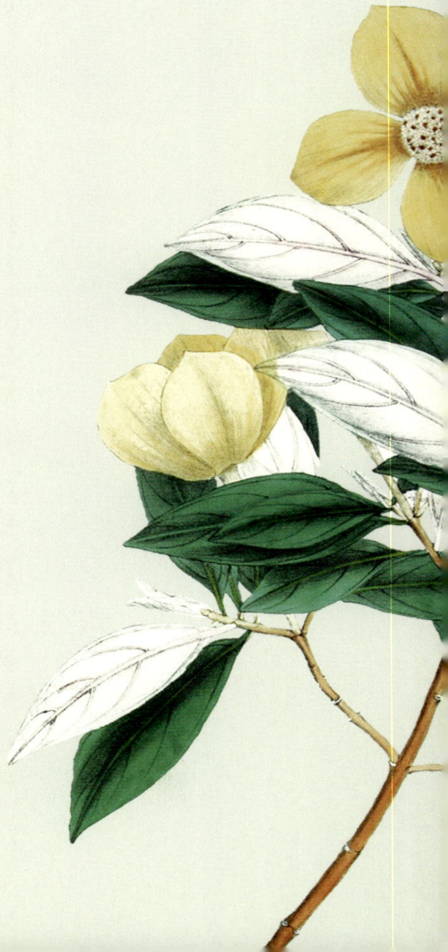

RUDRA AND WILLIAM CONNAL
Apna Cottage, Juhu-Tara Road,
Juhu, Mumbai 400049
(Page 31)

S.G. JHAVERI CONSULTANCY PVT. LTD.
Kamlesh Jhaveri, Managing Director,
1610/11 Prasad Chambers,
Opera House, Mumbai 400004
(Page 149)

THE SARASWAT CO-OPERATIVE BANK LIMITED
Ekanath K. Thakur, Chairman,
Mittal Court, "A" Wing, 1st Floor,
Vidhan Bhavan Marg,
Nariman Point, Mumbai 400021
(Pages 2, 95)

SARLA & SEVANTI PAREKH
Bhartiya Bhavan,
72, Marine Drive,
Mumbai 400020
(Page 75)

SATISH L. CHULLANI
"Mohini", 2nd Floor,
18th Road, Khar (W),
Mumbai 400052
(Page 32)

SHYAM HANSRAJ MARIWALA
Managing Director,
Marson Biocare Private Limited,
216, Vasan Udyog Bhavan,
Senapati Bapat Marg,
Lower Parel, Mumbai 400013
(Page 104)

SONAL & RATNESH DESAI
Mukund Desai & Co.,
75 Jeevan Vihar,
Mumbai Samachar Marg,
Mumbai 400023
(Page 49)

SOONOO TARAPOREWALA
Mumbai
(Pages 141, 162)

SUNEETA DATTATREYA JOSHI
9/199 Vijay Nagar Society, S.N. Marg,
Andheri (E), Mumbai 400069
(Page 100)

SURESH S. KOTHARI
155, Damarest Avenue,
Englewood Cliffs,
New Jersey, NJ 07632, USA
(Pages 56–57)

TATA CHEMICALS LIMITED
Bombay House,
24 Homi Mody Street,
Fort, Mumbai 400001
(Pages 76, 99, 107)

UNION BANK OF INDIA
Central Office, Union Bank Bhavan,
239, Vidhan Bhavan Marg,
Mumbai 400021
(Page 63)

UNITED PHOSPHORUS LTD.
Rajju D. Shroff, Chairman & Managing Director,
Uniphos House, 11th Road,
Khar (W), Mumbai 400052
(Page 178)

VENUS JEWEL
Ramniklal Premchand Shah, Partner,
902, Panchratna,
Opera House, Mumbai 400004
(Page 173)

VIJAY SALES
Nanu Gupta, Chairman,
384, Veer Savarkar Marg,
Opp. Siddhivinayak Temple,
Prabhadevi, Mumbai 400025
(Page 192)

YASHWANT DATTATREYA JOSHI
2/1 Chandra Shekhar Society,
S.N. Marg, Andheri (E),
Mumbai 400069
(Page 114)

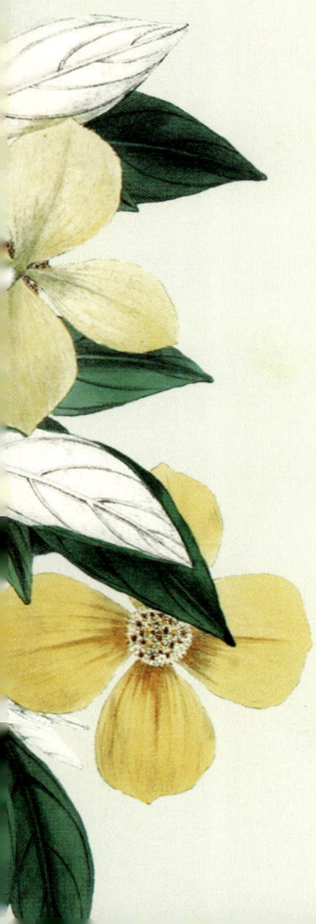

Detailed Captions for Jacket, Endpapers, and Preliminary Pages

FRONT COVER
Oriental Dwarf Kingfisher,
Black-backed Dwarf Kingfisher,
Three-toed Kingfisher
Ceyx erithacus
Synonyms: *C. tridactyla, C. erithaca*

A lovely, tiny, forest kingfisher with lilac-rufous head, rump, and tail, dark blue and black mantle, orange-yellow underparts, coral-red bill, and three-toed feet. It is found from Mumbai southwards; in the Himalayan foothills from Sikkim east to Arunachal Pradesh; Bangladesh and Sri Lanka. It keeps near shady, secluded streams in moist deciduous and evergreen forests, on the lookout for small fish, tadpoles, insects, and crustaceans. It makes a shrill call, *chichee-chichee*, while on the wing. During the nesting season (April–May in the north and northeast, July–September in the south) it builds a horizontal tunnel about 1 m long and 40 cm wide in a vertical earth bank by a forest stream or a roadside. It lays 4 or 5 white eggs.
From T.C. Jerdon, *Illustrations of Indian Ornithology*, 1847.
Sponsored by IndusInd Bank.

BACK COVER
Alpios carnea
Family: Fabaceae (alt. Leguminosae)
Synonym: *Cyrtotropis carnea*

A tall climbing perennial found from 1,200 to 2,400 m in Uttarakhand, Sikkim, Nepal, Bhutan, Myanmar, Tibet, Thailand, and Vietnam. It has large pinnate leaves with 5 ovate-acute leaflets, and long-stalked clusters of reddish flowers appear in August–September. The flowers are 2–2.5 cm long. The pods are 10–18 cm long and 6 cm wide, and contain poisonous seeds that are used as beads.
From N. Wallich, *Plantae Asiaticae Rariores*, 1832.
Sponsored by Deepak H. Shah, Heena Group of Companies.

ENDPAPERS
Front:
View from the top of the Bore Ghaut, drawn on the spot in 1803 by William Westall, ARA, coloured by J.B. Hogarth, engraved by T. Fielding.

Bhor Ghat – *Maa tujhe salaam* by Monica Bose
In red earth we rolled, Albert, Ashok, Rajani, Bella and I,
Scrambling for Jamun, Mulberry, Karvanda and Jaam;
Shielding from thunder, blue lightning and rain;
Melting into mist and cooling rain lashing our cheeks,
Sere hills, valleys and plains,
Bursting into verdure, bulbul song,
Jungle call and roaring waterfall.
Over jumping fish and slippery rock,
Amidst Katkaris drumming their mystic beats,
Jackals roamed and leopards prowled.
Maa Bhor Ghat, where have you gone?

Back:
The Mountain of Aboo, in Gujarat [now Rajasthan], drawn by William Westall, ARA, from a sketch by Capt. Grindlay, engraved by T. Fielding.
Courtesy Deepak H. Shah, Heena Builders.

Endpapers from *Scenery, Costumes and Architecture Chiefly on the Western Side of India* by Captain Robert Melville Grindlay, 1826.

FACING PAGE 1
Hodgsonia heteroclite (female plant)
Family: Cucurbitaceae

This gigantic woody climber with coiled wiry tendrils grows to a height of 30 m. Found in Darjeeling and Sikkim to an altitude of 1,400 m, extending to the Khasi hills of Meghalaya, Assam, and Arunachal Pradesh up to 1,000 m, it also grows in Nepal up to 400 m. The leaves are 15–24 cm across, and deeply lobed. The flowers are brown and hairy outside, white tinged with yellow within, 5 cm long, and fringed with 10-cm-long threads. The reddish-brown berry is roundish, 10–15 cm across, 12-grooved, hard, and bitter; it contains six or more flat, elliptic seeds. The seed kernel is eaten roasted. A paste made from the roots with ginger and lime is said to cure fever.
From J.D. Hooker, *Illustrations of Himalayan Plants*, 1855.
Courtesy Kavin C. Parikh, Parikh Foundation.

PAGE 1
Lesser Florican
Sypheotides indica

This species is resident and breeds in northwest India and Nepal, and winters in south India. The breeding male (illustrated here) has head plumes and black head, neck, and underparts. Females, non-breeding males, and immatures are buff or sand-coloured. Like other bustards, this species favours dry grasslands.
From T.C. Jerdon, *Illustrations of Indian Ornithology*, 1847.

PAGE 2
Orangebreasted Green Pigeon
Treron bicincta
Synonyms: *Osmotreron bicincta, Winga dicincta, Dendrophasia bicincta*

This orange-green pigeon has an orange breast bounded by a lilac band in the male. The tail feathers are slaty grey with black sub-terminal bands, except on the central pair. These birds, found in flocks, are resident in the Himalayan foothills from eastern Uttarakhand to the Assam hills and southern Arunachal Pradesh, the Eastern and Western Ghats south of Goa, Bangladesh and Sri Lanka. They feed solely on fruit. Their flight is very swift; their call soft and melodious. All green pigeons are difficult to spot when sitting quietly in trees, but are more easily seen when they start feeding.
From T.C. Jerdon, *Illustrations of Indian Ornithology*, 1847.
Courtesy The Saraswat Co-operative Bank Limited.

PAGE 5
Pink-headed Duck
Rhodonessa caryophyllacea

T.C. Jerdon writes: "I inserted this duck in my catalogue of Indian birds on the authority of some Shikarees of Jalnah, who described it accurately as being a rare visitor to that neighbourhood. ... I have obtained two specimens, one from the Nellore district and the other in the Madras Market. It however must be considered a very rare kind, even in the north of India, as Mr. Blyth has but seldom procured it. Latham says it is common in Oude, and lives generally in pairs, is often kept tame, and becomes tolerably familiar. ... I add a description – head, cheeks, sides and hind neck pale pink or rose colour; the rest of the plumage of a reddish brown, somewhat paler beneath; speculum and inner webs of quills pale reddish fawn; edge of the wing white; some of the tertiaries lengthened and glossy green – bill red – legs leaden. Length about 24 inches – wing 10.5 inches – tail 4 inches – tarsus 4.4 inches."
E.C. Stuart Baker (in *Fauna of British India*, Vol. VI, March 1929), writes about its distribution: "From Oude and Nepal to Assam and Manipur in the densest jungles running along the foothills of Himalayas. Elsewhere it has appeared as far west as Punjab, ... Its stronghold is probably Duars and Eastern Bengal, though even from there it has been pushed back by increasing cultivation .... Of all our Indian Ducks this is certainly the most shy and secretive, so that it is seldom seen except by chance, when a line of elephants are employed to beat through thick grass or forest when hunting for tiger or big game. In former days they were not rare in parts of Eastern Bengal .... Now, however, cultivation has beaten back the jungle and driven the birds to yet remoter and less trodden jungles, where if one could locate them, they still probably exist in some numbers. It feeds on both vegetable and animal food, remains of weed and crustacea having been found in its stomach. Its flight is powerful and fast and its voice is a musical edition of the Mallard's."
The last sighting of this bird in the wild was in 1935. According to Varad Giri, Curator, BNHS, the Society has in its collection *five* specimens of the Pink-headed Duck – collected from West Bengal by H.J. Walton in 1899, from Bihar by C.M. Inglis in 1903, from Burma by E.R. Jerdine in 1909, from north Kheri (U.P.), by J.B. Hearcey in 1921, and presented by the Indian Museum, Calcutta (date not known).
From T.C. Jerdon, *Illustrations of Indian Ornithology*, 1847.

PAGE 7
Barking Deer
Muntiacus muntjac
Painting by O.F. Tassart, from *JBNHS* Vol. 37, No. 4 (Supplement, *The Wild Animals of the Indian Empire and the Problem of their Preservation*, Part II).

FACING PAGE
Meadow Rue
Thalictrum cysticarpum
Family: Ranunculaceae
Synonym: *T. chelidonii*

This erect, slender perennial is found from Kashmir to southeast Tibet and Myanmar, in forests and shrubberies from 2,700 to 3,600 m. The leaves have rounded, heart-shaped, many-toothed leaflets, 2–3 cm long, pale beneath. The stems are usually much-branched. The flowers appear in July–September in branched hairless clusters; they are lilac or mauve, 1–2 cm across; the stamens are yellow.
From N. Wallich, *Plantae Asiaticae Rariores*, 1832.

BOOKMARK
Asian Paradise-flycatcher
Terpsiphone paradisi
From T.C. Jerdon, *Illustrations of Indian Ornithology*, 1847.

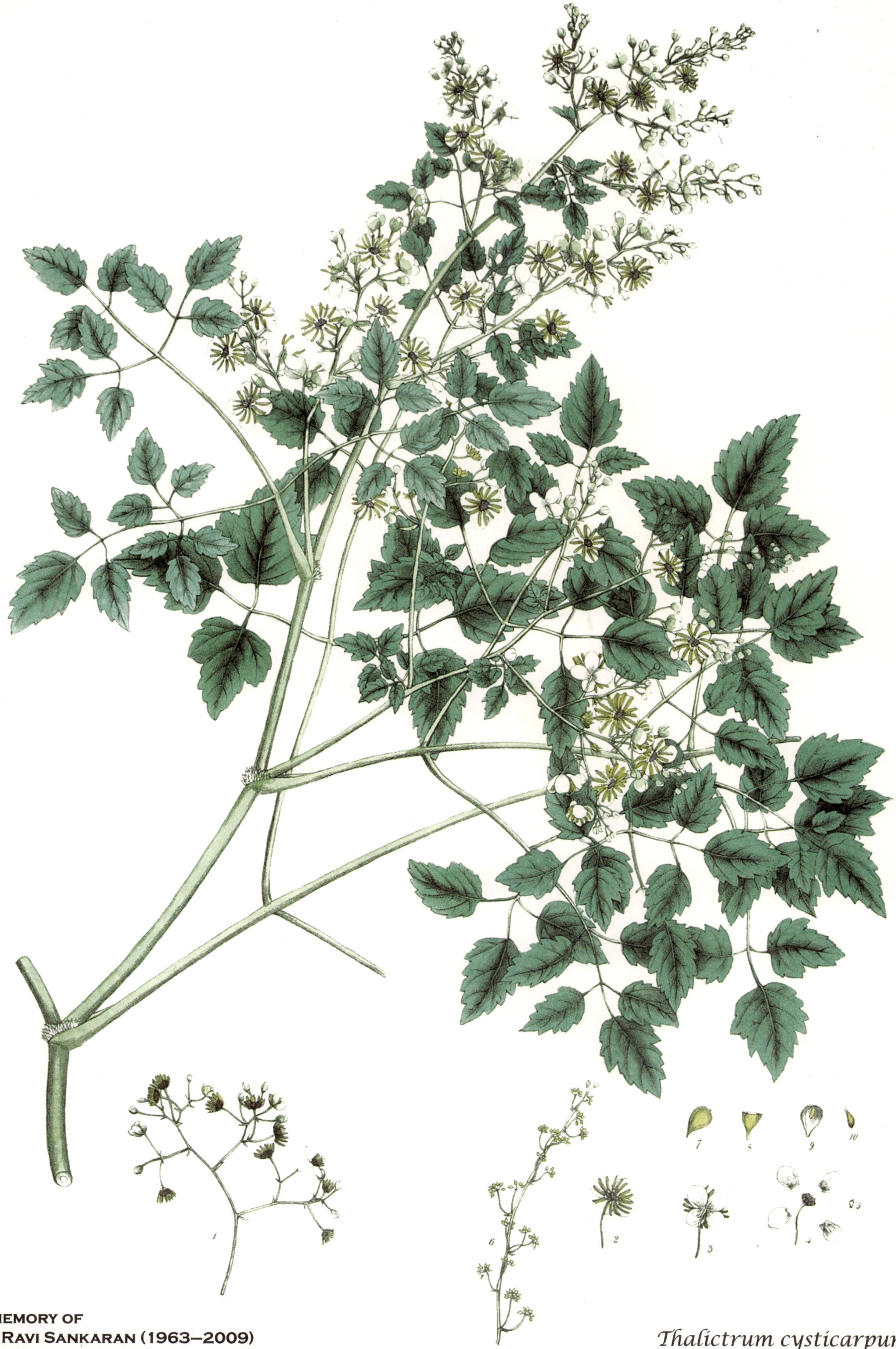

In memory of
Dr. Ravi Sankaran (1963–2009)
Former Director, Sálim Ali Centre for
Ornithology and Natural History (SACON)

Thalictrum cysticarpum

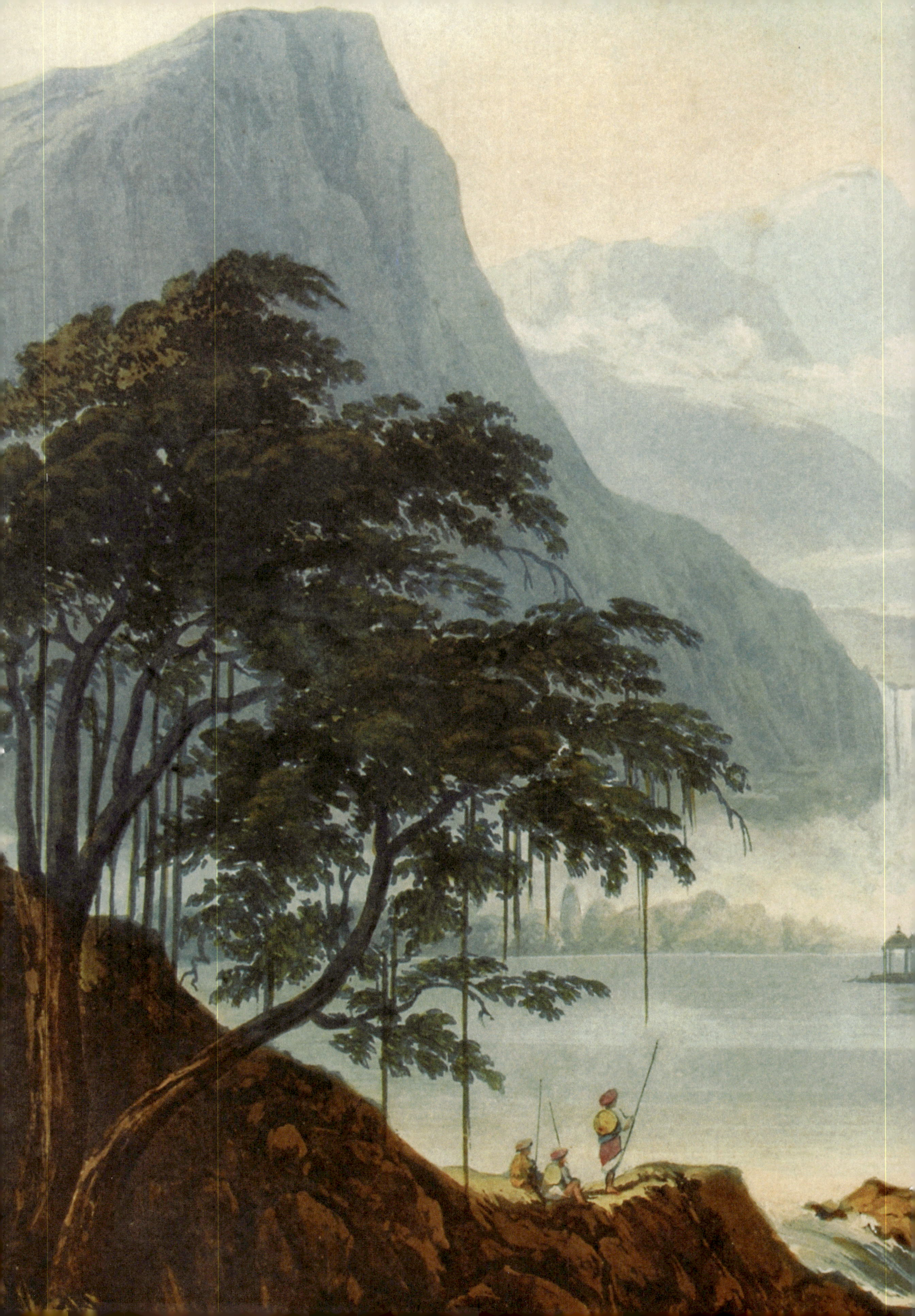